Deutsches Zentrum für Entwicklungstechnologien – GATE

Deutsches Zentrum für Entwicklungstechnologien – GATE – stands for German Appropriate Technology Exchange. It was founded in 1978 as a special division of the Deutsche Gesellschaft für Technische Zusammenarbeit (GTZ) GmbH. GATE is a centre for the dissemination and promotion of appropriate technologies for developing countries. GATE defines „Appropriate technologies" as those which are suitable and acceptable in the light of economic, social and cultural criteria. They should contribute to socio-economic development whilst ensuring optimal utilization of resources and minimal detriment to the environment. Depending on the case at hand a traditional, intermediate or highly-developed can be the „appropriate" one. GATE focusses its work on the key areas:
– *Dissemination of Appropriate Technologies:* Collecting, processing and disseminating information on technologies appropriate to the needs of the developing countries; ascertaining the technological requirements of Third World countries; support in the form of personnel, material and equipment to promote the development and adaptation of technologies for developing countries.
– *Environmental Protection:* The growing importance of ecology and environmental protection require better coordination and harmonization of projects. In order to tackle these tasks more effectively, a coordination center was set up within GATE in 1985.
GATE has entered into cooperation agreements with a number of technology centres in Third World countries.
GATE offers a free information service on appropriate technologies for all public and private development institutions in developing countries, dealing with the development, adaptation, introduction and application of technologies.

Deutsche Gesellschaft für Technische Zusammenarbeit (GTZ) GmbH

The government-owned GTZ operates in the field of Technical Cooperation. 2200 German experts are working together with partners from about 100 countries of Africa, Asia and Latin America in projects covering practically every sector of agriculture, forestry, economic development, social services and institutional and material infrastructure. – The GTZ is commissioned to do this work both by the Government of the Federal Republic of Germany and by other government or semi-government authorities.
The GTZ activities encompass:
– appraisal, technical planning, control and supervision of technical cooperation projects commissioned by the Government of the Federal Republic or by other authorities
– providing an advisory service to other agencies also working on development projects
– the recruitment, selection, briefing, assignment, administration of expert personnel and their welfare and technical backstopping during their period of assignment
– provision of materials and equipment for projects, planning work, selection, purchasing and shipment to the developing countries
– management of all financial obligations to the partner-country.

Deutsches Zentrum für Entwicklungstechnologien – GATE
in: Deutsche Gesellschaft für Technische Zusammenarbeit (GTZ) GmbH
P. O. Box 5180
D-65726 Eschborn
Federal Republic of Germany
Telephon: (06196) 79-0   Telex: 41523-0 gtz d   Fax: (06196) 7948 20

Albrecht Kaupp/John R. Goss

Small Scale Gas Producer-Engine Systems

A Publication of
Deutsches Zentrum für Entwicklungstechnologien – GATE
in: Deutsche Gesellschaft für Technische Zusammenarbeit (GTZ) GmbH

Springer Fachmedien Wiesbaden GmbH

The Authors:

Albrecht Kaupp, PhD, staff member of GTZ/GATE has been working in the fields of civil engineering, mathematics, and biomass energy conversion systems since 1972. Now project officer for biomass energy conversion systems since 1983. His field of expertise is gasification of biomass.

John R. Goss, M. S., Professor at the Department of Agricultural Engineering of the University of California, Davis. Mayor fields of research have been harvesting of agricultural crops, forestry, and gasification of agricultural residues.

CIP-Kurztitelaufnahme der Deutschen Bibliothek

Kaupp, Albrecht:
Small scale gas producer engine systems : a publ.
of Dt. Zentrum für Entwicklungstechnologien – GATE
in: Dt. Ges. für Techn. Zusammenarbeit (GTZ) GmbH /
Albrecht Kaupp ; John R. Goss. – Braunschweig ;
Wiesbaden : Vieweg, 1984.

NE: Goss, John R.:

ISBN 978-3-528-02001-9   ISBN 978-3-663-06868-6 (eBook)
DOI 10.1007/978-3-663-06868-6

Foreword

This monograph was prepared for the Agency for International Development, Washington D. C. 20523. The authors gratefully acknowledge the assistance of the following Research Assistants in the Department of Agricultural Engineering: G. Lamorey, E. A. Osman and K. Sachs.

J. L. Bumgarner, Draftsman for the Department, did most of the ink drawings. The writing of the monograph provided an unique opportunity to collect and study a significant part of the English and some German literature on the subject starting about the year 1900.

It may be concluded that, despite renewed worldwide efforts in this field, only insignificant advances have been made in the design of gas producer-engine systems.

Eschborn, February 13, 1984
Albrecht Kaupp

Contents

CHAPTER I: INTRODUCTION

Gasification of coal and biomass can be considered to be a century old technology. Besides gasoline and diesel oil, producer gas has been used to drive internal combustion engines almost since their invention. The generation of producer gas from wood and coal has been reliable and inexpensive compared to the use of gasoline and diesel oil for a long time but was generally only accepted during emergencies and war times. Although more than one reason accounts for this phenomena, the most significant factor has been the inconvenience and the required skill necessary to operate a gas producer-engine system.

The recent interest in gas producers has somehow diverted the attention away from the real problem of gasification. A gas producer itself is of little use. Gasification must be clearly seen as a whole system consisting of the gasification unit, the purification system and the final energy converter such as burner or internal combustion engine. The real difficulties are not so much to obtain a combustible gas, but to generate it in a physical and chemical state necessary for long-term internal combustion engine operation. Gasoline and diesel engines draw their fuel from a tank by natural suction or forced injection. These fuels are homogenous and do not change composition or physical properties over many months. It is therefore sufficient just to turn a key and start the engine. A gas producer driven power unit requires much more care and understanding. The gas producer generates the combustible gases as demanded by the engine with no storage container between the engine and the gas producing plant. Physical and chemical properties of the gas such as energy content, gas composition and impurities will vary widely, even within a few minutes or seconds. Their control is limited by the very nature of gasification, a complex sequence of partial combustion, distillation and reduction of lignocelluosic material under high temperatures and close to atmospheric pressure. The gas generated needs to be highly purified before it is used in an engine. The commercially available filter, condensing, and cooling components are not specifically designed to adequately handling the wide range of requirement for the many biomass fuels. In summary, a gas producer engine system, whether it is used for generating electricity, pumping water or driving an automobile must be custom tailored and the operator trained in the peculiarities of the system. No one would ever try to run a gasoline engine on diesel or vice versa. The same restriction applies to the gasifying unit of the system. It needs to be designed for a specific class of fuels. Variations in the physical and chemical composition of the fuel are tolerable within limits. For instance, a fixed bed gas producer designed to gasify wood blocks of a specific size and moisture content will not run as well on the same wood blocks with a much higher moisture content and will cease operation all together if fueled with straw. The claims sometimes found in papers and manufacturers' brochures of gasifiers operating on almost every type of waste product containing combustible carbon must be taken with extreme caution.

Although a gas producer-engine system is built as a unit and fine tuned for a successful operation, it is not necessary to develop special engines. The existing internal combustion engines can be used with little modifications. The usually unavoidable power drop, due to the lower energy density of the producer gas-air mixture is not a serious drawback. It can be recovered by turbocharging the engine or some other modifications described in Chapter VII. The most simple solution to this problem is to use a larger engine. A more serious problem has been the trend to build high-speed engines which are not as suitable as low-speed engines for operation with producer gas.

The design and construction of small units (5-100 hp) for power or electricity generation is a lost art. There are very few operational automotive units in the world today. Before and during the Second World War, over 1,000,000 portable units were in operation in European countries and their colonies. They were used in ships, on automobiles, tractors and in trains. An extensive search in the non-communist world came up with about a dozen operational units outside universities and research institutes and approximately 100 units used for research. Although the interest in this form of power generation has increased significantly and is growing fast there is a lack of functional units and off the shelf equipment. There are probably four or five companies world wide with enough experience that could deliver a small gas producer-engine system within a reasonable time span.

The same applies to published papers about the subject for the last decade. There is very little new concerning equipment or experimental results that has not been tried and published during the 1900-1950 period. However, the effect of these publications on the renewed interest in the subject, in particular, gasification of not so common fuels such as crop residues should not be underestimated. Although science hesitates to look back into the past, we simply can not ignore the fact that today's experience with small gas producer engine systems is insignificant and the little work that has been done in this field was closely related to previous experience. Moreover, there has been little concern about reliability and economics of the present test units, because of their specific status as learning systems.

The theoretical understanding of combustion and gasification of carbon fuels has made significant progress during the past decades. Its impact on new designs or better gas producers is minimal. There are no commercial systems today that can match the occasionally reported amazing reliability and long-term operation of some of the past systems. On the other hand, papers written about portable and stationary units of small and moderate size are in the thousands during the 1930-1950 period. As part of this report, at least 1200 papers about the subject have been located. Some of the information (over 600 publications) have been acquired, reviewed and incorporated into this report. Because gasification is a complex topic involving highly theoretical as well as purely practical matters, the reader will find such diverse topics as mathematical solution to the two dimensional heat transfer equation, CO poisoning, and how to start a gasifier at -20^{0}C in the reference list. In addition over 400 institutes, companies, consultants and private persons in 63 countries have been contacted. Our main interest was to receive information of existing units or previous experience with gasification on a broad basis. In doing this we have introduced our past and future projects to 250 of the contacts in form of an information letter. Although the information exchange resulting from this letter was limited to 130 responses, some conclusions and recommendations can be drawn:

1.   The scientific and practical data published during the 1930-50 period about small-scale, portable and stationary units should not be ignored and classified as old fashioned. Gasification is more an art and not so much a science when it comes to building and operating a gas producer-engine unit. The past knowledge documented in thousands of papers is therefore very helpful for the design of the gas producer and its auxillary equipment, as well as for its operation.

2.   The fuel situation must be critically examined and related to the social-economical condition in Developing Countries. There are little waste products in most Developing Countries that could be gasified on a large scale. In particular in arid zones the use of wood as a fuel even if it replaces much more expensive gasoline is out of question. The devastating long-term effects on the landscape and soil are too serious if wood is used even for a short period. The deforesting of whole areas for a quick profit or continuous supply of fire wood already shows its effects in Africa and has been a serious problem in Afghanistan and Pakistan for decades. On the other hand, in tropical countries such as Brasil and the Ivory Coast with fast renewable forests, the use of wood for gasification for small scale units will have very little, if any effect, on the overall wood situation. The present knowledge of gasification refers mostly to fuels such as wood, coal, charcoal and coke. This does not mean other perhaps more readily available biomass fuels such as nutshells, fruit pits or corncobs are unsuitable for gasification. Some of them are even superior. Their use as gasification fuels depends mainly on solving the logistic problems associated with their collection and processing.

3.   Any fuel for gasification should be processed and upgraded as little as possible. All biomass fuels need to be air dried before they can be gasified in a downdraft or crossdraft gasifier. Consequently facilities will be needed to store a few months supply of fuel. Besides drying, any further upgrading of the fuel is undesirable. In particular the charring of biomass is a highly wasteful process and densifying fuel to pellets, cylinders or cubes can be very costly and is only recommended for very large units. A hand operated densification unit may be justified under certain conditions for smaller units. Charring or densifying biomass fuels for use in gas producers does not always improve the gasification characteristics of the fuel. Adapting either method requires a careful evaluation of why the fuel can not be gasified in its original form and to what extent charring or densifying the fuel would improve its gasification characteristics.

4. The introduction of large biomass gasification units with automatic feed and ash-removal systems and units mounted on trucks and tractors should be undertaken at a later stage in a gasification development program.

Large units (above 200 hp) are considerably more expensive. Once built there is little room for modifications or improvements. The likelihood of failure and long-term technical problems are high and in most cases underestimated. Running a large plant requires skilled operators on a 24-hour shift. The automatic feeding and ash removal systems for large plants are sometimes more expensive and more difficult to control than the rest of the plant. The idea of portable units propelling trucks and tractors although rather attractive on first glance, lacks experience and reliability at this point. These units restrict possible fuels to wood, charcoal, coke, or anthracite. The necessary sophisticated cleaning equipment will not be available in most Third World Countries. The system is by no means fool proof and can be easily damaged through improper handling. Operating a producer gas driven truck requires considerably more skill than operating a diesel truck. There are some questions as to whether a gas producer has the ability to adjust its output to the need for fast changing engine speed. In fact the poor load following ability of gas producers has caused most of the problems in the past such as over heating, freezing of constituent gases, tar and dust burst, and poor gas quality. Our credibility in Developing Countries has been seriously undermined by our failure or inability to modify the transferred technology to local conditions. The usually high expectations of local government and their desire to set up large prestigious projects is a wide-spread phenomena in Third World Countries. Our present practical experience with automotive gas producers is insufficient and confined to a few running units, using a most suitable fuel such as charcoal or wood. Using Third World Countries as test locations to improve our lack of knowledge is not advisable and may further undermine our credibility. We do not disregard the sometimes reported amazing reliability of producer gas powered trucks that have travelled over 300,000 km without any operational problems, nor reported journeys over thousands of miles through the Middle East and desert areas by trucks run on producer gas. However, this was done 40 years ago by skilled personnel at a time when the technology was well developed and widely known. The only recent long distance journey by a producer gas fueled U.S. automobile known to us, was a trip from the East coast to the West coast through the Southern United States and a round trip from Southern U.S. to New York City (Figure 1). It is safe to say that very few people have the knowledge and theoretical expertise to set up a reliable system within a short time.

5. Our search for manufacturers of small gas producer engine systems in 49 countries was unsuccessful. There are no manufacturers known to us which could sell and install an off shelf unit and guarantee its performance. There are however some companies which do have the expertise and facilities to manufacture such units on request. A potential buyer of small gas producer-engine systems cannot expect to get any guarantees for the satisfactory operation, because of the well-known sensitivity of the gas producer to changes in the physical and chemical properties of the fuel. Any installment of a gas producer-engine system in Third World countries and elsewhere will therefore be a risk, and may require additional long-term testing to adapt the unit to local fuel properties.

6. The introduction of small scale producer-engine systems as replacement for diesel or gasoline driven power units and generators for small scale industries in urban areas, as well as on the village level, seems to be highly attractive and has a very good chance to be accepted. Ideal and most promising from an economical and social point of view are crop and wood processing industries with a need for power and electricity generation and a continuous output of residue products such as wood chips, sawdust, bark, corncobs, cotton gin trash and rice husks. These residues, although most of them are rather difficult to gasify with the present state of knowledge, are either a real waste product such

4

as about 50% of the world rice husk production or their use for gasification will not seriously interfere with established customs. We emphasize stationary or portable units for stationary applications, because successful application of producer gas will greatly depend on the purification system in the long run. There is a signficant difference in the design of a stationary purification system compared to a fully portable one. The latter system is much more sophisticated, expensive and built from material probably not available in most Third World Countries. We can see a possible use of gas producer units in the innumerable small rice milling industries around the world, provided the gasification of rice hulls can be satisfactorily done. The most commonly used 5-20 hp irrigation pumps in Third World Countries could be powered by producer gas as fuel for the existing engines. Most of these engines are old, low-speed engines. The low speed is an advantage for producer gas. The recent interest in the Humphrey pump, a simple device to lift water by combusting gaseous or liquid fuel, could be a promising application for two reasons. First, the design can handle gas impurities much better than internal combustion engines and second, the construction is possible in Third World Countries. In addition, power units in cotton gins and electrical generators in more remote areas are likely applications for producer gas. Another field for using producer gas which may not be as important in Developing Countries as it is in the U.S., is the artificial drying of crops.

7. Any further effort in gasification of biomass should therefore be more field experience in the long-term gasification of wood and charcoal wherever this can be justified. The gasification characteristics of both fuels are well known and the risk of failure of the system is greatly reduced. However, very few countries do have an excess of wood suitable for gasification or charcoal production and can afford to gasify large amounts without serious impacts on natural resources. The successful introduction of gas producers in the very short run is therefore limited to the few countries with a vast supply of wood or other proven gas producer fuels such as nutshells. In addition much more research is needed on the gasification of high ash fuels. This type of gas producer would most likely have a much better chance of acceptance because the unit could gasify many crop residues.

8. It can not be emphasized enough that the successful gasification of biomass can not be simply assessed on a global basis. A gas producer reacts quite sensitively to fuel parameters such as ash content, moisture content, ash composition and impurities. For instance, knowing the chemical analysis and the heating value of cotton gin trash is rather irrelevant in an assessment as to what extent this residue could be gasified. Seemingly unimportant factors such as climate, harvest pattern and further processing of cotton gin trash are much more relevant. The method of harvesting cotton has a considerable impact on the amount of soil in the cotton gin trash. Soil content quite clearly determines its potential and problems as a fuel for gasification. The same applies to other fuels in a different context. Wood usually considered an ideal fuel for gasification can be surprisingly difficult to gasify, in case its ash content is high, or it contains minerals in large amounts which lower the natural ash melting point considerably. The first stage of gasification development should be seen as a careful evaluation of the fuel available, and to what extent and for what periods it can be used. The fuel ash content and composition should be known. Based on the above information a conservative decision can be made as to whether it

5

is technically feasible to gasify it and what type of system should be used. The examples where gasifiers have been built for a fuel assumed to be suitable and could not be put into operation are not rare. Whether it is feasible to upgrade unsuitable fuels in order to gasify them is a purely economical question and depends on the specific case. For instance, cotton gin trash could be screened and most of the dirt removed, or sawdust may be densified to cubes or pellets and therefore essentially upgraded to wood blocks. The so-called doping of unsuitable or less suitable fuels is a well-established technology and its widespread use is only limited by economic factors.

9.   The construction of a small gasifier including the purification system does not require sophisticated equipment or highly skilled mechanics. It can be built in workshops comparable to the auto repair workshops found in most Third World Countries. The understanding and the skill to repair the innumerable old trucks in those countries are on the average high. In summary the construction of the gasifier and the modifications on the engine do not require foreign help.

However, the design of a prototype and the testing should be done at well-established institutions with the necessary equipment and knowledge, particularly if problematic fuels are planned as the feed material. Although a small gas producer is a most simple machine, not much different from a stove, its sensitivity to a change in its design parameters and fuel properties are notorious. To fine tune a unit so that it can gasify the desired fuel is not an easy task. It requires a continuous net of temperature and pressure measurements inside and outside the gasifier. There is always the danger to seriously damage the gas producer or the internal combustion engine during the testing period. This is due to high temperatures in the gas producer and unknown impurities in the gas. On the other hand, once the mode of running and the geometry of the plant has been established, a highly reliable operation can be expected.

A program set up with prospective collaborators in Third World Countries should as a first stage include at least one person from this country at the test site during the testing period. Although theoretical knowledge about gasification is desirable and helps in understanding the overall process and identifying solutions to the sometimes startling behavior of a gasifier, it does not automatically lead to an ability to design and build a gas producer in a responsible fashion. It is therefore important to have collaborators at the earliest stage of the project. Providing collaborators with plans to build a well-tested unit or even ship a complete commerical unit will require technical advisors for a long time.

10.   No attempts have been made in this report to incorporate new trends or describe in detail some of the hardware such as steam injectors or automatic temperature control devices associated with some plants. In principle it is quite feasible to automate the entire system even on a small scale. It is rather questionable whether all this is necessary and does actually improve the operation characteristics of the plant. A classical example for "over designing" gas producers were the units sold for a short period during the 1930's. Their air blast injectors were distributed at the wall of the gas producer as well as in the middle of the partial combustion zone. All this was done to ensure a complete and thorough heat penetration in the partial combustion zone. Later

it was recognized that a careful design of air blast inlet and partial combustion zone could guarantee a homogeneous, hot, partial combustion zone with only one set of air injectors (tuyeres). To what extent a small-scale gas producer with all kinds of technical hardware attached to it such as automatic fuel bed stirrers, automatic ash removal-fuel feed system and protective layers of high temperature alloys or refractories; or simple devices built out of oil barrels or home-made clay bricks are a better solution, is an open question.

Engineering ingenuity came up with about 400 granted patents during one single year in the later 1930's in England. This may indicate how much space for either improvement or freedom in the design of a gas producer is available. In any case one should carefully examine what technical aids are necessary to improve operation and which ones are only boosting the convenience of running the unit. The trend to automation has mainly economical reasons. 24-hour attention to the plant and the labor involved in feeding the fuel and removing the ash by hand may be too expensive in the U.S. However, in Third World Countries the situation is totally different and speaks against automation at any price.

11. Our information letter mailed to 250 institutions in 36 countries has revealed a considerable interest in the subject and that some amazing units exist, such as one on the island of Bora Bora in New Guinea, which is run with coconut husks and supplies the electricity for several villages. Gas producers on a village level are operating in Tanzania to provide power for a corn mill. The large colonial empires of the European countries were equipped with their technically advanced gasification systems from 1900-1945. Consequently, gasification is not new to Developing Countries. However, the information received by us indicates that these units have been put out of operation and the knowledge and information is mostly lost.

Chapter II: History of Small Gas Producer Engine Systems

The history of gasification can be dated back far earlier than usually stated. In 1669 Thomas Shirley conducted crude experiments with carbureted hydrogen and 30 years later Dean Clayton obtained coal gas from pyrolitic experiments. The first patents with regard to gasification were issued to Robert Gardner and John Barber in the year 1788 and 1791. Robert Gardner suggested the application of waste heat of furnaces to raise steam, by combusting the heated products in a boiler. John Barber's patent mentioned the use of producer gas to drive an internal combustion engine. However, the first confirmed use of producer gas from coal was reported in 1792. In this year Murdock generated gas from coal and used it to light a room in his house. For many years, after Murdock's development, coal gas was one of the principal fuels used for lighting purposes in England. Its use declined in favor of electricity but the use of producer gas still continued and became increasingly important for cooking and heating. Experiments to gasify wood or at least use the gases obtained from charring of wood started surprisingly early in the year 1798, when Lebon tried to gasify wood and make gas out of it. In 1801 Lampadius proved the possibility of using the waste gases escaping from charring of wood. The process of generating water gas by reaction of water with a hot carbon bed was mentioned by Fourcroy in 1804. It took five more years before it was realized by Aubertot that the stack gases of blast furnaces can be combusted and used to roast ore and burn lime. He received a patent for this process in the year 1812. The first gas producer built used oil as a fuel and the patent was given to J. Taylor in 1815 who designed and operated the unit. Between the years 1815 and 1839 many patents were issued for utilization of waste heat and stack gas from blast furnaces. However, the first commercially used gas producer can be attributed to Bischof who built a large unit at the iron works of Audincourt, France in 1840. During the next 20 years many researchers and engineers tried to improve the technology. They already used low grade fuel and combusted the gases in gas fired furnaces. The real breakthrough came in 1861 with the Siemens gas producer which is considered to be the first successful commercial unit. Before the turn of the nineteenth century there are three more important events to mention. First, the introduction of the Dowson gas producer in 1878 which was the starting point of the modern gas producer - engine system. This was the first producer that was successfully used for stationary power engines. Second, the introduction of the Mond by-product process on a large scale in 1889. And third, the introduction of the Bernier suction gas producer in 1895, which was the beginning of the use of gas producers in small, compact units. The Mond by-product process proved for the first time that other valuable products such as ammonia could be obtained via gasification. The residual gas from this process was low in heating value but still could be used for industrial heating purposes. This process was also adapted to gasify high volatile fuels such as peat and brown-coal and several plants were in operation in Japan, the United States and Europe.

As far back as 1819 a portable gas producing apparatus comprising of a gas producer and a gas vacuum engine were patented in England. No record can be found that it was ever fitted on a vehicle. The task to actually operate a passenger vehicle with producer gas for the first time ever must therefore be credited to J. W. Parker who covered over 1000 miles with his $2\frac{1}{2}$ and 25 hp automotive gas producers in Scotland during 1901 to 1905. It is interesting to note that the inadequate protection Bernier got for his patented gas producer-engine system, permitted other enterprising engineers with the opportunity of getting something for nothing. Many competing designs were put on the market in increasing numbers for the next 15 years. One such make is the Brush Koela plant that was first introduced as a patented device in 1901 and was actually designed for import to India and other Developing Countries. The name Koela is the Indian word for charcoal. The oil engines used during this time period were actually replaced by producer gas engines. Some companies in England did a brisk business selling producer-engine sets to generate electricity throughout the country for lighting mansions. The necessity to stay ahead of competitors lead some companies to utilization of the waste heat and the CO_2 generated in the process. However, these early attempts of co-generation were not very successful, although the general ideas behind it are no different from today's principles of co-generation. The first decade of the 20th century was also full of attempts to spread the new concept of suction gas producer-engine systems to other applications.

The Duke of Montrose convinced the British Admirality to introduce some of the new compact suction plants on ships, because similar experimental units were already in use on barges for channel and river transport in Germany and France. A small gas producer carried by four men and used for disinfection purposes was manufactured by J. Pintsch. The gas, rich in carbon monoxide, was used for killing mice, rats, or other vermin on farms and ships. The technology of gasification of wood and charcoal was stepped up, mostly to provide the colonies of the British and German Empires with gas producers that did not depend on scare anthracite coal. H. A. Humphrey had considerable success with operating huge pumps on producer gas. Several types of these 1000 hp waterpumps were built in Alexandria (Egypt), Berlin (West Germany) and Chingford (England). Some enthusiasts considered producer gas the future fuel for internal combustion engines. On the other hand a talk given by Ade Clark for the Institution of Mechanical Engineers, London, in which he discussed industrial applications of the diesel engine signaled, in 1904, the increasing interest in this new technology. The manufacture and operation of producer gas plants was in no way restricted to European countries and their colonies. In fact the United States Geological Survey had for several years investigated the economical value of coals and lignites as gas producer fuel. The early tests done with a pilot plant erected at the Louisiana Purchase Exposition in 1904 were very encouraging and demonstrated the use of many coals that could not be combusted in the existing steam-power plants. The fact that the technology of large updraft gas producers became more and more reliable encouraged gas engine manufacturers to build larger and larger units. Before the wide spread use of producer gas only small gas engines up to 75 hp were found economical to operate with town gas. However the cheap producer gas led to the operation of huge gas engines. The first 600 hp engine was exhibited in Paris in 1900. Larger engines, up to 5400 hp were put into service in the U.S. shortly thereafter. The results of a survey of 70 plants out of the 376 existing plants in the U.S. in the year 1909 are published in United States Geological Survey, Bulletin #416.

Figure 1. The ECON wood gas producer resulting from a privately funded development program started in 1978. The
compact, modular gas producer
system weighing 350 pounds is
conviently mounted in the pick-
up bed. Commerical production is
planned for 1981. Courtesy ECON
(The Energy Conservation Company),
P.O. Box 828, Alexander City,
Alabama 35010.

With regard to the present situation, this report is important because it states
for the first time the many difficulties caused by lack of knowledgeable engineers,
lack of knowledge and confidence in the technology on the part of the public,
inexperienced salesmen not familiar with the details of the engine and the gas
producer concept, lack of types of gas producers that could gasify inferior fuel
and the large number of unsuccessful or only partly successful installations made
during the experimental period of this development. One of the key problems
with gas producer systems that has persistently remained to the present is quoted
from the bulletin:

"It can not be denied that many of the difficulties charged to producer-gas power plants are due entirely to incompetent operators. Some plants have been put out of commission temporarily by the prejudices or the lack of ability and training of the operators or engineers in charge. A few of these failures are due to the impossibility of finding men competent to operate the plants, but many of them have undoubtedly been the result of a short-sighted policy on the part of some manufacturers, who are not willing to give proper and necessary information about the design, construction, and operation of the plants made by them. The possibility of a sale at the time is apparently the only interest they keep in mind, and the future is allowed to take care of itself."

Sales brochures from many countries and personal contacts indicate the situation is very much the same today. The demand for better education of the designers and builders of gas producer plants and furnaces, drivers of automotive gas producer vehicles, the existence of special schools teaching gasification and the demand for higher wages for drivers of automotive gas producer vehicles can be found throughout the entire literature covering the 100 years of commercial gas producers.

Further development of the automotive gas producer was done by Porter and Smith in England during the First World War. The impetus for this work was the possibility of disruption of gasoline supplies which had become the dominant fuel for motor transport. Although most of the early development of automotive gas producers was done in England, wide spread application during and after the First World War was crippled by the British taxation system that assigned taxes to cars according to their weight which included the gas producer. The 1919 special report of the Inter-Departmental Committee on the employment of gas as a source of power which dealt at considerable length with the automotive gas producers and its advantages was not followed by any government action to put the automotive gas producer in a more favorable tax situation.

A totally different situation prevailed in France. There the use of wood and charcoal as a fuel had a long history and the French government was actively encouraging the development of automotive gas producers after 1919. Further public awareness of this method to drive an automobile was greatly increased through ralleys organized each year since 1926 by the Automobile Club de France. The distances that had to be covered were between 1600 and 3000 km. One of the greatest names in the development and manufacture of automotive gas producers was the Frenchmen, Imbert. He filed its first patent for a downdraft gas producer in 1923 and many successful designs including the recently built small automotive gas producers are based on this design. The interest in the automotive gas producer faded in France during the 1930s and most of the development in this field continued in Germany. In fact the Imbert Company is still manufacturing small portable gas producer-engine systems in West Germany. Although the automotive gas producer never played any role in the development of gasification in the U.S., more than 12,000 stationary gas producers were in operation during the 1920 and 1930 decades in the U.S. and Canada. In addition, over 150 companies in Europe manufactured small and large gas producers for various applications. The gas producer concept was especially appealing for applications in remote areas or Developing Countries which had bush or timber. For instance, the British company, Crossly, sold gas producers for remote mines in Australia and the Tulloch Reading 50 hp truck developed in England was mostly purchased by the Empire Cotton Growing Cooperation for use in Nigeria.

The next decade from 1930 to 1940 can clearly be considered as a development decade for small automotive and portable gas producers that reached its peek during World War II. New concepts and designs such as downdraft and crossdraft gas producers were developed or improved. Efforts were undertaken to build the automotive gas producers lighter and improve the gas cleaning system which was the vulnerable part of the units. New units, capable of gasifying more readily available fuels such as bituminous coal, anthracite and wood, were developed and tested in small numbers. The British gasification efforts were still more directed to their overseas markets and not so much for domestic use. There were signs of an increasing critical view toward the automotive gas producer in France. It was claimed that at least one new gas producer mounted on a truck was more expensive to run and operate than a comparable gasoline truck despite all government grants and subsidies. It is of interest to recall the official position of the French and British governments during the early '30s. Authorities in both countries felt at that time that the automotive charcoal gas producer was more suitable for their colonies where the supply of gasoline was scarce, and wood that could be charred to charcoal at very low labor costs was readily available. The emerging gas producers using wood and low grade coal were not given much of a chance for general use. History has proven that assessment to be correct.

The first well reported conversion of internal combustion engines, in this case tractors, to producer gas drive under economical pressure happened during the 1931 to 1934 period in Western Australia. The large quantities of wood available, the neglible oil resources at this time and the collapse of the wheat prices during 1930 set the scenario for a rather hasty, uncoordinated conversion of kerosene tractors to producer gas drive. Many farmers, in order to avoid bankruptcy had to consider all alternatives, including producer gas, although it was well known that the power loss of the tractors would be considerable. What happened during these years until the recovery of the wheat prices was just a small part of what happened later during World War II on a much broader basis. Many gas producers were failures from the start. Others deteriorated rapidly owing to faulty construction. Several firms were interested in the manufacture and sale of such units, but had neither the money nor time to do the necessary research and development engineering. As a consequence, there were often totally dissatisfied customers, who after a short trial, resolved they would never again have anything to do with gas producers.

On the other hand, a small number of farmers having ingenuity and mechanical skill, operated their units very satisfactorily for a number of years. In this context it should be mentioned that there has never been an automotive engine especially designed and built for producer gas, although the technology was wide spread for over 100 years. With plentiful fossil fuels available during peaceful and stable economical times, there was no need for the producer gas concept. During emergencies and war times the concept of producer gas engine systems was always so hastily recalled that there was simply not enough time and money available to develop a specially designed producer gas, internal combustion engine for automotive use. This explains in part the difficulties some farmers had to convert their kerosene tractors to producer gas drive. The interest in gas producers faded quickly after the 1930 depression was over. Only 62 producer gas tractors out of 4548 tractors in Western Australia were operating at the end of 1937.

Figure 2. UCD Laboratory Downdraft Gas Producer. Air blown and mounted on platform scales to determine fuel rate. The fire box is one foot in diameter and will produce enough gas when cleaned and cooled to operate a 35 Hp engine from about 60 to 65 pounds of air-dry wood per hour.

In late 1930 the effort of Nazi Germany to accelerate the conversion of vehicles to producer gas drive was the beginning of a world-wide effort to use the gas producer concept as part of a plan for national security, independence from

13

imported oil and acceleration of the agricultural mechanization. A typical example was the Soviet Union. The build-up of the military as well as rapid expansion of heavy industry necessitated a major change in the mechanized agricultural units. The change was directed toward the fuel used. It became apparent that despite a high priority for the agricultural sector, the transport of the fuel was becoming a problem. The big agricultural areas were far from the large oilfields and the distribution of the fuel even when plentiful was one of the biggest problems. The introduction of gas producer powered tractors and trucks to the Rusian farmers can therefore not be viewed as an emergency measure to reduce the consumption of gasoline and diesel oil. Instead it was viewed as an alternative to use fuels available locally and ease the transportation and distribution problem. Almost all early Russian tractors were powered by gasoline engines which required extensive rebuilding of the engine to avoid a severe power reduction. (A later model the Stalinez C65 tractor and the Kharkov caterpillar tractor were equipped with diesel engines). From the design of the gas producer and its gas cleaning system, it seems most likely that various German gas producers were used as the basic design for this final model. Despite some criticism about the gas producer concept, its economics and future, new advanced crossdraft gas producers were built in France. In particular the Sabatier and Gohin Poulence plant showed an astonishing performance, equal to most gasoline powered vehicles. However, it became more and more obvious that good gas producer performance was closely connected to the quality of the fuel. Plants like Sabatier or later, the Swedish Kalle model were highly reliable and worked well only with specially manufactured charcoal having carefully controlled quality. In 1938 most European countries stimulated the use of producer gas through subsidies for conversion, favorable tax or even edicts such as in France that required all public transport companies to change at least 10% of their vehicles to producer gas. The Italian government was even more strict, requiring all buses in public service to use home produced fuel, wood charcoal, alcohol or home produced petrol and oil. These various measures led to 4500 gas producer vehicles in France, 2200 in Germany and over 2000 in Italy by the early part of 1939. England, the country that did most of the pioneer work in the beginning, however, saw its producer gas program entangled in politics, resulting in very little conversion to producer gas for vehicles. This situation can be read in an article written by the Coal Utilization Council appearing in the Fuel Economist in July 1938. The Director of this organization complained bitterly about the stubborness of the British government in this matter and his arguments for producer gas vehicles in England were similar to what is said about today's energy situation in the United States. Nevertheless, some British bus companies ran their city buses on producer gas quite successfully and on schedule.

What happened to the development of the automotive gas producer after 1939 must be seen in the context of the World War II. From the numbers of articles published about gasification in German journals each year and the work of several national committees on the subject it was obvious that Germany was much better prepared to deal with the logistic problems associated with the operation of hundreds of thousands of automotive gas producers. However, the most drastic development took place in Sweden, which experienced a most severe fuel shortage. Other countries delayed the conversion to producer gas drive, because there was simply no need for it. For instance, not too many automotive gas producers were seen in Australia in the year 1940, compared to a considerable larger number in New Zealand which was much earlier affected by the fuel shortage. The United States coped with gasoline shortage by means of rationing but nevertheless automotive and stationary gas producers were manufactured in Michigan. They were not available for domestic use and most of them were sold to China under Lend-Lease terms. "Woman Who Fled Nazis Makes Gas Producers in Michigan Plant for Export to China" was one of the headlines of several articles that appeared in the National Petroleum News and Chicago Tribune about this activity.

The development of the European gasification activities was closely monitored by the Forest Service of the United States Department of Agriculture and some of the findings have been published. At the end of 1944 it was concluded that wide spread commercial adoption of gas producers in the United States would not be promoted. Only under special circumstances in remote areas, gas producer operation might be acceptable.

Even after the outbreak of the war, the British government was in no hurry to regulate or require the use of automotive gas producers. One of the reasons was the unsuitability of most existing gas producers for the soft and brown coals of England which had little anthracite. Nevertheless, a so called government emergency crossdraft gas producer was developed especially for the British coals and low temperature coke and it was planned to manufacture 10,000 units. The government developed producer worked reasonably well but in 1942 it became increasingly difficult to obtain the necessary low ash coal to run the gas producer and plans to mass produce the unit were given up. The conversion of vehicles to producer gas drive was therefore mostly restricted to bus companies and some private companies that installed the stationary Cowan Mark 2C gas producer as an emergency power supply to factories affected by air bombing. Therefore, large scale conversion of vehicles took place in Sweden and the countries occupied by Germany during World War II.

In December, 1939, about 250,000 vehicles were registered in Sweden. At the beginning of 1942 the total number of road vehicles still in service was 80,000. About 90% of which were converted to producer gas drive within $1\frac{1}{2}$ years. In addition, almost all of the 20,000 tractors were also operated on producer gas. 40% of the fuel used was wood and the remainder charcoal. Dried peat was used to some extent. This fast and almost complete conversion was accompanied by the drastic decline of imported petroleum from 11 million barrels in 1939 to 800,000 in 1942.

It is far more interesting to recall the logistic difficulties associated with the conversion of gasoline vehicles on a large scale during World War II, because the technical advances made after 1940 were not significant and dealt mostly with the improvement of gas cleaning systems and better alloys for the gas producer shell.

Schläpfer and Tobler, who conducted extensive tests with various gas producers during the 1930 to 1939 period in Switzerland, pointed out the human element involved. They argued that most of the converted post buses running on producer gas in Switzerland did not perform well because drivers had difficulties getting used to the new driving style and certainly rejected the additional work involved. Most troublesome was the required daily cleaning of the entire gas-purification

15

system and the preparations for the next run, which included the clearing of the fuel hopper, because overnight storage of the fuel in the fuel hopper caused considerable starting difficulties. It also became apparent that neither the manufacturers nor the general public really understood the problems associated with a gas-producer operated bus.

Figure 3. Swedish farm tractor fueled with wood gas. Four cylinder, naturally aspirated diesel engine, dual-fueled do operate with about 10 percent diesel as the pilot fuel. Gas producer is mounted at the left-front corner of operator cabin. Ahead of it is the hot gas filter with the cooler mounted in front of the tractor radiator. Note bags of air-dry wood chips on top of cab. Development by the National Machinery Testing Institute, Uppsala, Sweden which began in about 1958. Photograph taken in 1976. The WW II Imbert gas producer served as the starting point for the gas producer on the tractor.

One company in France had a mixed fleet of gasoline and gas producer driven vehicles. The drivers had to carry out refueling and making repairs after the day's work at regular rates of pay. The producer vehicles were constantly having to go to the shop for repairs which the drivers alledged were beyond what they could do. After the company decided to pay overtime for the time spent to clean and refuel the gas-producers, the producer vehicles became at once as trouble free as the gasoline vehicles. This situation was not new in connection with gas producers. In the early 20th century, owners of large stationary gas producers talked about sending their engineers to special schools on gasification and paid them higher wages to ensure that the gas producer was properly operated. The German government finally agreed after many years of rejection,

16

to pay their drivers of automotive gas producers higher wages, which improved the situation. However, the uninformed private driver remained a persistent problem. At the beginning he was faced with hundreds of makes of gas producers and no manufacturer's guarentee about the performance. Although one could not prove that some manufacturers actually sold equipment they knew would not work, it cannot be denied that many of them did not know much about the performance of their units or could only prove reliable performance with high quality fuel having carefully controlled physical and chemical properties. Large numbers of unsatisfied customers finally led to government action in Germany and Sweden as well as in the occupied countries. The number of manufacturers of gas producers was significantly reduced to about 10 with models that had been proven to be successful. However, the fuel supply and the quality of gas producer fuel was still a problem that actually was never solved. Until the end of 1941, wood and charcoal were the fuels most widely used in Germany. The collection and preparation of gas producer fuel was handled by the Gesellschaft fur Tankholzgewinnung und Holzabfallverwertung which kept a tight control over the size, shape and moisture content of the fuel. The fuel could be purchased at over one thousand official filling stations all over the country. This service was more or less operated and organized like today's oil companies and gasoline stations. It soon became apparent that at the prevailing wood consumption rate and the tendency of drivers to use charcoal, there would not be much forest left within a few years. The construction of charcoal gas-producers was therefore forbidden in France and Denmark after July 1st, 1941 and greatly restricted in Germany and Sweden. The new policy was to encourage the use of brown-coal, peat coke, anthracite and low temperature coke made from bituminous coal. Problems associated with the use of these fuels will be discussed in subsequent chapters. It however can be concluded that their use was plagued by problems with the quality of the fuel, such as high sulfur content, too much volatile matter, poor physical shape of the various cokes sold, too expensive production methods and improper handling of the fuel bags. Most customers did not understand the differences among the various fuels they could buy or their influence on the gas producer. The situation today is about the same and any introduction of small stationary or portable gas producers on a broad basis would likely lead to the same difficulties. Some users of automotive gas producers even produced their own fuel out of brush wood collected in the national forests.

A slightly different situation prevailed in Sweden with its vast supply of wood. At the beginning the unrestricted use of charcoal led to various designs of high performance gas producers, which operated very well as long as they were fired with the specially prepared charcoal they were designed for. The tar oils from wood carbonization were also not wasted and used for heavy engine fuels and as lubricant. Over 3000 furnaces producing charcoal were in operation in 1944, to provide the necessary fuel for metallurgical operations and the fleet of gas producers. Although the officially produced fuel was strictly classified and controlled, not all of the fuel related problems could be solved. For instance first grade low volatile fuel of less than 3% volatiles turned out to be medium volatile fuel with over 8% volatiles that could not be gasified in most gas producers. Hard, high grade charcoal leaving the factories with a low moisture content of 10% and only a 10% fractions of fines, reached the consumer broken up and crumbled with a moisture content of over 20% and was therefore rendered useless. Although the emergency situation was on everybodys mind, the temptation was high to buy and operate the very convenient, high performance gas producers which depended on special fuels.

Figure 4. Scania Vabis, 6 cylinder, naturally aspirated, diesel engine, dual-fueled to operate on wood gas with about 10 percent diesel as the pilot fuel. Truck is used by a Swedish machinery dealer to service his district and has been driven nearly 200,000 kilometers. The engine has not been overhauled during its service life. Development by the National Machinery Testing Institute, Uppsala, Sweden. Photograph taken in 1976.

It's obvious that an automotive gas producer that can be started within 2 minutes, and does not require much cleaning sounded much more appealing for the private customer than one with more flexibility with regard to the fuel needed to operate the unit. The tendency to modify the fuel for a gasifier in question instead of investing the time and money to design and construct a gas producer for a fuel in question can be found throughout the entire history of gasification. This approach was not changed during the first 100 years of gasification and present signs indicate that there will be slow progress toward designing gas producers for specific fuels.

Although the number of accidents related to the use of automotive gas producers was considerably higher than with gasoline vehicles, most accidents were due to negligence of the driver. The increasing numbers of accidents caused

18

by operators not familiar with their equipment was of much concern to the Swedish government and the manufacturers. This was reflected in very detailed operation manuals and the introduction of a special driver's license for the operation of an automotive gas producer. Of concern were simple operational mistakes such as not ventilating the unit after a day's use which resulted in a gas built up in the gas producer that could explode while the owner was checking the fuel level next morning. Other operaters had the opinion that as long as the engine was running on the produced gas everything was fine and switched too early to producer gas drive during the startup period. In most cases this led to totally tarred up manifold and valves, because the initially produced gas, although of high heating value was rich in higher hydrocarbons that condensed out in the engine. More serious and not so easily controlled is the danger of long term carbon monoxide poisioning which occurred frequently according to Swedish reports. The problems in the past with automotive gas producers, should be viewed in the light of the enormous task that was undertaken in Europe to convert hundreds of thousands of gasoline vehicles to producer gas drive within three years in a difficult time. An automotive gas producer must be also viewed as the most advanced gas producer, much more difficult to design and operate than a stationary unit.

Shortly after World War II, automotive gas producers as well as all the large stationary units were put out of service because of abundant, cheap supplies of gasoline, diesel oil and natural gas. The change away from producer gas operation was also drastically reflected in the research done in this field. The number of publications listed in major engineering indexes dropped sharply from several hundreds a year to less than 10 a year during the 1950 to 1970 period. It can be said with one exception, gasification and in particular small portable gas producers were a forgotten technology during this time period. The only research done in this field which can be called a considerable contribution to the advancement of automotive gas producers took place in Sweden during the 1957 to 1963 period. This research was initiated by the Swedish Defense Department during the Suez Crisis and undertaken by the National Machinery Testing Institute. The research made considerable contributions to the improvement of the gas cleaning system and the modifications of diesel engines for gas producer drive.

The 1970s brought an increasing renewed interest in this form of power generation and a more general look at the complexity of gasification. Some of the present work concentrates on the revival of the old ideas and designs and their modification and expansion to fuels different from wood and coal. Our worldwide search for small scale gas producers in operation and researchers working on the subject as well as the increasing number of daily inquieries about gasification received, show a considerable interest and demand in small gas producers. However it can also be noted that, in the public opinion, gas producers still have the image of a simple stove like energy conversion system easy to design and operate. The present demand is therefore also stimulated by the belief that gasifiers can convert almost any carbonecous material to useful mechanical and electrical energy. This image of a gasification system is far removed from any reality and in particular the history of gasification has shown that a fixed bed gasifier providing fuel for an internal combustion engine is a very selective energy conversion system with little flexibility with regard to the fuel it was designed for. A further handicap is the little knowledge we have about the behavior of various biomass fuels under thermal decomposition. This knowledge is certainly basic for any further optimization of gas producers and cannot be obtained within months. On the other hand, amazing performances of gas producer-engine systems have been reported and verified throughout the history of gasification. It is not just an assumption but confirmed reality that trucks have been operating on producer gas for over 300,000 km with no major repair and less engine wear than obtained from diesel fuel. Large Italian rice mills have gasified their rice husks and used the gas to drive the power units used for milling for decades prior to World War II. The number of quite satisfied owners of small and large gasifiers is certainly not small and there is lots of evidence that it can be done. The history of gasification has also shown that it is not one of the most convenient technologies, but in a time with less fossil fuel available and costing more each year, convenience will be a luxury that cannot be afforded very much longer.

Figure 5. 100 kW mobile farm power plant. Powered with a 8.8 liter, turbo-charged and inter-cooled diesel engine that has been dual-fueled to operate on producer gas generated from corn cobs. The unit was designed and constructed in 1978 by the Agricultural Engineering Department, University of California, Davis under contract for the John Deere Harvester Works, East Moline, Illinois. The unit was given to the Department by Deere and Company in 1981.

Chapter II

1. Allcut, E. A., Producer Gas for Motor Transport, Engineering Journal, v 25, n 4, 1942, pp 223-230.

2. Anderson, M., Case for the Encouragement of the Producer-Gas Vehicle in Britain, Fuel Economist, v 14, July, 1938, pp 245-246, 256-257.

3. Anonymous, A New Gas Producer-Gas Plant for Road Transport, The Commercial Motor, January, 20, 1933, pp 787-788.

4. Anonymous, Alternative Fuels for Wartime, Gas and Oil Power, October, 1939, pp 235-238.

5. Anonymous, Emergency Gas Supplies for Factories, Power and Works Engineer, v 36, June, 1941, pp 137-139.

6. Anonymous, Forest Gas for Traction, Engineer, v 166, n 4311, 1938, pp 230-231.

7. Anonymous, Gas as a Substitute for Gasoline Part 1, Petroleum Times, v 42, n 1073, 1939, pp 169-170, 189.

9. Foster Wheeler Energy Corporation, Gas from Coal: A Volatile Solution, Energy Guidebook, 1978, pp 108-110.

11. Anonymous, Gas Producer for Road Vehicles, Engineering, May 26, 1939, pp 631-632.

12. Anonymous, Gas Utilization for Automobiles, Gas Age, December 7, 1939.

13. Anonymous, German Portable Gas Producer Practice, Engineering, v 155, May, 1943, pp 423-424.

14. Anonymous, Improvements in the "Brush Koela" Gas Producer, Engineering, v 169, n 4398, 1949, pp 395.

15. Anonymous, Official Specification for Portable Gas-Producer Fuels, Engineering, February, 1940, p 150.

16. Anonymous, Producer Gas Plant Manufacture, Gas and Oil Power, v 37, n 443, 1942, pp 147-150.

17. Anonymous, Producer Gas: Present and Future, Gas and Oil Power, v 40, n 473, pp 49-50.

18. Anonymous, Producer Gas versus Petrol Operation in Germany, Petroleum Times, v 47, n 1193, 1943, p 190.

19. Anonymous, The P.S.V. Gas Producer, Bus and Coach, November, 1942, pp 228-230.

20. Anonymous, The Soviet Producer-Gas Tractors, Gas and Oil Power, March, 1945, pp 89-95.

21. Anonymous, The Tulloch-Reading Gas Producer for Motor Vehicles, Engineering, v 127, May, 1929, pp 641-644.

22. Bailey, M. L., Gas Producers for Motor Vehicles: A Historical Review, Department of Scientific and Industrial Research, Chemistry Division, Report CD 2279, New Zealand, 1979.

23. Branders, H. A., Producer Gas is the Motor Fuel of Finland, Automotive Industries, May, 1941, pp 482-485, 522-523.

24. Breag, G. R. and A. E. Chittenden, Producer Gas: Its Potential and Application in Developing Countries, Tropical Products Institute, Report G130, London, England, 1979.

25. Brownlie, D., Producer-Gas Driven Vehicles, The Iron and Coal Trades Review, January, 1940, pp 121-123.

26. Campbell, J. L., Gas Producers: An Outline of the Compulsory Government Tests in Australia, Automobile Engineer, v 32, n 422, 1942, pp 156-158.

27. Clarke, J. S., The Use of Gas as a Fuel for Motor Vehicles, Institute of Fuel Journal, v 13, n 70, 1940, pp 102-117.

28. Dimitryev, A. P., Automotive Gas Generators Used in USSR, Automotive Industries, v 83, n 10, 1940, pp 534-535, 551.

29. Dunstan, W. N., Gas Engine and Gas Producer Practice in Australia, Engineer, v 180, n 4688, 1945, pp 400-401.

30. Egloff, G. and M. Alexander, Combustible Gases as Substitute Motor Fuels, Petroleum Refiner, v 23, n 6, 1944, pp 123-128.

31. Egloff, G., Fuels Used in Sweden, Petroleum Engineer, v 18, n 5, 1947, pp 86-88.

32. Forbes, W., Experiments with Gas Producer Vehicles in Cardiff, Passenger Transport Journal, November, 1939, pp 201-205.

33. Fowke, W. H., Operating Results with Producer Gas, Bus and Coach, v 10, n 2, 1938, pp 84-86.

34. Freeth, E. E., Producer Gas for Agricultural Purposes, Journal of the Department of Agriculture of Western Australia, v 16, n 4, 1939, pp 371-414.

35. Gall, R. L. and J. D. Spencer, Caking Coal Behavior in Gas-Producer Tests, Coal Age, v 71, n 2, 1966, pp 128-130.

36. Goldman, B., Fuels Alternative to Oil for Road Transport Vehicles, Fuel Economist, v 14, July, 1938, pp 248-252.

37. Goldman, B. and N. C. Jones, The Modern Portable Gas Producer, Institute of Fuel, v 12, n 63, 1939, pp 103-140.

38. Goldman, B. and N. C. Jones, The Modern Portable Gas Producer, The Engineer, v 166, December, 1938, pp 248-252.

39. Goldman, B. and N. C. Jones, The Modern Portable Gas Producer, The Petroleum World, v 36, n 460, 1939, pp 3-5.

40. Greaves-Walker, A. F., The Design and Construction of a Producer-Gas House for Clay Plants, Transactions of American Ceramic Society, v 18, 1916, pp 862-866.

41. Hurley, T. F. and A. Fitton, Producer Gas for Road Transport, Proceedings of the Institution of Mechanical Engineers, v 161, 1949, pp 81-97.

42. Kralik, F., Rail Car with Charcoal Gas Producer, The Engineers' Digest, December, 1943, pp 24-25.

43. Lang, W. A., Alternative Fuels for Motor Vehicles, Engineering Journal, v 26, n 8, 1943, pp 449-454.

44. Langley, F. D., The Revival of Suction-Gas Producer, Gas and Oil Power, v 37, n 446, 1942, pp 236-240.

45. Lindmark, G., Swedish Gas Producer Buses, Bus and Coach, April, 1944, pp 266-269.

46. Littlewood, K., Gasification: Theory and Application, Progress in Energy and Combustion Science, v 3, n 1, 1977, pp 35-71.

47. Lustig, L., New Gas Producer for Dual Fuel Engines, Diesel Progress, v 13, n 5, 1947, pp 42-43.

48. Mellgren, S. and E. Andersson, Driving with Producer Gas, National Research Council of Canada, RP 15/43, Ottawa, Canada, 1943.

49. Miller, R. H. P., Gasogens, U.S. Department of Agriculture, Forest Service, Forest Products Laboratory, Madison, Wisconsin, 1944.

50. Overend, R., Wood Gasification: An Old Technology with a Future? Sixth Annual Meeting, Biomass Energy Institute Symposium, Winnipeg, Manitoba, Canada, October, 12, 1977.

51. Pavia, R. E., Woodgas Producers for Motor Vehicles, Institution of Engineers Journal, Australia, v 14, n 12, 1942, pp 279-292.

52. Rambush, N. E., **Modern Gas Producers,** Van Nostrand Company, New York, 1923.

53. Renton, C., Producer Gas Tests in the Queensland Railway Department, Institution of Engineers Journal, Australia, October, 1940, pp 274-278.

54. Ridley, C., Temporary Fuels, A Consideration of the Prospect of Their Permanency, Automobile Engineer, v 34, n 446, 1944, pp 63-67.

55. Roberts, R. P., Producer Gas Equipment on Tractors in Western Australia, Journal of the Department of Agriculture of Western Australia, v 15, n 4, pp 391-402.

56. Ruedy, R., Wood and Charcoal as Fuel for Vehicles, National Research Council of Canada, n 1157, Ottawa, Canada, 1944.

57. Skov, N. A. and M. L. Papworth, **The Pegasus Unit,** Pegasus Publishers Inc., Olympia, Washington, 1975.

58. Telford, W. M., Some Notes on the Design of Mobile Producer Gas Units, Institute of Engineers Journal, Australia, v 12, n 11, 1949, pp 299-304.

59. Telford, W. M., Some Notes on the Design of Mobile Producer Gas Units, Gas and Oil Power, v 36, September, 1941, pp 179-181.

60. Tookey, W. A., Suction Gas Plant Development Fifty Years Ago, Engineer, v 193, n 5028, 1952, p 754.

61. Twelvetrees, R., Paving the Way for Producer Gas Operation, Bus and Coach, February, 1944, pp 104-107.

62. Walton, J., Alternative Fuels, Automobile Engineering, v 30, March, 1940, pp 91-92.

63. Woods, M. W., An Investigation of the High-Speed Producer Gas Engine, Engineer, v 169, n 4401, 1940, pp 448-450.

64. Woods, M. W., Producer Gas Vehicles, Institution of Engineers Journal, Australia, v 10, n 3, 1938.

65. Wyer, S. S., **A Treatise on Producer Gas and Gas Producers,** Hill Publishing Company, 1906.

CHAPTER III: CHEMISTRY OF GASIFICATION

The essence of gasification is the conversion of solid carbon to combustible carbon monoxide by thermochemical reactions of a fuel. Complete gasification comprises all the processes which convert the solid fuel into a gaseous and liquid product leaving only parts of the mineral constitutents of the fuel as a residue. Complete combustion takes place with excess air or at least 100% theoretical air; whereas, gasification takes place with excess carbon. The gasification of solid fuels containing carbon is accomplished in an air sealed, closed chamber under slight suction or pressure relative to ambient pressure. The fuel column is ignited at one point and exposed to the air blast. The gas is drawn off at another location in the fuel column as shown in Figure 6.

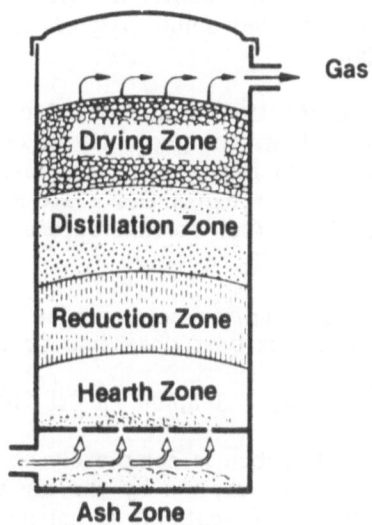

Figure 6. Updraft Gasification (16).

Incomplete combustion of the fuel with air is the initial part of the gasification of lignocellulose material. The process oxidizes part of the carbon and includes distillation and reduction zones, which are separated from the partial combustion zone in a physical and chronological sense.

The research that has been done in this field for the last 140 years can be categorized in three major topics:

1. Design and construction of plants for commercial purposes, utilizing observations and information obtained from existing plants.

2. Basic research about the energy balance, gas composition and chemical reactions in gasification on a macroscale.

3. Research on a microscale under laboratory conditions. Most of this work concentrates on three major questions:

a. Where do the basic chemical reactions take place and in what chronological order?

b. What type of model best fits certain chemical reactions and transport phenomena observed in the gasification of carbon?

c. Can gasification be optimized for a particular objective function?

This chapter will discuss in some detail topics 2. and 3. simultaneously. Topic 1. is discussed in the remaining chapters.

The understanding of the chemical and physical processes in a gasifier is not completely known and the gap between observed data obtained from practical operations and data obtained under controlled laboratory conditions is still being investigated, despite the fact that some progress has been made to explain the discrepancies (9,10,14,15).

In discussing the chemical reactions that take place in a gasifier, the reader is referred to Figure 6 which shows the geometry of one of several modes in which a gasifier can be operated. In this Figure, combustion air is introduced at the bottom of the reactor vessel through a flat grate and the generated gas stream penetrates through the entire fuel column before leaving the producer at the very top.

The heterogeneous chemical reaction between the oxygen in the combustion air and the solid carbonized fuel is best described by the equation:

$$C + O_2 = CO_2 + 393,800 \text{ kJ (at } 25^{\circ}C, 1 \text{ atm).}$$

In this reaction 12.01 kg of carbon is completely combusted with 22.39 standard cubic meters (SCM) of oxygen supplied by the air blast to yield 22.26 SCM of carbon dioxide and 393,800 kJ of heat. It is important to observe that the fuel reaches the oxidation zone in a carbonized form with all volatile matter driven off while passing through the reduction and distillation zones. Therefore, in a theoretical sense only carbon and mineral matter are present in the combustion zone. If complete gasification takes place all the carbon is either burned or reduced to carbon monoxide, a combustible gas, and some mineral matter is vaporized. The remains are mineral matter (ash) in several forms such as friable ash and clinkers. In practice, some char (unburned carbon) will always be present in the ash. The combustion of part of the carbon is the main driving force of gasification and supplies almost all the heat necessary to sustain the endothermic reactions that take place in the reduction and distillation zones. The reader is cautioned that the above equation does not describe the physical and chemical processes on a microscale. Several authors (4,6,7,9,12,13,15,17,18,19,20) have put a great deal of effort into examining combustion on a microscale. The results are not presented because of the highly theoretical nature of these observations and the apparent disagreements.

The introduced air contains, besides oxygen and water vapor, the inert gases in air such as nitrogen and argon. Nitrogen and argon are for simplicity assumed to be non-reactive with the fuel constituents. However, the water vapor reacts with the hot carbon according to the heterogeneous reversible water gas reaction:

$$C + H_2O = H_2 + CO - 131,400 \text{ kJ (at } 25^\circ C, 1 \text{ atm).}$$

In this reaction 12.01 kg of carbon reacts with 22.40 SCM of water vapor to yield 22.34 SCM of hydrogen, 22.40 SCM of carbon monoxide and 131,400 kJ of heat is absorbed in this chemical reaction.

A schematic temperature distribution through a vertical cross section of an updraft gas producer is shown in Figure 7. The highest temperature reached is not shown in the diagram and depends on the design, fuel gasified and mode of operation. Prevailing gas temperatures in the oxidation zone are in the range of $1000^\circ C$ to $1600^\circ C$.

In order to understand the sometimes confusing results and observations, the overall reaction can be divided into two basically different partial processes. The physical process is referred to as mass exchange or mass transport which transports one reactant to the other. This process is certainly a necessary condition to trigger the second chemical process, the reaction itself. The mass transfer is by diffusion and convection and therefore, depends mainly upon factors characteristic of the gas flow and the fuel such as, fuel surface, particle size and bulk density. The overall process described by the chemical equations previously mentioned is limited by either the mass transport or the chemical reaction rates. For instance, the combustion of carbon to carbon dioxide is a very fast chemical reaction and the process is probably limited by insufficient mass transport. The immensely high chemical reaction speed cannot be fully effective because it is not possible for the relatively slow oxygen transport to not even roughly keep pace (10).

Principal reactions that take place in the reduction and distillation zone are:

a. The Boudouard reaction: $CO_2 + C = 2 CO - 172,600 \text{ kJ (at } 25^\circ C, 1 \text{ atm).}$

This highly endothermic reaction generates 44.80 SCM of combustible CO out of 12.01 kg of carbon and 22.26 SCM of noncombustible CO_2 while absorbing 172,600 kJ of energy.

b. The water shift reaction: $CO_2 + H_2 = CO + H_2O + 41,200 \text{ kJ}$
$$\text{(at } 25^\circ C, 1 \text{ atm).}$$

This reaction relates the water gas reaction and the Boudouard reaction and is weak exothermic.

c. The simplified form of methane production:

$$C + 2 H_2 = CH_4 + 75,000 \text{ kJ (at } 25^\circ C, 1 \text{ atm).}$$

This, also weak exothermic reaction generates 22.38 SCM of methane out of 12.01 kg of carbon and 44.86 SCM of hydrogen while releasing 75,000 kJ of heat.

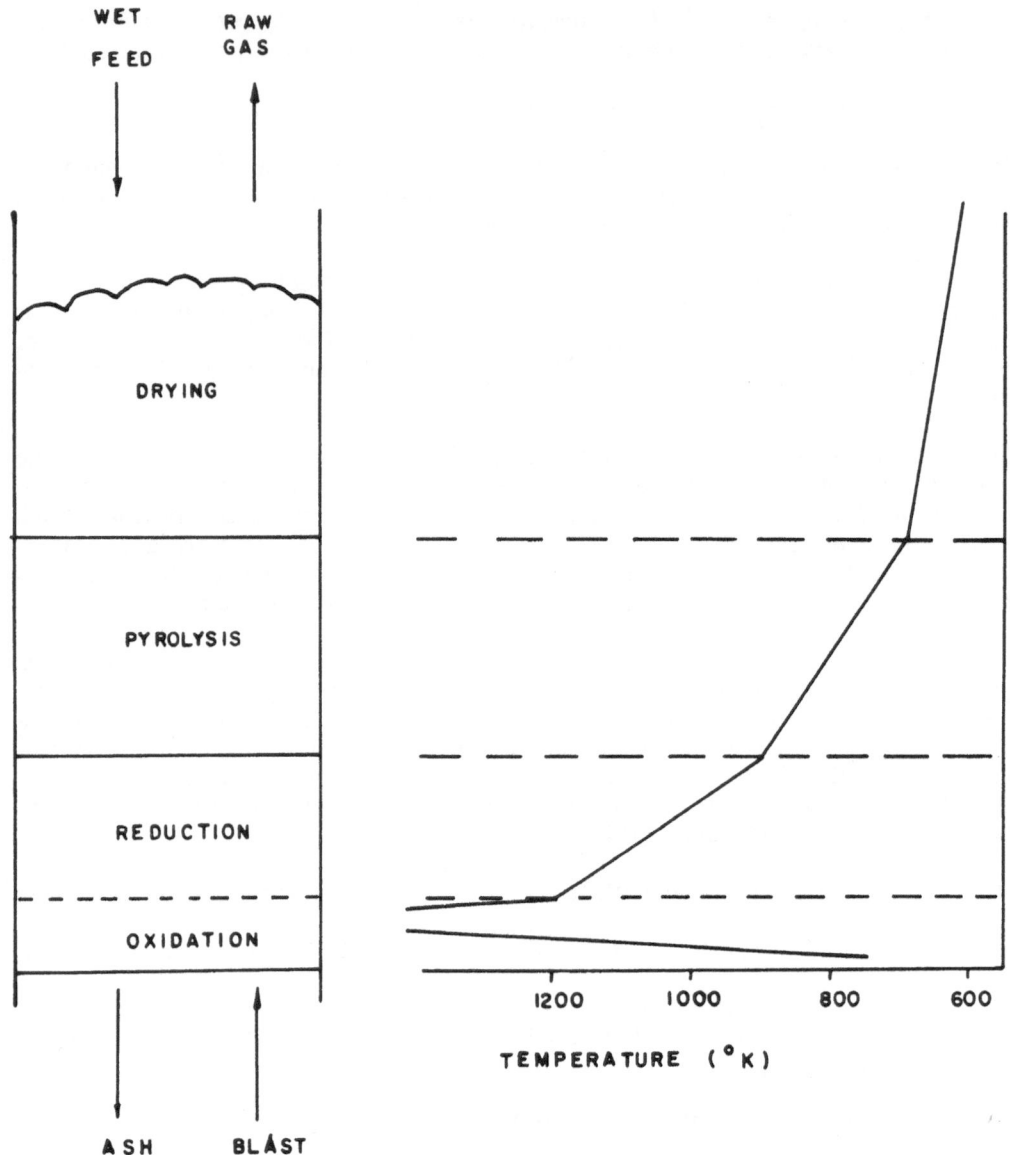

Figure 7.   Temperature Distribution in an Updraft Gas Producer (14).
Oxidation and Partial Combustion are used as synonomous terms.

Obviously the distillation, reduction and partial combustion zones are overlapping and not strictly separated in a physical sense. The previously described five equations, although the major ones, do not represent gasification as a whole. For instance, the mineral matter in biomass fuels and coal reacts as well. Some

28

of it becomes vaporized and oxidized and leaves the gas producer in gaseous form. Moreover, the gaseous products and vapors from the distillation zone are an extremely complex conglomerate of at least 200 constituents. They will mix freely with the gaseous products from the other reaction zones and make any comparison of actual data with calculated data a rather tedious undertaking. The steady decrease in temperature through the vertical cross section of an updraft gas producer raises the question: Why doesn't the exothermic methane formation provide the reduction zone with a temperature floor? However, at the prevailing temperatures of $200^{\circ}C$ to $500^{\circ}C$ in the distillation zone, the methane generation is too slow to control the temperature (14). In general, at low temperatures of $500^{\circ}C$ the chemical reaction speeds are insufficient for obtaining an equilibrium under conditions present in gas producers. At temperatures above $700^{\circ}C$ the physical reaction resistance caused by slow mass transport compared to the increasingly high chemical reaction speed will control the process. The question whether the five main reactions listed below will attain their equilibrium state in a gas producer has been the main issue in attempting to relate actual gasification data to calculated data.

(Combustion) $\qquad C + O_2 = CO_2 + 393,800$ kJ/kg mole

(Water gas) $\qquad C + H_2O = CO + H_2 - 131,400$ kJ/kg mole

(Water shift reaction) $\quad CO + H_2O = CO_2 + H_2 + 41,200$ kJ/kg mole

(Boudouard reaction) $\quad C + CO_2 = 2\ CO - 172,600$ kJ/kg mole

(Methane reaction) $\qquad C + 2\ H_2 = CH_4 + 75,000$ kJ/kg mole

In this context, equilibrium may be defined as the limit state toward which the reaction proceeds when given enough time. The nature of the chemical reactions and the definition of the equilibrium state certainly excludes any oscillatory movement around the equilibrium and in addition does not imply the existence of such a state in all cases or reaching it in finite time. This seemingly philosophical statement explains some of the discrepancies reported by several authors. The question of equilibrium is more related to physical properties of the fuel and gasifier design such as depth of fuel bed, size and grading of the feed material and gas velocity and not so much to the temperature. Estimated data for the depth of the oxidation zone range from 0.1 cm to 15 cm. This wide range is easily explained, as the depth of the partial combustion zone strongly depends on the fuel size. Experiments have shown that the depth of the zone can be predicted as being equal to (2.66)(average particle size) (14). The depth of the reduction zone can be assumed to be 80 to 100 cm in large plants. The fact that for the Boudouard reaction to reach equilibrium under laboratory conditions needs hours of time, whereas, the residence time of the gas in a gas producer is only a fraction of a second, has led many authors to prematurely conclude that an approach to equilibrium cannot be expected. Figures 8 and 11 show the original curves obtained by Boudouard. The conversion of CO_2 into CO at $800^{\circ}C$ and $650^{\circ}C$ using carbon in the form of wood charcoal is much faster at $800^{\circ}C$ and reaches its equilibrium after one hour under the given laboratory conditions. At $650^{\circ}C$ no asymptotic behavior of the curve seems to be apparent after 12 hours. The degree of conversion of CO_2 into

CO is also much lower at 650°C. The influence of the fuel on the conversion of CO_2 into CO is demonstrated by Boudouard's second set of curves. In the case of wood charcoal with its high porosity and large accessible surface area, the equilibrium is obtained much faster than in the case of high temperature coke with few pores and small accessible surface area. With today's knowledge of gasification kinetics it is of course easy to verify Boudouard's experimental results in much more detail.

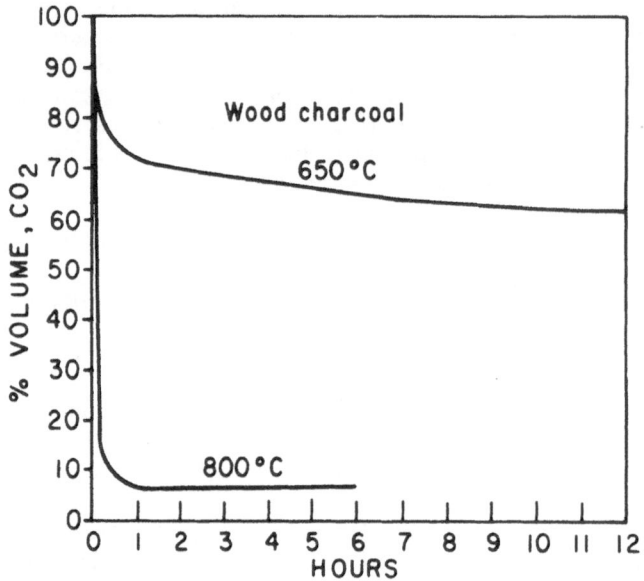

Figure 8. Influence of Temperature on the Conversion of CO_2 to CO (3).

The conflicts of opinion or interpretation are mostly based on an uncritical application of laboratory tests to commercial scale gas producers and the misinterpretation of temperature measurements in gas producers. In order to have some justification as to why a mathematical treatment of gasification is highly valuable in understanding the chemical processes, it seems worthwhile to look into some of the common mistakes made in comparing data.

1. Law of similarity: CO_2 conversion into CO under laboratory conditions can not be compared to actual data as long as the law of similarity is disobeyed as it has been done quite often in the past. For instance, reduction of CO_2 to CO with carbon particles of average size 5 mm in a 15 mm tube has no relevance whatsoever to actual gas producer practice. Such an experiment would roughly represent the gasification of 60 cm coke nuts in a gasifier of 2 m diameter.

2. Misinterpretation of temperature measurements: The reactants in the gas phase are assumed to have a "finite" reaction time and consequently require a specific path length or reaction space within the fuel column in order to reach

30

the equilibrium state. Only after passing through the needed reaction space can they reach equilibrium. The temperature that corresponds to this state is obtained through the energy balance under conditions which represent this final state. Temperature is clearly a function of time and location and the temperature change of the gas phase is much more drastic than those of the solid phase, due to the endothermic reactions which mainly influence the gas phase. There will also be a significant temperature change at the phase boundaries. This phenomena is illustrated on a microscale in a proposed double film model of the boundary layer around a carbon particle, as shown in Figure 9. The temperature difference between the phases on a macroscale as a function of the location in the fuel column is shown in Figure 10, where the reaction temperature, T_R, is arbitrarily defined as the equilibrium temperature at the end of the reduction zone which also is identical with the surface temperature of the fuel particles.

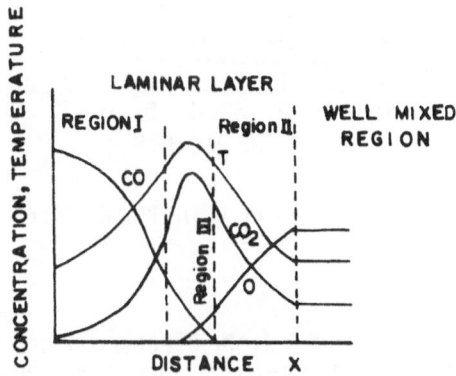

Figure 9.   Schematic Concentration and Temperature Profiles in the Double Film Model (2).

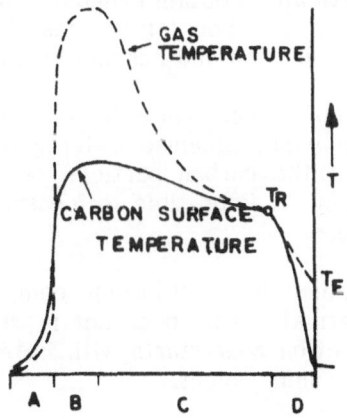

Figure 10.   Temperature of Solid and Gas Phase in a Gas Producer (10). A - ash zone, B - partial combustion zone, C - reduction zone, D - distillation zone, T_R - reaction temperature, and T_E -exit gas temperature.

31

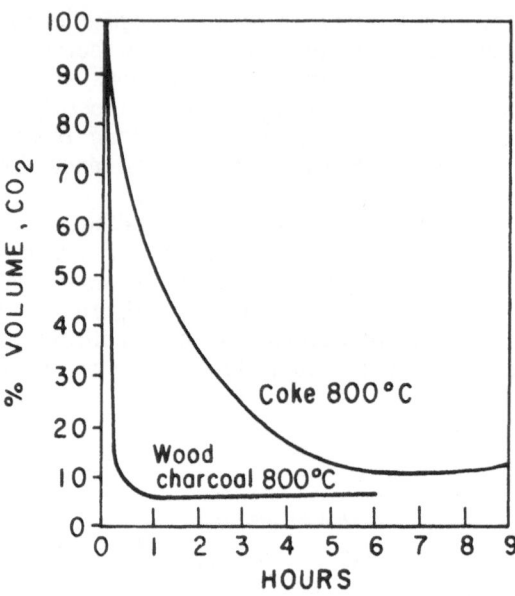

Figure 11. Conversion of CO_2 to CO With Wood Charcoal and Coke (3).

The analysis of gas samples taken from a gas producer at various heights and simultaneous measurements of the temperature when compared to the computed equilibrium curve at this temperature may or may not agree. The results are in no way any contribution to answer the question whether equilibrium is reached. The gas may have been sampled at points where the chemical reaction is still in process and not completed. Moreover, even with today's advanced measurement techniques it is extremely difficult to obtain reliable "true" temperature measurements. Temperatures obtained are those of the gas phase altered by the usual errors caused by radiation, convection and conduction for the temperature probe.

Where the assumed equilibrium temperature in heterogeneous reactions occurs and how to measure it are unsolved problems. Heterogenous gasification reactions take place at the surface of the carbon particle, or in the vicinity of a very thin boundary layer which makes it impossible to measure this temperature under actual gasification conditions.

3.   Experiments to determine the equilibrium composition under laboratory conditions are mostly isothermal. This does not represent the conditions in a gas producer. Here the reduction zone starts with initial high temperatures and high concentration of the reducing agent.

At the present state of knowledge it seems justified to postulate that the equilibrium state of the four major chemical reactions in a gas producer are reached to a high degree. This is particularly true for updraft gas producers that develop a sufficient depth in the reduction zone. Consequently it is beneficial and illustrative to present a mathematical treatment of gasification based under the assumption of equilibrium of the four major reactions in the overlapping reduction and partial combustion zone. However, this description can not take into account reactions occurring in the distillation zone which are highly unstable and complex and do not tend toward an equilibrium. The reader should keep in mind that these products mix with the products of gasification and will show up in the overall gas analysis.

The two most common methods to describe the physical reaction and the equilibrium composition of the four major reactions are: (1) the equilibrium curves calculated under the assumption of no dissociation and (2), the use of the mass action coefficient curves. The total differential, dG, of the Gibbs function $G = H - TS$ equals zero at this state. This also means that the graph of G attains its minimum at this point as shown in Figure 12.

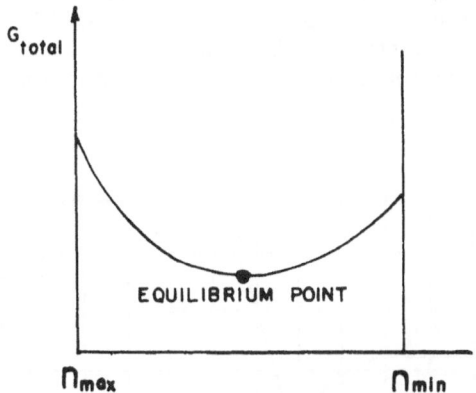

Figure 12. Behavior of Gibbs Function at Equilibrium (11). n_{max} -initial moles of a reactant and n_{min} - final moles of same reactant.

Figure 15 shows the calculated equilibrium curve for the Boudouard's reaction at 1 atm. This Figure indicates that at a temperature of 650°C only about 40% of the CO_2 is converted into CO, a result that agrees with Boudouard's experiment shown in Figure 8. The graph also shows that high temperatures favor CO generation, but one has to keep in mind that this highly endothermic reaction is mostly sustained by the heat released through combustion of some of the carbon. Consequently, the temperature drop of the gas phase will be considerable through the reduction zone and in practice not all of the CO_2 generated in the partial combustion zone will be converted. Large stationary gas producers which usually come reasonably close to an equilibrium state have very little CO_2 in the raw gas (less than 1% under favorable conditions) because of their extended reduction zone, the long residence time of the gas and the gradual decline of temperature. However, small, portable units, especially downdraft gas producers, can yield considerable amounts of noncombustible CO_2 in the raw gas. This is mainly due to the extremely short residence time of the gas, and moderate temperatures combined with a small reduction zone.

33

The amount of CO_2 in the raw gas does not represent the fraction that escaped the reduction process. The distillation products in the raw gas also contain CO_2 and as shown in Tables 40 to 42 this can be considerable. A CO_2 content of more than 3% in the raw gas of updraft gas producers has been usually attributed to a poorly constructed or carelessly operated producer. It is either an indication that CO_2 is not well reduced or CO has been oxidized through air leaking into the reactor vessel. Not only is the CO_2 a diluent, but the additional oxygen required for its formation will increase the amount of inert nitrogen in the gas and thus further reduce the heating value per unit volume of producer gas.

Figure 13 shows the calculated composition of gasification of carbon with dry air assumed to contain 21% oxygen and 79% nitrogen. The only combustible gaseous product is CO. Although this graph is only of theoretical interest since there is usually plenty of hydrogen in the fuel as well as in the air blast, it shows quite clearly the importance of high temperatures for conversion of carbon into CO. As indicated in the Figure, not more than 35% CO can be evolved and in practice the CO content of the raw gas is well below this figure, due to the formation of H_2 in the water shift reaction and the water gas reaction.

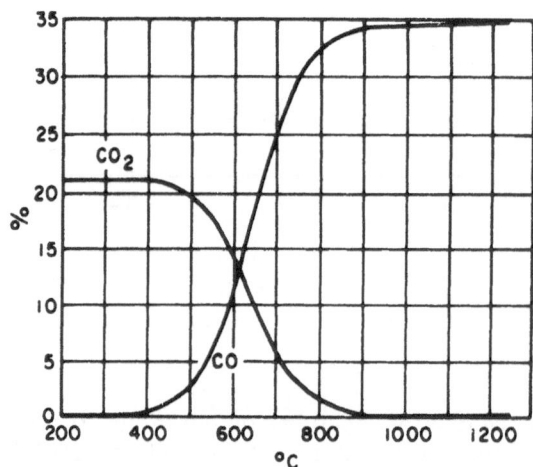

Figure 13. Air-Gas Composition of Gasification of Carbon at 1 atm (10).

In considering the hydrocarbon component of producer gas, especially that generated from biomass, the notion of cracking these components in the very hot carbon bed is introduced. No technical literature was found that substantively dealt with this notion. Thus, that it happens will be left as "art" of the gasification process.

Contrary to the CO formation, the exothermic methane formation:

$$C + 2\ H_2 = CH_4 + 75,000\ kJ\ (at\ 25^{\circ}C,\ 1\ atm).$$

is favored by low temperatures as shown in Figure 14. The above equation does not really describe the actual formation of methane, because methane could be as well formed according to the equations:

$$CO + 3 H_2 \rightleftharpoons CH_4 + H_2O \quad \text{or} \quad CO_2 + 4 H_2 \rightleftharpoons CH_4 + 2 H_2O$$

Although the latter two reactions are less likely to occur at low pressure and require a catalyst in order to be important, it is nevertheless wrong to assume that CH_4 is only a product of the distillation zone. CH_4 has been found in parts of gas producers where no volatile matter could any longer exist. Moreover laboratory experiments show evidence that at sufficiently high hydrogen partial pressure, virtually all of the carbon not evolved during distillation can be gasified quickly to methane. Unfortunately the very low pressure, around 1 atm in air-blown gas producers, is not suitable for a high methane yield. The present state of knowledge does not provide any final answer to where and how methane is formed.

Besides the usual assumption of CH_4 formed as a product of distillation simultaneously with the rest of the distillation products, one could as well postulate a distillation stage followed by a rapid rate methane formation and a low rate gasification. The fact that the methane formation occurs at a much slower rate than devolatilization justifies this approach. However, at temperatures above $1000°C$ methane cannot exist.

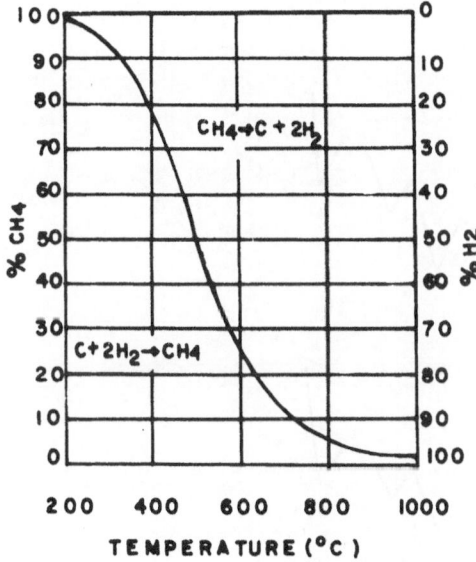

Figure 14. Equilibrium Curve for Methane at 1 atm (10).

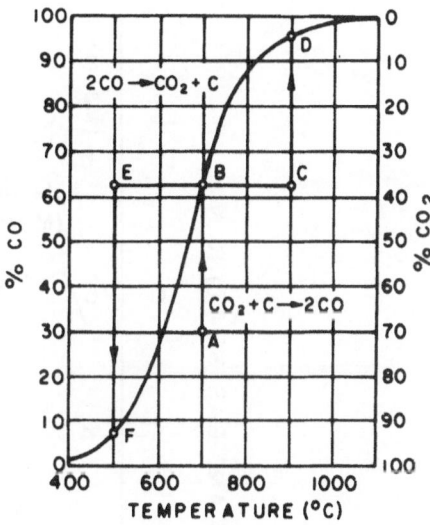

Figure 15. Equilibrium Curve of Boudouard's Reaction at 1 atm (10).

In practice a high CH_4 content in the raw gas is most desirable because of the high heating value of methane. From Table 1, which lists the higher heating values of the main combustible products in producer gas, it can be seen that even small amounts of methane in the gas can considerably raise the heating value.

Table 1.   Higher Heating Values of the Constituents of Producer Gas (11).

Gas	Higher heating value kJ/kg mol at 25°C
Hydrogen, H_2	285,840
Carbon monoxide, CO	282,990
Methane, CH_4	890,360
Ethane, C_2H_6	1,559,900

How well the CH_4 formation and the heating value follow the temperature is illustrated in Figure 16 which shows the continuous gas analysis of a downdraft gas producer fueled with densified waste paper cubes and municipal sludge at the University of California, Davis. During start up time when temperatures are low throughout the gasifier and during the batch fuel load period when the air blast is shut off, the methane formation and with it the heating value of the gas increased considerably.

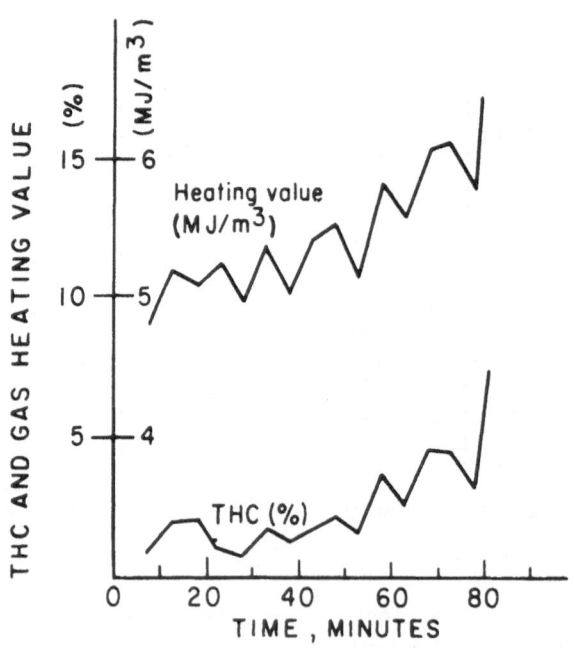

Figure 16.   Gas Composition and Energy Content of Producer Gas as a Function of Time.

It should not be concluded that low temperature gasification may be a method to increase the heating value of the gas. Low temperatures in the partial combustion zone prevent a downdraft gasifier from cracking the tarry products and therefore generate an unsuitable gas for further use in internal combustion engines or burners fed with ambient primary combustion air. In fact to run an engine on gas produced during start up time is one of the most serious operational mistakes.

In practical calculations, the amount of noncondensable hydrocarbons in the gas is measured as THC (total hydrocarbons) where practice has shown that 95% CH_4 and 5% C_2H_6 is a good approximation for total hydrocarbons in the raw gas. The amount of THC in the gas may be as low as 0.1% and occasionally above 10% on a dry basis depending on the type of gas producer and its thermodynamical state. Tests at the University of California, Davis with 26 crop and wood residues in a downdraft gas producer yielded the lowest THC value of 2.9% for peach pits and the highest value of 9% for olive pits (8).

Although the term dry gasification usually refers to gasification without additional steam injection, there will be plenty of moisture in the air and biomass fuel to trigger the water gas reaction:

$$C + H_2O = CO + H_2 - 131,400 \text{ kJ (at } 25^{\circ}C, 1 \text{ atm).}$$

This strong endothermic reaction together with the water shift reaction balances the CO and H_2 formation. With respect to the heating value of the gas from the reduction zone, only the sum of H_2 and CO in the raw gas is of interest, because both contituents have roughly the same heating value as shown in Table 1. Pure water gas can be practically obtained by alternately blowing with

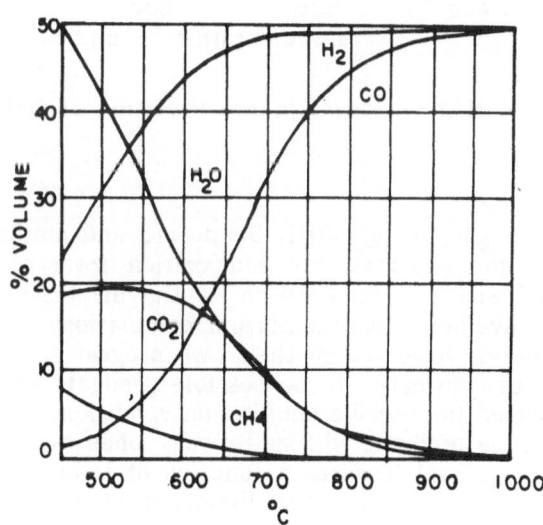

Figure 17. Composition of Water Gas from Carbon at 1 atm (10).

air and steam to produce enough heat for the subsequent steam injection. Figure 17 shows the calculated composition of pure water gas from carbon at various reaction temperatures on a wet basis. In this context, reaction temperature refers to the equilibrium state where the CO_2 concentration in the gas phase equals that of the phase boundary and in addition, no temperature differential exists across the phase boundaries. This state may be physically realized and defined as the end of the reduction zone.

The graph illustrates the drawback of gasification with too much moisture provided either by the fuel or through the air blast. The strongly endothermic reaction will quickly lower the fuel bed temperature and consequently a considerable amount of undecomposed steam will be present in the gas which makes it hard to ignite and lowers the heating value of the raw gas.

How much undecomposed steam leaves the gas producer depends on the temperature. Figure 18 shows this dependence and how much H_2 is generated.

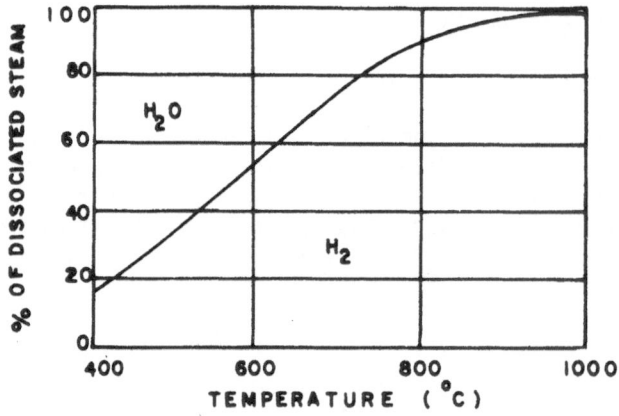

Figure 18. Decomposition of Steam in the Presence of Hot Carbon (16).

Combining the water gas, water shift, Boudouard and methane reaction allows a precalculation of the expected gas composition from an ultimate chemical analysis of the fuel and the composition of the air blast. There are a few computer programs available for equilibrium calculations and many researchers dealing with gasification have set up their own programs. The programs differ in the kind of species considered to be possible products and reactants and the basic equations assumed to describe equilibrium conditions. Although the quantitive analysis of possible products will surely vary when using different programs, they all describe the general trend as a function of temperature, or equivalence ratios of various reactants. It seems of little importance what kind of program is used in the design of a gas producer and prediction of limits of the various constituents of the raw gas. Particularly, if one keeps in mind the assumption of equilibrium which is underlaying all programs known to us. All equilibrium curves should be treated with great caution below 500°C. In general at lower temperatures an effect called "freezing the gas composition" takes place. For instance, cooling down the gas from an equilibrium state at 700°C to 500°C should result in heavy soot formation according to the reaction:

$$2 \ CO \rightarrow CO_2 + C$$

In practice this soot formation reaction has not been observed to occur to a great extent since the chemical reaction becomes very slow and stops altogether. Figure 19 shows the calculated gas composition and the energy content as a function of temperature. Equilibrium is assumed for the water gas, methane and water shift reaction in an adiabatic reactor. Computations are carried out within the H-C-O-N system disregarding the chemical composition of the fuel and assuming an H/O ratio of less than two.

A slightly different approach to equilibrium calculations is the equivalence ratio:

$$ER = \frac{weight \ of \ oxidant/weight \ of \ dry \ fuel}{oxidant/fuel \ (stochiometric \ weight \ ratio)}$$

This rather arbitrary definition is more significant than the temperature as a parameter when evaluating gasification processes. Its usefulness lies in the fact that gas composition, heating value, adiabatic flame temperature, "useful" chemical energy and "not so useful" sensible energy in the gas can be viewed as a function of variables such as temperature and air to fuel ratio or ER. When expressing the above properties of a gas as a function of the equivalence ratio, which is a normalized, dimensionless parameter, one can show that maxima and minima as well as inflection points of the various curves occur all at about ER = 0.255 in the case of wood gasification. This establishes the ER as a more natural parameter. ER = 0 corresponds to thermal decomposition without external oxygen introduction (pyrolisis or distillation). The other extreme of ER = 1 or larger corresponds to complete combustion with 100% theoretical air or excess air. The ER for gasification processes as they take place in practice lies between those two extrema and within a range of 0.2 to 0.4 for steady state operation. This range refers to the partial combustion zone of the gasification process. One should keep in mind that there is also a distillation zone in gasifiers which cannot be avoided. Occasionally a high ER, close to 1, is noticed in gasifiers which means that the unit is malfunctioning due to bridging of the fuel or clinker formation. A sharp increase in temperature in the lower fuel zone, indicates complete combustion within parts of the unit. Figure 20 shows the various chemical processes and the adiabatic flame temperature as a function of the equivalence ratio for wood gasification.

The total energy in producer gas is the sum of its sensible heat plus the chemical energy. In most applications the sensible heat is lost because the gas has to be cooled down and water, tar, and oil vapors are condensed out of the gas stream. The total energy is therefore no practical indicator to what extent the gas if useful in practice. This is shown in Figure 21 for dry gasification of dry wood at 1 atm. The chemical energy of the gas reaches its maximum at an ER = 0.275.

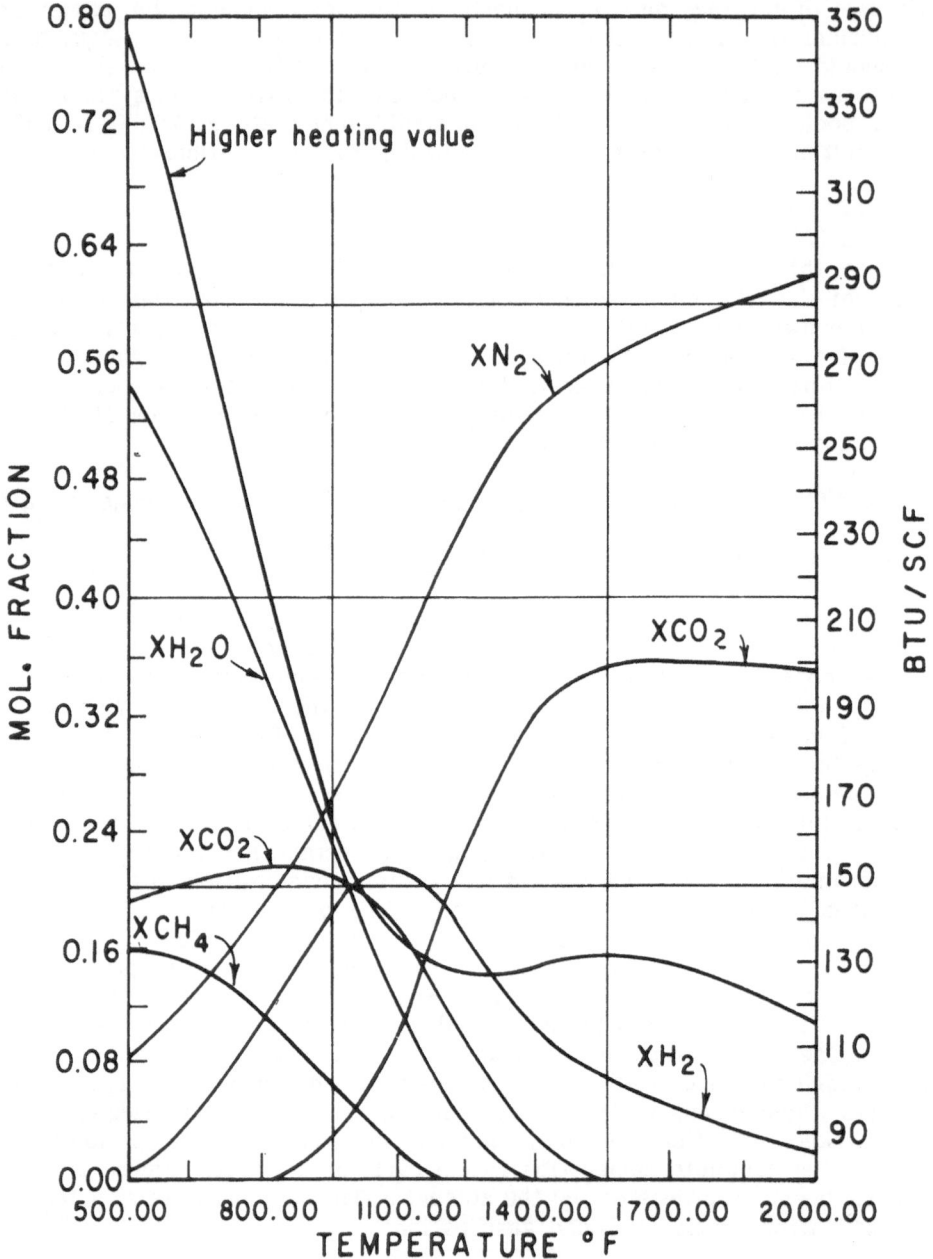

Figure 19. Equilibrium Composition and Energy Content of Producer Gas at 1 atm (1).

40

Increasing the ER closer to combustion will cause a rapid decrease in useful chemical energy and an increase in temperature and sensible heat of the gas. The total energy however will roughly be the same.

Unfortunately the ER is not adjustable to as wide a range as desired. Two obvious ways to change the ER are to change the combustion air rate or the cross sectional area of the tuyeres. Both methods will physically expand or contract the partial combustion zone and also influence the temperature, therefore, nullify part or all of the additional oxygen available in this zone. In addition, temperature is a function of the ER and in practice the control of the temperature is important for operational reasons and has priority.

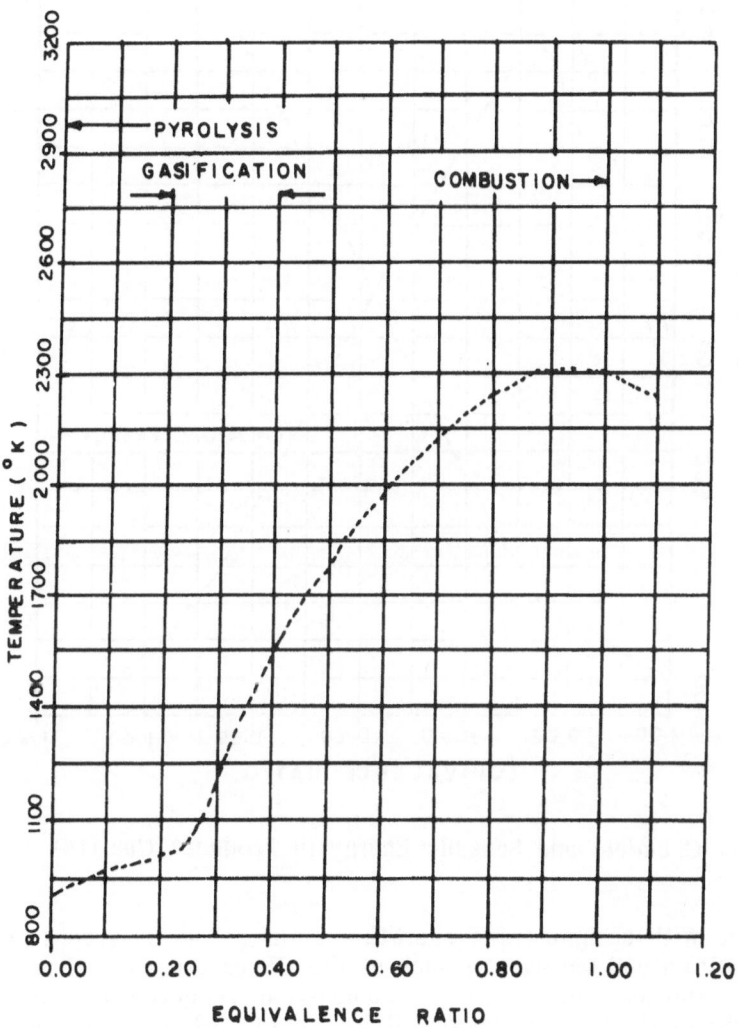

Figure 20. Equivalence Ratio and Adiabatic Flame Temperature for Wood Gasification (14).

A typical computed composition of wood gas at 1 atm is shown in Figure 22. It can be seen that the highest CO content and lowest CO_2 content are obtained at an ER of 0.255.

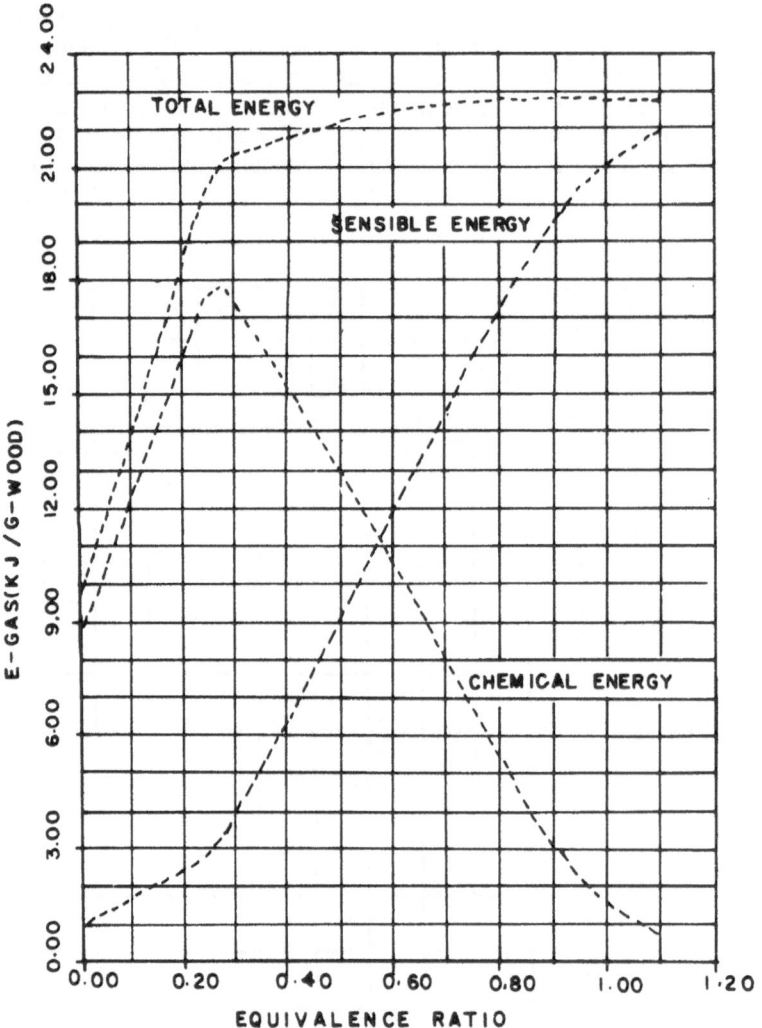

Figure 21. Chemical and Sensible Energy in Producer Gas (14).

When faced with designing and operating a gas producer, graphs predicting the gas composition and parameters such as the ER or the air-fuel ratio as well as arbitrary definitions such as partial combustion, reduction and distillation zone are helpful in setting limits and visualizing the overall complex process. Combining this purely physical knowledge with more detailed data about the chemical nature of the products obtained from the various zones in a gasifier, can show that a generalized optimization of the process is theoretically impossible. This

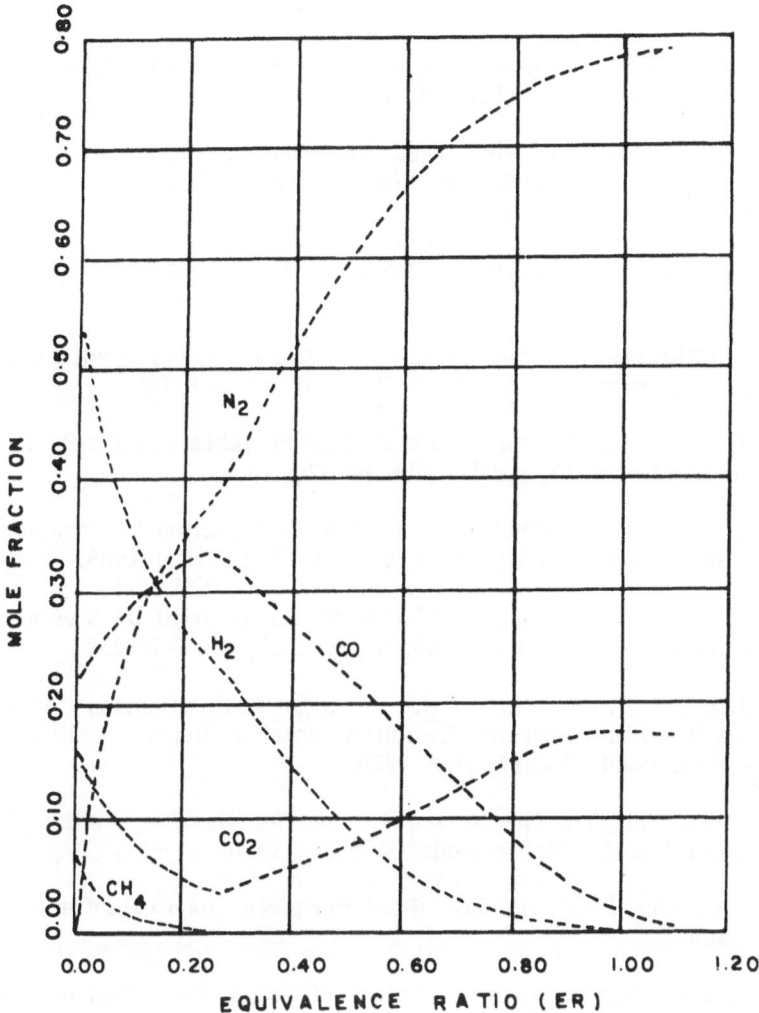

Figure 22. Computed Composition of Wood Gas at 1 atm (14).

leaves the designer of a gas producer with the difficult task to decide which parameters are most important for a particular fuel and use of the gas and how they can be controlled. Based on the chemical composition of the fuel and air blast one can always predict a gas composition although the input data tells very little about the suitability of the feed material for gasification. In fact it will be shown later that seemingly unimportant physical properties of the feed material are even more important with regard to a successful gasification.

43

Reference Chapter III

1. Baron, R.E., **Chemical Equilibria in Carbon-Hydrogen-Oxygen Systems**, MIT Press, Cambridge, Massachusetts, 1977.

2. Caram, H.S. and N.R. Amundson, Diffusion and Reaction in a Stagnant Boundary Layer about a Carbon Particle, Part 1, Ind. Eng. Chem. Fundam., v 16, n 2, 1977, pp 171-181.

3. Dowson, J.E. and A.T. Larter, **Producer Gas**, Longmans Green and Co., London, 1907.

4. Ergun, S., Kinetics of the Reaction of Carbon Dioxide with Carbon, The Journal of Physical Chemistry, v 60, 1956, pp 480-485.

5. Goldman, B. and N.C. Jones, The Modern Portable Gas Producer, Institute of Fuel, London, v 12, n 63, 1939, pp 103-40.

6. Goring, G.E., et al., Kinetics of Carbon Gasification by Steam, Industrial and Engineering Chemistry, v 44, n 5, 1952, pp 1051-1065.

7. Goring, G.E., et al., Kinetics of Carbon Gasification by Steam, Industrial and Engineering Chemistry, v 45, n 11, 1953, pp 2586-2591.

8. Goss, J.R., An Investigation of the Down-Draft Gasification Characteristics of Agricultural and Forestry Residues: Interim Report, California Energy Commission, P500-79-0017, Nov 1979.

9. Groeneveld, M.J., The Co-Current Moving Bed Gasifier, Thesis, Technische Hogeschool Twente, Netherlands, May 29, 1980.

10. Gumz, W., **Gas Producers and Blast Furnaces**, John Wiley and Sons, New York, 1950.

11. Holman, J.P., **Thermodynamics**, Third Edition, McGraw-Hill Book Co., New York, 1980.

12. Howard, J.B., et al., Kinetics of Carbon Monoxide Oxidation in Postflame Gases, 14th Symposium on Combustion, Penn. State Univ., August 1972, pp 975-986.

13. Johnson, J.L., Kinetics of Bituminous Coal Char Gasification with Gases Containing Steam and Hydrogen, Symposium on Coal Gasification, Division of Fuel Chemistry, Meeting of A.C.S., Dallas, Texas, April 1973.

14. Reed, T.B., A Survey of Biomass Gasification, Volume 2, Principles of Gasification, Publication # SERI/TR-33-239, Solar Energy Research Institute, Golden, Colorado, 1979.

15. Schläpfer, P. and J. Tobler, **Theoretische und praktische Untersuchungen über den Betrieb von Motorfahrzeugen mit Holzgas**, Schweizerische Gesellschaft für das Studium der Motobrennstoffe, Bern, Switzerland, 1937.

16.   Skov, N.A. and M.L. Papworth, **The Pegasus Unit**, Pegasus Publisher Inc., Olympia, Washington, 1975.

17.   Wen, C.Y. and J. Huebler, Kinetic Study of Coal Char Hydrogasification Rapid Initial Reaction, Industrial and Engineering Chemistry Process Design and Development, v 4, n 2, 1965, pp 142-147.

18.   Wen, C.Y. and J. Huebler, Kinetic Study of Coal Char Hydrograsification Second-Phase Reaction, Industrial and Engineering Chemistry Process Design and Development, v 4, n 2, 1965, 147-154.

19.   Zielke, C.W. and E. Gorin, Kinetics of Carbon Gasification, Industrial and Engineering Chemistry, v 47, n 4, 1955, pp 820-825.

20.   Zielke, C.W. and E. Gorin, Kinetics of Carbon Gasification, Industrial and Engineering Chemistry, v 9, n 3, 1957, pp 396-403.

CHAPTER IV: GAS PRODUCERS

This chapter deals only with small and medium sized fixed bed gas producers with the oxygen for partial combustion supplied from ambient air. There has been no significant development in the design of these gas producers for the last 50 years. Today's gas producers are built out of better heat resisting material such as high temperature alloys and longer lasting refractories but the design itself has shown very little change over the past century. The dramatic advancement in understanding combustion and transport phenomena in gases has certainly not changed the engineering principles of gasification nor contributed anything important to the design of a plant. However, it has provided a microscale understanding of the gasification process and its sensitivity to minor changes in the gas producer geometry, fuel size and general operation. Its sensitivity, known quite well during the booming years of gasification, has resulted in detailed operating manuals in particular for large plants, where a shut down is much more serious than in smaller or portable plants. The general rule was that a well-designed gasifier is as good as the man who operates it and this principle seems to still be valid.

A small-sized gas producer is a very simple device, consisting usually of a cylindrical container filled with the fuel, an air inlet, gas exit and a grate. It can be manufactured out of fire bricks and steel or concrete and oil barrels (6, 12, 22). If properly designed and operated the plant is highly reliable and does not require maintenance other than the periodical removal of ash, char and clinkers. The design of a gas producer depends mainly on whether it is stationary or portable and the fuel to be gasified. Portable gasifiers mounted on trucks and tractors need to operate under a wide range of temperatures and load conditions, whereas stationary units used for heating, generation of electricity or pumping water operate under a steady load in most cases. It is in any case highly desirable to generate a clean gas leaving the producer at a moderate temperature and containing as little moisture as possible. These conditions, which guarantee a high efficiency and reliable operation are difficult to achieve. Moreover, the choice of fuel dictates the mode of running the gas producer and greatly influences the type of difficulties to be expected. Gas producers are mainly classified according to how the air blast is introduced into the fuel column. Most gas producers have been downdraft or updraft. Their evolution for the last 140 years has been guided by typical gas producer fuels such as coal, wood charcoal and wood and the use of the plant for propelling an automobile or generating electricity. Unfortunately basic thermodynamic laws prevent designing a gas producer that is optimal in all respects. In practice a decision has to be made as to what the most desirable property of the gas and the plant should be. High efficiency, tar free gas and excellent load following capabilities are desirable properties that contradict each other thermodynamically and cannot be simultaneously optimized. A gas producer mounted on an automobile should have a good load following capability and generate a tar free gas which leaves the gas producer as cold as possible. Producer gas combusted in a burner can have a high temperature and a high tar content as long as it is burned at a gas-combustion air temperature above the condensing point of the tar vapors.

The Updraft Gas Producer: An updraft gas producer has clearly defined zones for partial combustion, reduction and distillation. The air flow is countercurrent to the fuel flow and introduced at the bottom of the gas producer. The gas is drawn off at a higher location as shown in Figure 23.

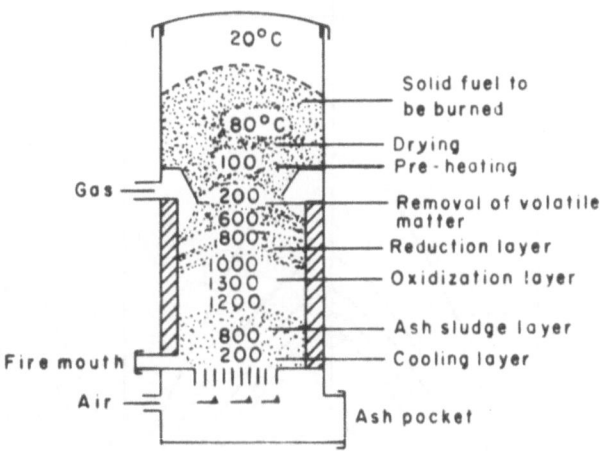

Figure 23. Diagram of Updraft Gasification (27).

The updraft gas producer achieves the highest efficiency because the hot gas passes through the entire fuel bed and leaves the gas producer at a low temperature. The sensible heat given up by the gas is used to dry and preheat the fuel before it reaches the reduction zone and is therefore not lost. A typical temperature profile of a small updraft gas producer which has the gas exit at the very top is shown in Figure 24. Products from the distillation and drying zone consists mainly of water, tar and oil vapors and are not passed through an incandescent hot carbon bed. They therefore leave the gas producer uncracked and will later condense at temperatures between $125^\circ C$ -$400^\circ C$. A common updraft gasifier with the gas outlet at the very top is therefore unsuitable for high volatile fuels when tar free gas is required. To overcome this handicap updraft gasifiers have been built with funnels to draw off the gas at the middle of the gas producer. Other methods like recycling the distillation gases through the hot carbon bed at the bottom or burning in an external combustor and feeding the products back into the air blast will be discussed in Chapter VI. Most updraft gas producers are operated with a wet air blast to increase the gas quality and keep the temperature below the melting point of the ash. Important points in the design of an updraft gas producer are:

1. The method of the air feed
2. The position of the gas exit
3. The type and size of the grate
4. Means of vaporizing water for the wet air blast
5. Fire box lining
6. The expected specific gasification rate
7. The height of the fuel bed

There have been very few updraft gas producers on the market for propelling an automobile because of the excessive amount of tar in the raw gas and the poor load following capabiblity. All successful commercial updraft units drew off the gas right above the reduction zone and in most cases were fired with

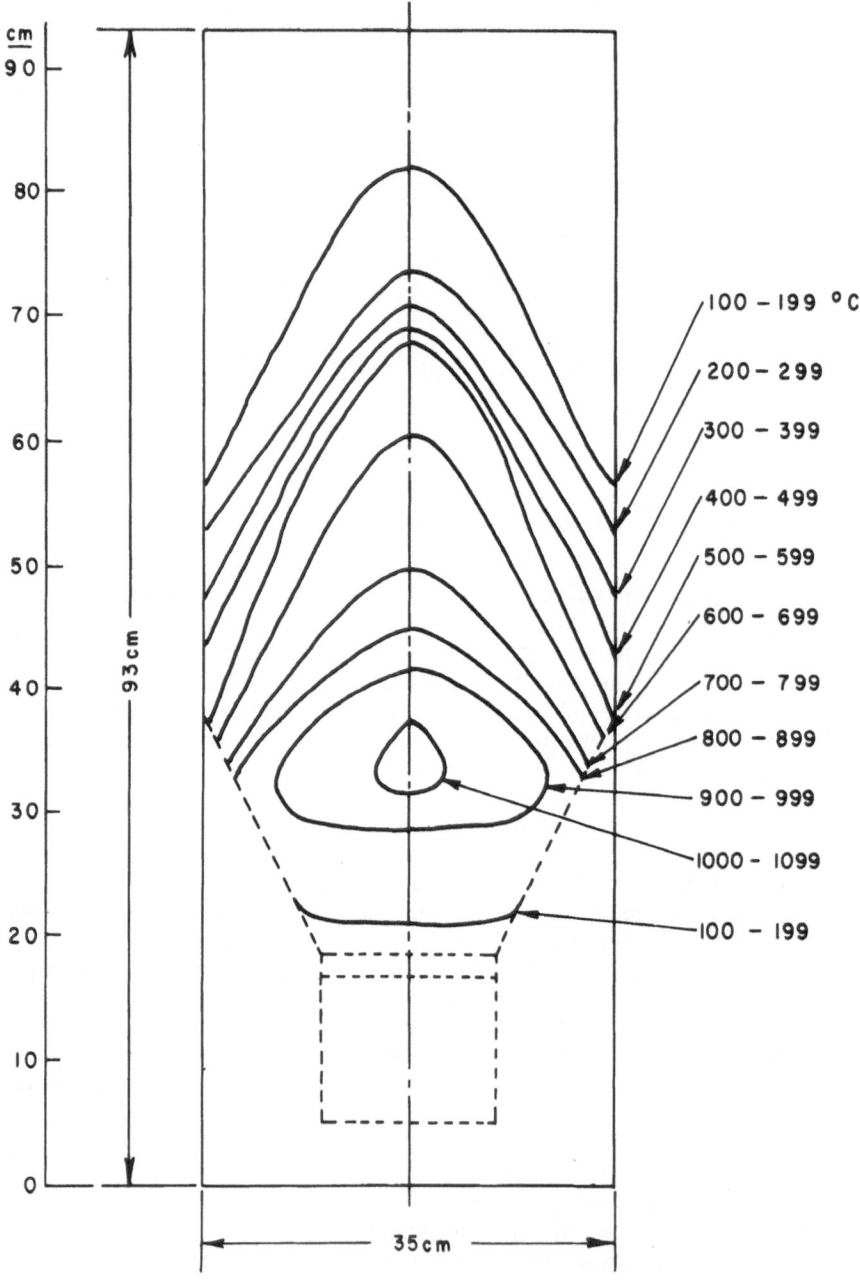

Figure 24. Temperature Profile in an Updraft Gas Producer (27).

low volatile fuels such as charcoal and coke. In these units the air blast is most commonly introduced through or around the grate as shown in Figures 25 and 26.

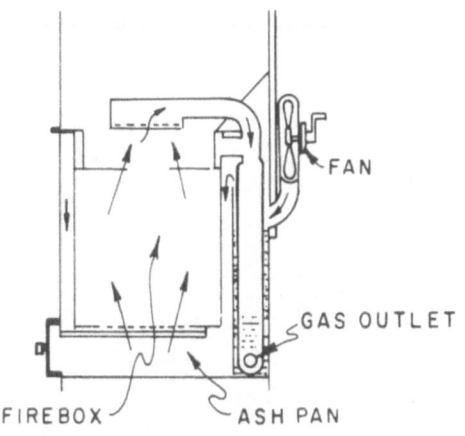

Figure 25. Hearth zone of C.G.B. Producer (10).

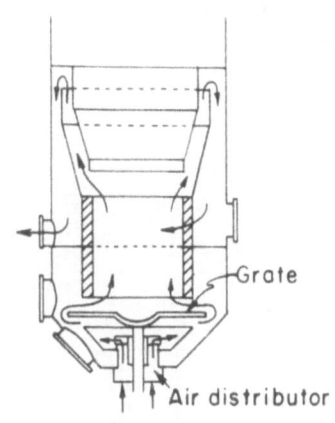

Figure 26. Hearth zone of Malbay Producer (10).

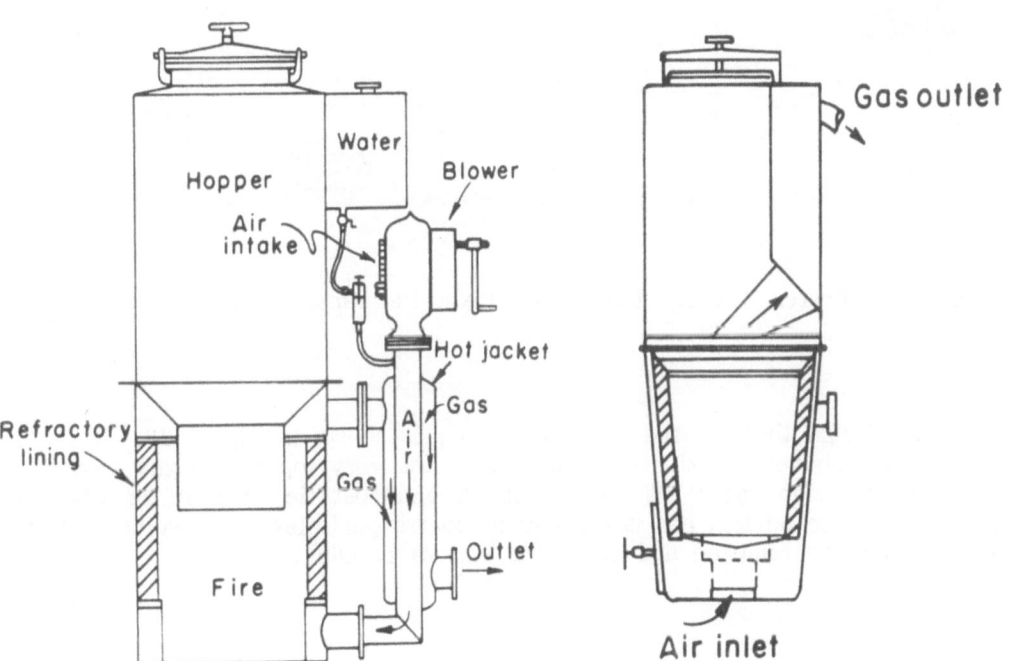

Figure 27. Koela Updraft Gas Producer (10).

Figure 28. Hansa Gas Producer (10).

Drawing off the gas above the reduction zone has the beneficial effect of obtaining a more tar free gas but results in high exit temperature and decreased overall efficiency. To recover some of the sensible heat in the gas, a simple parallel heat exchanger to heat up the incoming air with the sensible heat of the gas was used as shown in Figure 28; or the gas exit funnel was extended through the entire fuel column above the reduction zone and therefore serves as a heat exchanger inside the gasifier, Figure 27.

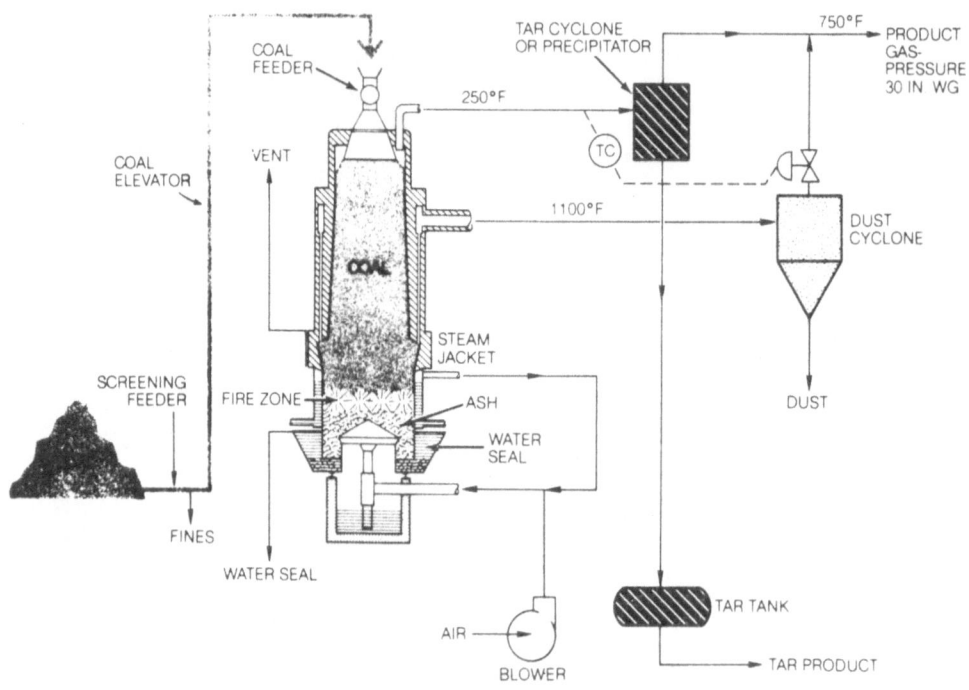

Figure 29. Foster Wheeler Two-Stage, Fixed-Bed Gasifier (9).

A unique design that is used to fire a boiler is shown in Figure 29. The gas is drawn off above the oxidation zone and at the very top of the gas producer. The tar-laden top gases are cleaned in an electrostatic precipitator whereas the hot tar-free bottom gas is cleaned of coarse particles in a cyclone. Both gas streams are reunited and enter the boilers at 400°C.

Figure 30 shows a test unit at Mie University, Japan, with the gas exit at the very top. This design guarantees a high overall efficiency, extremely low gas exit temperature of 20-80°C and a high degree of reaching the desired equilibrium states in the reduction zone but suffers from heavy tar formation if unsuitable fuel is used. The unit, fired with charcoal or coke, drives a 5 hp engine.

Another important point concerning the gas exit is the space between the top of the fuel column and the gas exit. All large updraft gas producers provide a space free of fuel below the gas exit that allows the gas to expand, cool down and decrease its velocity before it reaches the outlet pipe. Consequently, coarse fuel particles entrained in the gas current are allowed to settle down and do not reach the gas exit. This should be taken into consideration, in particular, when fuel with a high content of fine particles is gasified under a high specific gasification rate. In such a case the energy loss in particles carried away by the gas current can be unacceptably high. Moreover, a continuous high dust content in the gas requires cumbersome cleaning equipment and frequent maintenance. Figure 31 shows this general principle in the case of one of the original gas producers.

An updraft gasifier may be designed and operated under high temperatures (above 1300°C) to liquify the ash or it may be operated under controlled low temperatures below the softening point of the ash. These two modes of operating a gasifier require different grate designs. In the slagging type, the hearth zone must be kept continuously above the melting point of the ash and, in order to improve the viscosity of the molten ash some flux such as limestone,

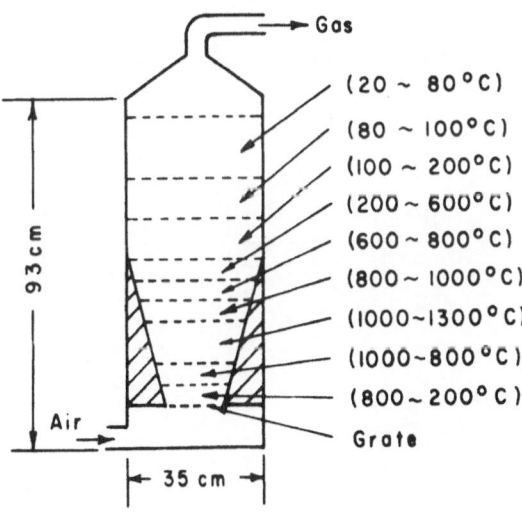

Figure 30. Pilot Gas Producer, Mie University (26, 27).

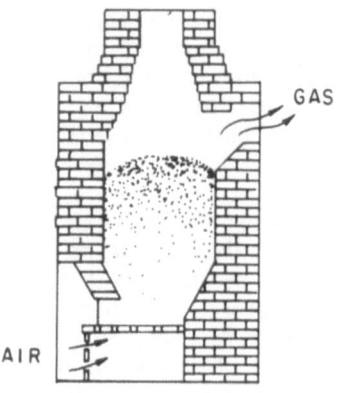

Figure 31. Bischof's Flat Grate Produer (22).

sand or iron furnace slag ranging from 10-25% of the fuel is sometimes added. The molten slag is then tapped off through slag notches as shown in a typical design in Figure 32 and 33. Due to the intensity of the heat around the tuyeres and in the lower part of the gasifier a rapid wearing out of the fire lining takes place and in most cases the tuyeres must be water cooled or specifically protected by refractories. The amount of flux added to the fuel must be determined by experience. The fluxibility of the ash does not increase with the amount of limestone or any other flux added because the minerals in the ash together with the flux form an eutetic mixture with one or more lowest melting points (see Chapter VI). Another method to keep the ash in a liquid state is shown in Figure 34. In this case a gas fire was maintained below the brick crown at the bottom.

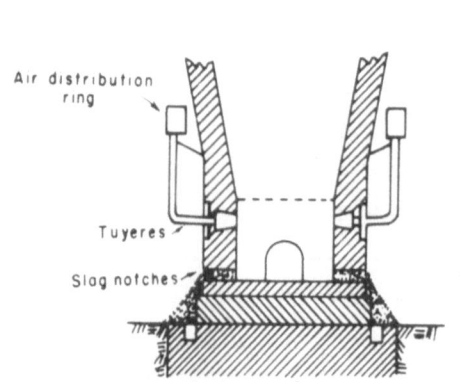

Figure 32. Liquid Slag Gas Producer (22).

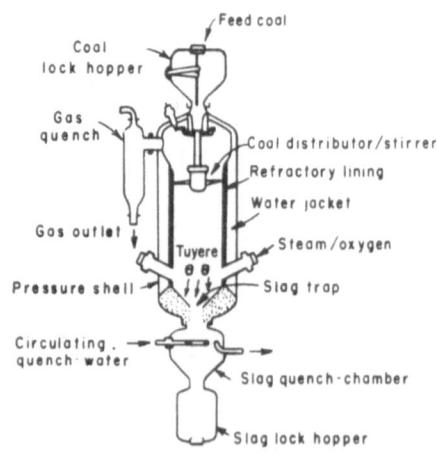

Figure 33. Liquid Slag Gas Producer With a Central Tap Hole (23).

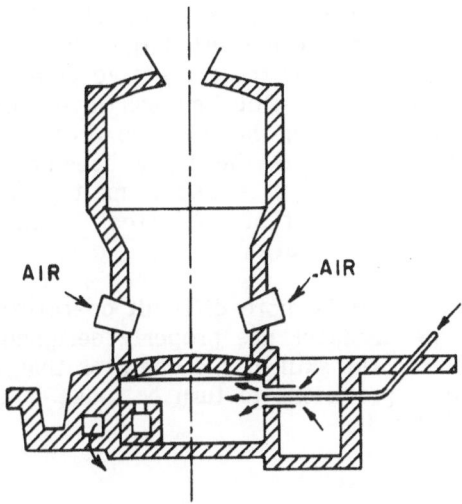

Figure 34. Liquid Slag Gas Producer with Heated Bottom Crown (22).

Obviously, variation of the rate of gasification will seriously interfere with the melting and fluidity of the ash and slagging medium. Low specific gasification rates of 100-150 kg/m^2-h are undesirable for obtaining a temperature high enough for melting of the slag. It is therefore unlikely that liquid slag updraft gas producers will ever be employed to drive internal combustion engines when a frequent change in power output is required. One may conclude that liquid type updraft gas producers work at a higher rate of gasification, which has the advantage of smaller capital outlay, no mechanical parts required to remove the ash and no carbon in the ash. On the other hand, the high rates of gasification limit the grading of fuel that can be employed due to the dust which is carried away from the producer in the hot gas. A continuous and steady load rate is essential to satisfactory operation. The gas leaves the producer at a high temperature, and thermal losses are great when the gas is used in its cold state. Upkeep charges become higher due to the repeated renewals of the necessary special brick lining in the lower part.

Updraft gasifiers designed to operate under temperatures below the melting point of the ash differ from the previously discussed type in that they all have a grate at the bottom of the plant. The grate separates the ash bin from the partial combustion zone and supports the entire fuel column. It was soon recognized that the grate is the most vulnerable part of an updraft gas producer because of the several functions it performs. Its design must allow for the ash to move freely through it into the ash bin and at the same time prevent carbonized fuel from falling through it. Although the plant is designed to operate at temperatures well below the melting point of the ash, in most practical cases the formation of clinkers can not be avoided. This applies in particular to plants used for power generation under unsteady conditions and fuels with a high ash content as outlined in Chapter V and VI. Therefore, it is desirable to construct the grate so that it can crush large clinkers. Another important point in the design and operation is the protective layer of ash that should be maintained above the grate. Too thick an ash layer seriously interfers with the operation due to an increase in the pressure drop across the gas producer and a lower gasification rate. If the layer becomes too thin the partial combustion zone may reach the grate and a melt down of the grate takes place when the grate is made out of mild steel or another material with a low heat resistance or heat conduction.

A simple grate does not result in more difficult operation of the plant as long as all other parts of the producer are properly designed. Moreover the life of a grate depends more on the skill of the operator than on the actual design. A fixed flat grate with no provisions to turn or shake it is one of the simplest designs as shown in Figure 35.

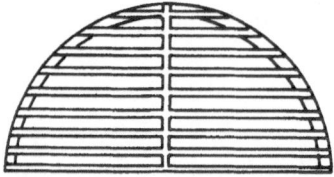

Figure 35. Flat grate (22).

It is used when very limited clinker formation is expected and no large amounts of ash are produced, which is the case for fuels with an ash content below one percent. In most other cases, means must be provided to periodically or continuously shake the ash through the grate and crush any clinkers above the grate which may obstruct the air flow. Several designs have been proposed which are working more or less successfully depending on the particular case. Figures 36 and 37 show two representative cases of shaker grates which facilitate the detachment and separation of the ash. However, these types of grates are ineffective in crushing clinkers.

For fuels with high ash content and the tendency for clinker formation, a continuously slowly rotating grate that has a milling effect on the clinker is usually employed. Two general principles are most common: The star grate which allows the air to enter through slots as shown in Figure 38 or a rotating eccentric grate, Figures 76 and 77. The eccentric grate discharges the ash through horizontal slots into the ash bin.

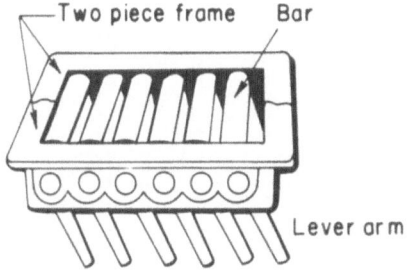

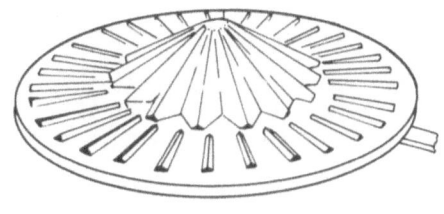

Figure 36. Shaker Grate (25).   Figure 37. Imbert Shaker Grate (24).

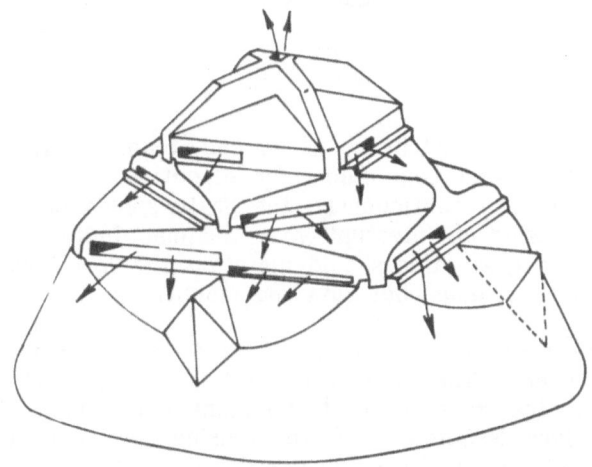

Figure 38. Star Grate (22).

Besides the flat grate, several different types of inclined step grates have been built as shown in Figure 39. The advantage of the inclined grate over the flat grate is the fuel bed is more accessible and can be stirred easier if necessary. The ash is discharged through the grate into a water sink at the bottom of the plant. Some steam is raised in the water by radiation of the grate and quenching the hot ashes.

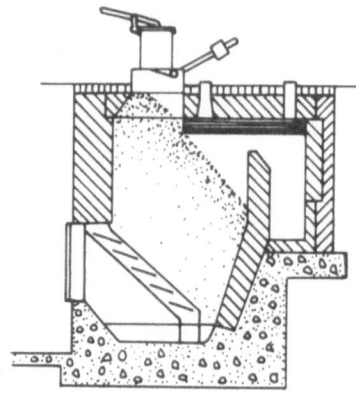

Figure 39. Inclined Step Grate Producer (22).

A well designed grate should distribute the air and steam evenly over the entire grate area and at the same time allow for effective ashing and clinkering. For producers that operate continuously, the grate should be one that allows for ashing without causing an interruption in the manufacture of the gas. In the case of a rotating grate this should maintain the lower part of the fuel bed in a steady and continuous, but slow movement.

In most updraft gas producers, steam is injected or evaporated into the hot partial combustion zone. The procedure has a beneficial effect on the gas quality and prevents the lower part of the plant from overheating. A very large number of various designs of self steam-rising devices have been used in gas producers. Most small and medium-sized plants utilize the sensible heat in the gas or the radiative heat emitted from the gas producer shell to generate the necessary amount of steam. The general principle is to build a water jacket around the plant and conduct the generated steam through a pipe into the gas producer below the grate where it is mixed with the incoming air blast.

Figure 42 shows the Dowson and Mason Self Vaporizing Suction Gas Plant. The water jacket is located at the upper part of the gas producer. The sensible heat of the gas together with the heat of radiation from the reduction and distillation zone is used to generate the steam. A slightly different design with the water jacket around the partial combustion zone is shown in Figure 29. This plant built by Foster Wheeler Energy Corporation has some other unique features. Here only the radiative heat is used to generate the steam. This design also protects the combustion zone walls from over heating because of the large heat sink in the water jacket.

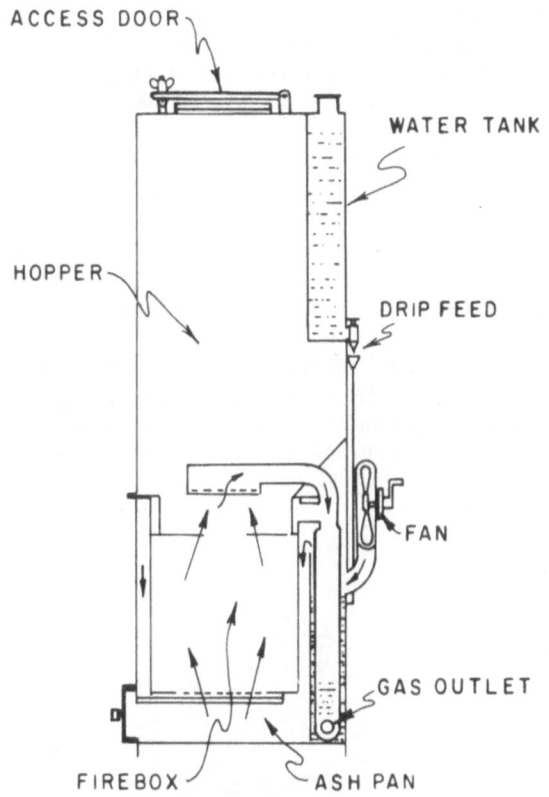

Figure 41. C.G.B. Producer (10).

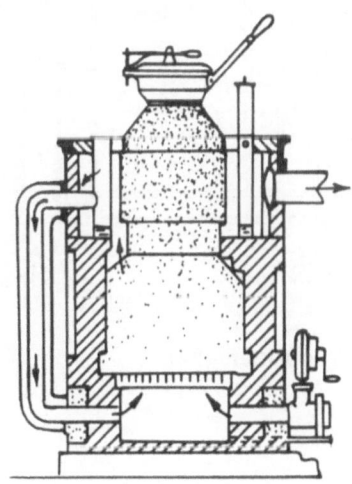

Figure 42. Dowson and Mason Suction Gas Plant (22).

Small portable units employ the water jacket principle or inject the vapor directly into the air stream. Figure 41 shows the C.G.B. Producer where steam is generated by the sensible heat of the hot exit gas. Both water and incoming air surround the gas exit pipe as shown in the sketch.

A German type vaporizer is shown in Figure 43. The vaporizer and distillation zone consists of four concentric shells. The hot gases are passed through the central annular space, A', before leaving at B. The air entering at C is passed over to the top of the boiling water surface before being admitted to the grate. D is a water supply funnel and E the sight gauge glass. This design differs from the previously described system insofar as the steam or vapor is not injected into the air stream but picked up by the air through convective transfer. This old system is of interest because the water vapor in saturated air at 50° - 70°C should be sufficient to generate a high fraction of hydrogen in the raw gas. One of the most simple steam introducing devices is a pan filled with water at the bottom of the gas producer as shown in Figure 39.

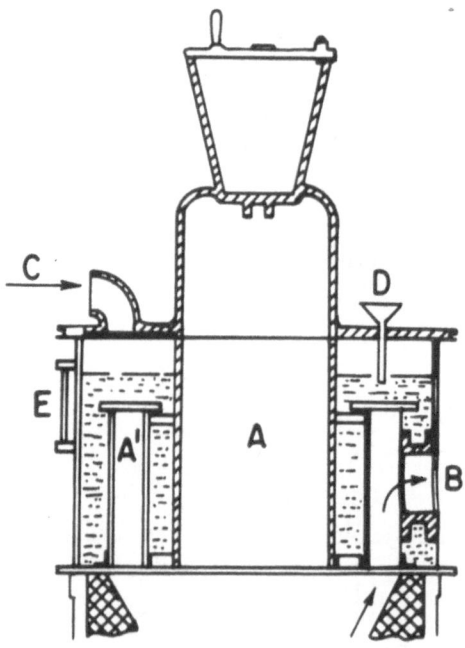

Figure 43. German Type Vaporizer (22).

From the previous examples one can see that some systems generate steam for the sole purpose of increasing the heating value of the raw gas through the generation of hydrogen, whereas others use the steam generation for the additional purpose of cooling down the fire box walls and grate and quenching the ash. It is questionable whether the injection of steam or water is of any advantage in small portable updraft gas producers. A dry gasification simplifies operation and equipment, and no water has to be supplied. If steam is used, the temperature of the exit gas that generates the steam may drop below the dew point of the water and tar vapors contained in the raw gas. These tar and moisture condensates will clog the cleaning equipment. The difference in the heating value of the raw gas and efficiency of the plant are actually very slight since the heating value of the stoichiometric gas-air mixture is the important factor in driving an internal combustion engine and not the gas heating value. However, in stationary plants with a steady load and controlled conditions, the injection of steam can be beneficial as outlined in Chapter III.

In most updraft gas producers the air blast is distributed over the entire grate area and consequently the combustion zone extends to the wall of the fire box. For this reason most updraft gas producers have fire brick lining in the combustion zone that protects the outer shell of the plant. In the case of a sufficiently large water jacket surrounding the lower part of the producer a fire brick lining is not necessary. In addition, the air inlet through the grate can be confined to a smaller circular area than the grate itself which allows a protective layer of carbonized fuel between the walls and the fire zone.

The gas output of an updraft gas producer is limited by the specific gasification rate; i.e., the amount of fuel that can be gasified per square meter of grate area in one hour. This number should be given on a dry fuel basis, because most biomass fuels contain a considerable amount of moisture which is driven off in the distillation zone. Specific gasification rates from 100 kg/m^2-hour to 300 kg/m^2-hour are considered normal for coal gasifiers. The specific gasification rate of a gas producer depends on the fuel, the design and the mode of running. Rotating and fixed grate plants are usually operated from 100-200 kg/m^2-h; whereas slagging type gas producers require a higher rate close to 300 kg/m^2-h in order to keep the temperatures high enough. Rates above 300 kg/m^2-h have occasionally been reported, but a prolonged operation under such high loads results in excessive wear of the fire lining and the tuyeres. The considerable loss of fuel particles entrained in the gas current must also be taken into consideration at high specific gasification rates.

Among all types of gas producers for immediate combustion of the producer gas in a fire box, the updraft gasifer achieves the highest efficiency. Because of the "natural" upward sequence of partial combustion, reduction and distillation zone and the countercurrent flow of air and fuel it is most suitable for high moisture or high ash fuels. The limiting factor using high ash fuel is the design of the grate and the ash discharge mechanism. However, it must be emphasized that updraft gasifiers cannot crack tar and oil vapors generated in the distillation zone. Consequently, they are unsuitable for portable units mounted on automobiles where the cleaning equipment has to be compact and light. In the modified updraft form with the gas taken off above the reduction zone and fired with low volatile fuel such as charcoal or anthracite the tar generation is not as severe.

The Downdraft Gas Producer: Because the tar vapors leaving an updraft gas producer in uncracked form seriously interfer with the operation of internal combustion engines, the next step in the evolution of gasifiers was taken toward downdraft gas producers. In this type, the air is introduced into a downward flowing packed bed of solid fuel and the gas is drawn off at the bottom as shown in Figure 44.

The general idea behind this design is that the tarry oils and vapors given off in the distillation zone are highly unstable at high temperatures. In order to reach the gas outlet they must pass through the partial combustion zone where a high amount will be cracked and reduced to noncondensible gaseous products before leaving the gasifier. Although the general principle behind this idea seems convincing, in practice it requires some testing and skill to come up with a downdraft gas producer capable of generating a tar free gas under equilibrium conditions.

Points of importance in regard to the design of downdraft gas producers include:

1. The design of the combustion zone
2. The air feed
3. Design of grate.

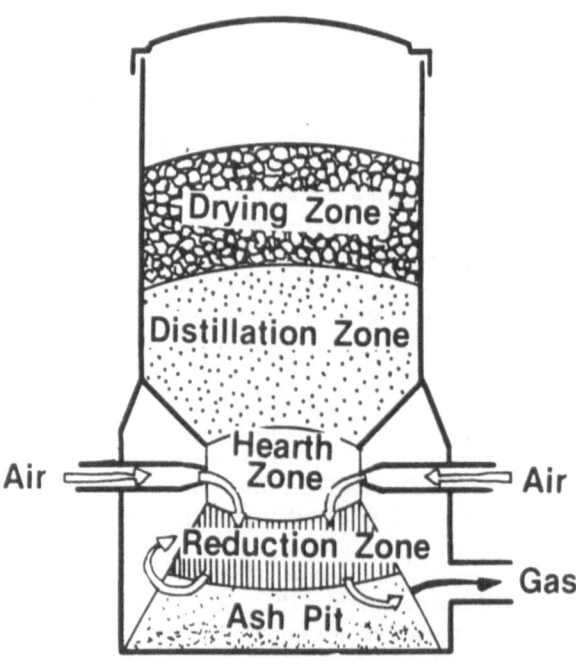

Figure 44. Downdraft Gasification (25).

Two parameters which determined to a great extent proper tar cracking are the methods of air injection and the geometry of the partial combustion zone. Downdraft gas producers have a reduced cross-sectional area above which the air is introduced. This so-called throat ensures a homogenous layer of hot carbon through which the distillation gases must pass. Figures 45 to 50 show some of the many designs that have been successful.

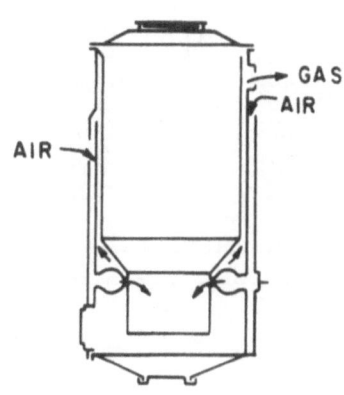

Figure 45.   Wall Tuyeres
             Central Air Feed
             (24).

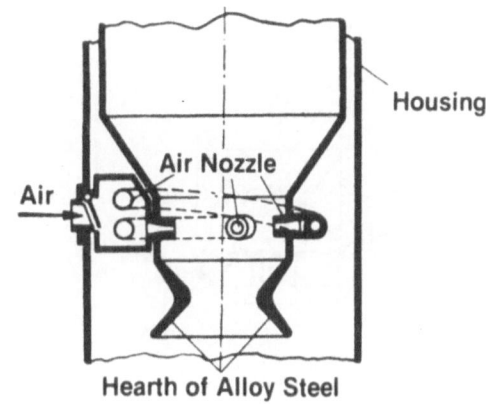

Figure 46.   Wall Tuyeres
             Individual Air
             Feed (25).

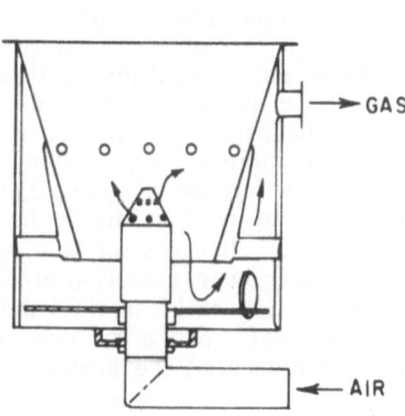

Figure 47.   Central Air Inlet
             From Below (24).

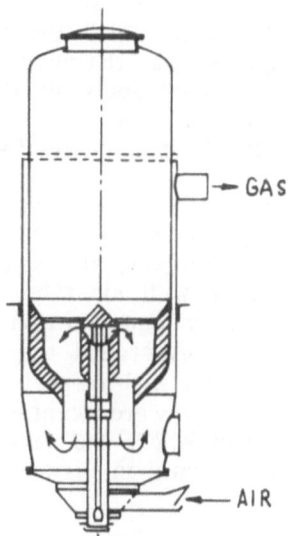

Figure 48.   Central Air Inlet
             From Below and
             Refractory Lining
             (24).

61

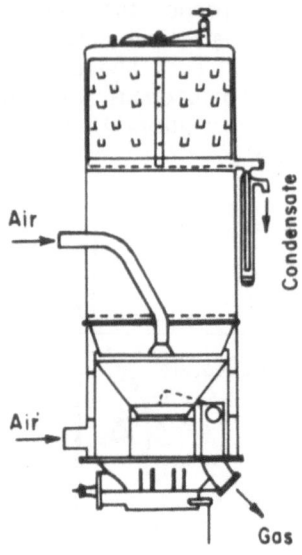

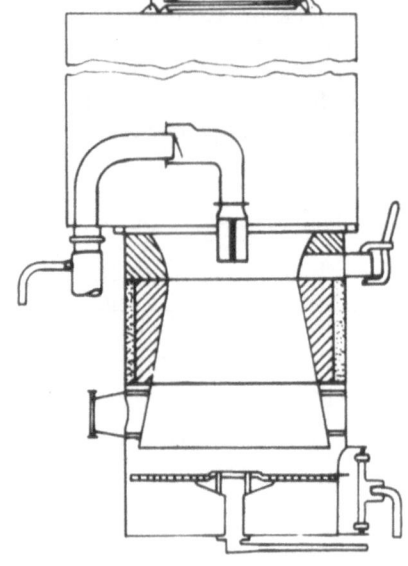

Figure 49. Central Air Inlet
            From Above and
            Below (24).

Figure 50. Central Air Inlet
            From Above and Re-
            fractory Lining

One of the most successful gasifiers, the Imbert type, had initially a central annulus from which the tuyeres were fed with air (Figure 45). It was soon recognized that the unavoidable pressure drop between the tuyeres resulted in hot and cold spots in the partial combustion zone because of unequal air distribution. The design was later changed as shown in Figure 46. Here the tuyeres are individually fed with pipes connected to one central air inlet port. Figure 47 shows a typical design of a downdraft gas producer with a middle air inlet and in addition, tuyeres in the wall of the partial combustion zone. It was believed that this design would result in a more equally heated fire zone. However, the same effect can be achieved through a well-designed middle or wall air feed system and a combination of both seems to be an unnecessary complication of the air inlet system. Because of the high temperatures around the tuyeres or middle air inlet, some models protected the air inlet with refractories as shown in Figure 48. Figure 50 shows a model with a downward pointed middle air inlet. Some units had a built-in heat exchanger where the sensible heat of the raw gas preheated the incoming air blast as shown in Figure 28.

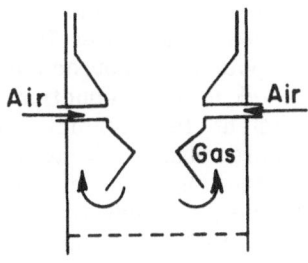

Figure 51. Wall Tuyere and
Conventional Throat (12).

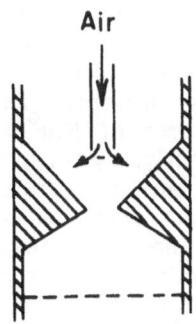

Figure 52. Middle Tuyere
Pointed Downward
and Conventional
Throat (12).

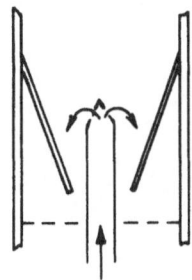

Figure 53. Middle Tuyere Pointed
Upward and Convention-
al Throat (12).

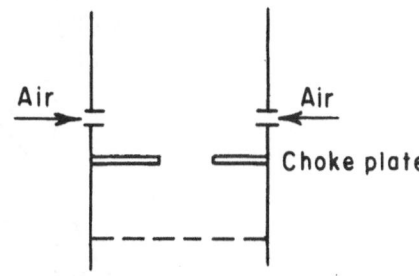

Figure 54. Wall Tuyere and
Choke Plate.

In general, four different types of downdraft gas producers have evolved over
the past 50 years:

1. Wall tuyeres and conventional throat (Figure 51)
2. Middle tuyere, pointed downard and conventional throat (Figure 52)
3. Middle tuyere pointed upward, and conventional throat (Figure 53)
4. Wall tuyeres and choke plate (Figure 54).

One should not underestimate the problems associated with the high temperatures
around the tuyeres and throat area. Cracking of the metal or ceramic throats
as well as melt downs of the tuyeres have been frequently reported. The
choke plate design in the UCD laboratory gas producer, Figure 54, seems to
be one solution to the thermal stresses occurring in the throat area. Because
the throat and the position of the air inlet determine how well the distillation
products will be cracked before they leave the gas producer, care must be
taken in their design. Figure 55 shows schematically the oxidation zone formed

in front of wall tuyeres. Between the nozzle near the wall and in the center
are spots which are not reached by oxygen resulting in lower temperatures.
All distillation products passing through these spots are unburned because no
oxygen is present. They may be partly cracked in the reduction zone but
temperatures of 650°C are by no means sufficient for complete reforming.
Consequently, systems with wall tuyeres are more susceptible to release tar
vapors than the models with middle tuyeres that achieve a better, more
homogenous oxidation zone at the throat. However, one has to keep in mind
that the throat creates a barrier to the downward fuel flow and systems with
middle tuyeres pointed upward tend to increase the bridging problem and yield
a too loose bed of incandescent fuel that also hinders the tar conversion.

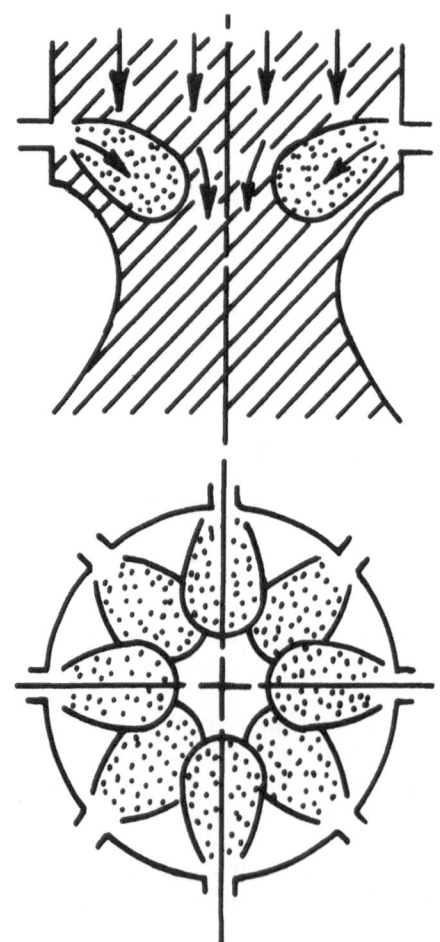

Figure 55. Fire Zone in Front of Wall Tuyeres (13).

As already pointed out, slight changes in the diameter of the throat or choke plate and position of the air inlet can change drastically the gas composition and the tar yield. The best configuration will depend on physical parameters of the fuel and the load factor and consequently must be found by trial and error. Rough guidelines about the relative dimensions of tuyeres, diameter of the throat or choke plate and height of the tuyeres above the throat are given at the end of this chapter. One of the most extensive tests concerning downdraft gas producers with middle air inlet has been presented by Groeneveld in his thesis: "The Co-Current Moving Bed Gasifier" (12). Readers interested in the modelling of downdraft gas producers as shown in Figure 52 are referred to this paper.

From Figure 44 it can be seen that downdraft gas producers are not well suited for high ash fuels, fuels with high moisture content or the tendency to slag. The fuel moisture, usually driven off by the sensible heat in the hot gas stream passing through the distillation zone in an updraft gas producer, will not get into contact with the hot gas in a downdraft unit. A lower overall efficieny and difficulties in handling moisture contents higher than 20% were common in small downdraft gas producers. Any slag formed in the partial combustion zone will flow downward, quickly cool and solidify in the reduction zone and finally obstruct the gas and fuel flow. A well-designed rotating grate and operation below the ash melting point are therefore essential if fuels with high ash contents are used in a downdraft gasifier. The unit tested with eleven crop residues at UCD could not gasify high ash fuels (15% to 20%) such as cubed cotton gin trash or rice straw. An upper limit of 5% ash content was established and fuels with higher ash contents could not be gasified over a prolonged period.

Additional steam or water injection is uncommon in downdraft producers. The combined moisture in the fuel and the humidity of the air are sufficient for the generation of hydrogen. To fire line the partial combustion zone seems also unnecessary, since the position of the tuyeres generates a natural protective layer of carbonized fuel between the fire zone and the walls of the gas producer, except at the throat, where such a layer would be highly undesirable. It should be mentioned that the throat is one of the most vulnerable spots in a downdraft gas producer. The high temperatures in this area lead to metal fatigue, melt downs and cracking. Why some units work extremely well and others have material problems at the throat and the tuyeres is more or less due to the specific gasification rate and the desire to achieve high temperatures for the tar cracking purposes. During three years of testing under no slagging conditions, the UCD laboratory gas producer has never shown any damage to the tuyeres and the choke plate. The end section of the tuyere was a stainless steel nut and the choke plate was made from A 515 steel. The highest specific gasification rate ever reported was 5020 kg/h-m^2. The fraction of tar cracked was low at low specific gasification rates.

Downdraft gas producers can be operated at a considerable higher specific gasification rate. An upper limit of 1 Nm^3 of gas per hour per square centimeter throat area has been established for small portable downdraft gasifiers. This corresponds to 2900 kg/h-m^2 to 3900 kg/h-m^2 of dry fuel depending on the heating values of fuel and gas. Consequently, the downdraft gas producer can utilize the available grate area much better than updraft gas producers and this has been one more reason why they are used in automobiles. The time needed to ignite the fuel and bring the plant to a working temperature with good gas quality and little tar in the gas is shorter than for the updraft gas producer, but still inconveniently long in the range of 15-30 minutes. Variables such as weight, start up time as well as load following capability of the plant are important from a driver's point of view. The load following capability of a gas producer, in a physical sense, is its ability to extend the partial combustion zone to produce more gas per minute without a decrease in the heating value. This determines the acceleration behavior and the hill climbing capability of the engine. Consequently, the next step in the evolution of gasifiers was toward the crossdraft type that could much better meet the desired fast start up time, high load following capability and compact design.

The Crossdraft Gas Producer:

Crossdraft gas producers, although they have certain advantages over updraft and downdraft types, are not the ideal gasifier. Unsatisfactory performance of a unit can be overcome by replacing it with another unit better suited to the particulars of the situation at hand. Figure 56 shows the schematic design of a crossdraft gas producer. It is obvious that certain disadvantages such as high gas exit temperatures, poor CO_2 reduction and high gas velocity with extremely short gas residence time are the consequence of the design.

In almost all cases the ash bin, fire and reduction zones are not separated by a grate as in updraft or downdraft gas producers, which limits the type of fuel suitable for operation to low ash fuels such as wood, wood charcoal, anthracite and coke. The load following ability of a crossdraft unit is quite good due to the concentrated partial combustion zone which operates at temperatures up to $2000^{\circ}C$. Start up time is in general much faster than those of downdraft and updraft units (5-10 minutes). The desired concentrated combustion zone is best achieved by one single tuyere which is in most cases water cooled and only rarely air cooled. The shape of the air jet exit and the air jet velocity determine the extent of the combustion zone. Although there have been a few crossdraft gas producers with fire lining, most units operate without it and confine the partial combustion zone to the center of the bottom part. However the danger of quickly burning out the vertical grate in front of the gas exit is always present, since an extension of the partial combustion zone can be much faster and is easily achieved by increasing the air velocity and amount of air blown into the oxidation zone. Of specific interest are the various tuyeres, their shape, cooling systems and in some cases built-in steam injection devices. Figure 57 shows one design which has a flat rectangular orifice. This design is believed to avoid turbulence and unnecessary eddies and the wide flat stream of air does not diffuse as quickly, thus causing a comparatively small oxidation zone at very high temperatures (1).

Figure 58 shows the water cooled tuyere of the South African High Speed Gas Producer (H.S.G. Plant). In addition, this design had a steam injection channel to boost hydrogen production and cool down the partial combustion zone if necessary.

66

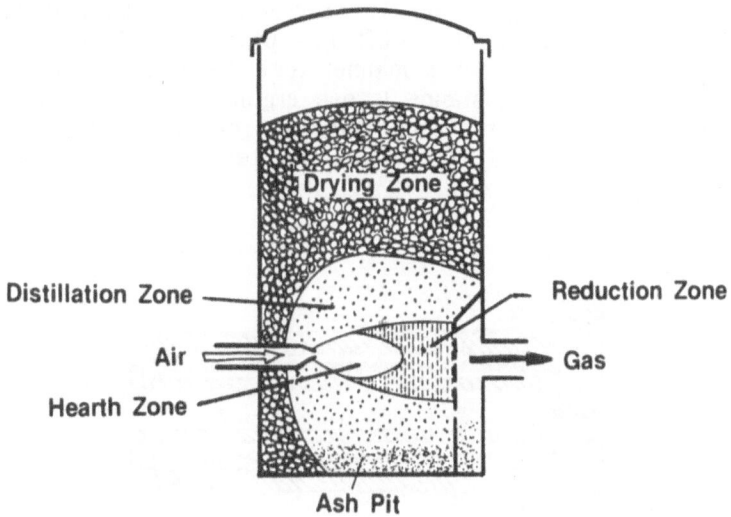

Figure 56.   Crossdraft Gas Producer (25).

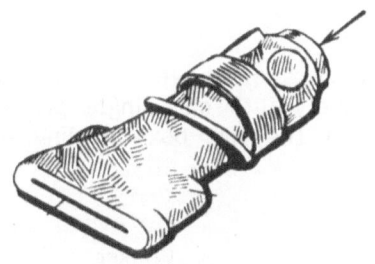

Figure 57.   Watercooled Rectangular Tuyere for Crossdraft Gas Producer (25).

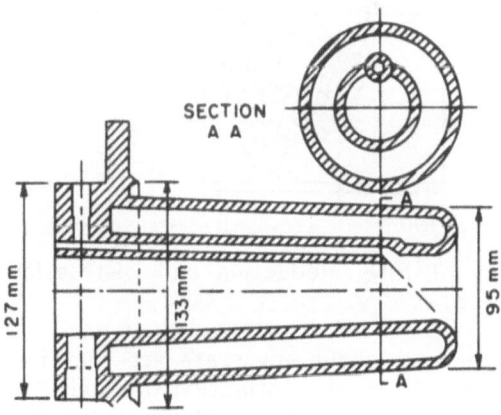

Figure 58.   Tuyere of H.S.G. Plant (3).

Another design is shown in Figure 59. This tuyere is air cooled and was originally used in the French Sabatier crossdraft gas producer. It consists of three concentric tubes arranged in such a manner that the entering air is the cooling agent. The increased air resistance is one argument against this design. Its advantages are no breakdowns through blockage of the cooling water and the degree of cooling is directly related to the temperature in the fire zone.

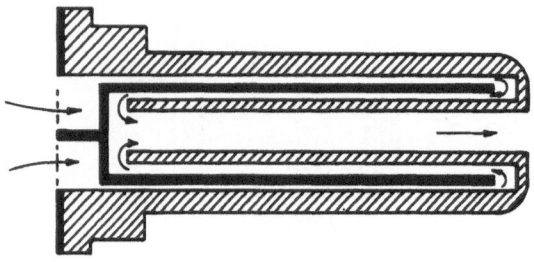

Figure 59. Sabatier Air Cooled Tuyere (31).

At the prevailing temperatures of above $1500^{\circ}C$ in a crossdraft gas producer, the ash will fuse most likely into one single piece of clinker which will be deposited at the bottom of the fire box or cling to the tuyere and walls as shown in Figure 60.

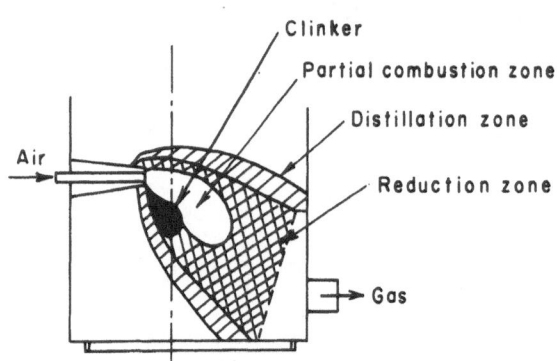

Figure 60. Diagram of Fire, Reduction and Distillation Zone in a Crossdraft Gasifier (14).

Crossdraft gas producers without any grate are therefore not suitable for high ash fuels. For low ash fuels, the formation of clinker is insignificant within a reasonable time period and does not obstruct the gasification process. Some gasifiers such as the British Emergency Producer featured inclined tuyeres to prevent molten slag from clinging to the outer surface, (Figure 61).

68

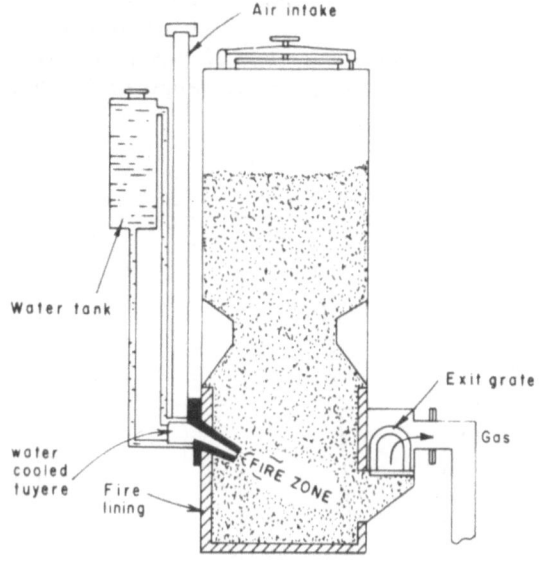

Figure 61.   Gas Producer with Inclined Tuyere (1).

Crossdraft gas producers are very sensitive to changes in the fire length, transit time of the gas and amount of water injected.   In this context the fire length is defined as the distance from the tip of the tuyere to the exit grate in front of the outlet pipe.   Transit time of the gas is by definition the fire length divided by the air velocity at tuyere exit.   Although both definitions do not have any real physical meaning, in a crossdraft gas producer their usefulness is established through experimental results presented in Figures 63 to 65.   The test unit used was a Wishart crossdraft gas producer with an adjustable exit grate and provisions for water injection as shown in Figure 62.

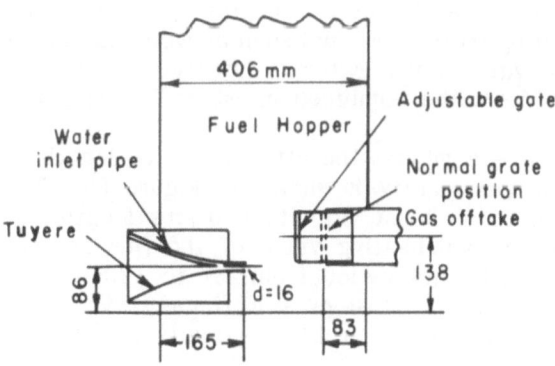

Figure 62.   Wishart Crossdraft with Adjustable Grate (2).

Figure 63 shows the lower heating value of the gas as a function of the fire length. The test was conducted with charcoal. Water admitted to the partial combustion zone amounted to 53 g per m^3 of gas generated. As expected the curve attains a maximum within the possible range of 63 mm - 180 mm for the fire length.

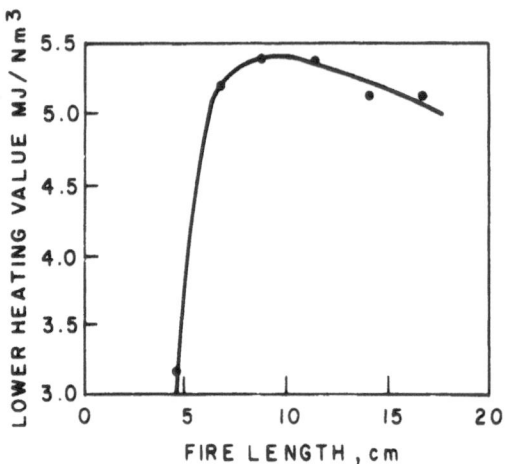

Figure 63. Lower Heating Value of the Gas Versus Fire Length (2).

It is interesting to note that increasing the fire length to the practical largest value and consequently expanding the reduction zone did not yield a better gas.

The considerably higher temperature in crossdraft gas producers has an obvious effect on the gas composition such as high CO content and low hydrogen content when dry fuel such as charcoal is used. The CH_4 generated is also negligible. This is shown in Figure 64. It can be observed that the injection of water into the air blast has some effect on the heating value of the gas and a considerably greater effect on the composition of the gas. One should note the increase in the heating value of the gas of 12% at optimal water injection of 10 g per minute, which amounts to an increase in engine power of at most 6%. Crossdraft gas producers operating on less dry fuel such as wood do not show this pronounced increase in heating value when water is injected. This is due to the already high H_2 generation from the combined moisture in the fuel.

Finally the transit time combines the effects of fire length and air velocity and its influence on the gas quality is shown in Figure 65. This curve represents various combinations of air blast velocity and fire length. It can be seen that at 0.009 seconds a maximal heating value of the raw gas is obtained. The air velocity in these experiments varied from 2.7 to 12.5 m/s which is rather low for crossdraft gas producers. Most crossdraft gas producers for cars and trucks operated on considerably higher air blast velocities of up to 100 m/s.

Extensive tests of the same type as just described with two different crossdraft gas producers and dry air blast are published in Reference (16). In addition to guidelines on how to size the tuyeres, the tests revealed that seemingly un-

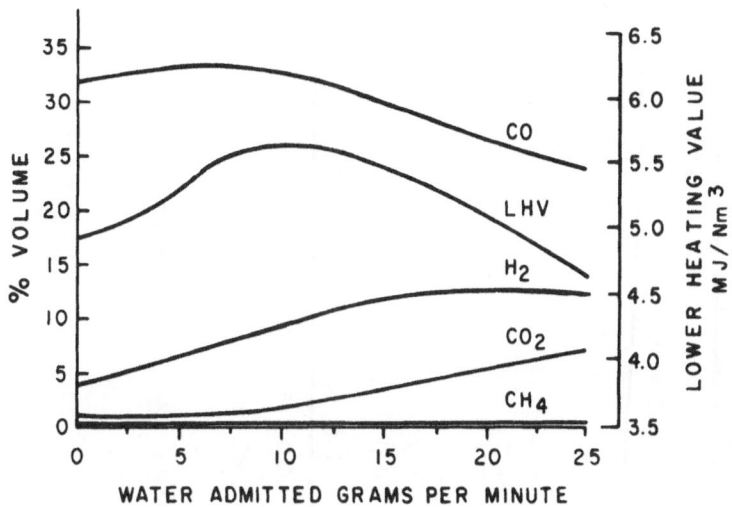

Figure 64. Gas Composition Versus Water Injected in a Crossdraft Gas Producer (2).

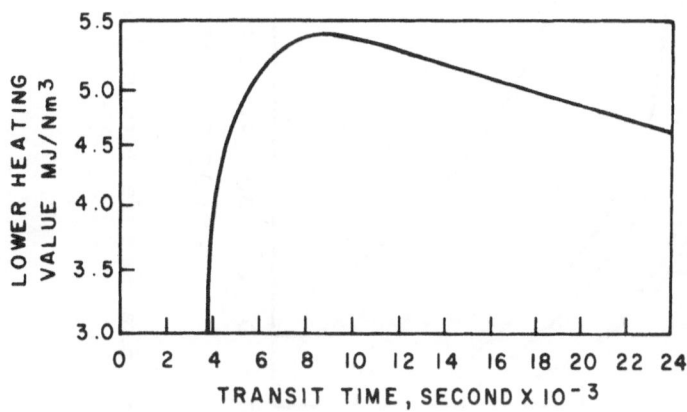

Figure 65. Heating Value of the Gas Versus Transit Time in a Crossdraft Gas Producer (2).

important design differences such as refractory lined tuyeres versus water cooled tuyeres make a difference in the performance of crossdraft gas producers. Figure 66 shows the temperature of the partial combustion zone in front of the tuyere for various tuyere diameters as a function of the air blast velocity.

It can be observed that at air blast velocities of 30 m/s, the temperatures are high enough to induce slagging and evaporation of mineral vapors in any kind of biomass or coal ash.

Crossdraft gas producers operated on dry air blast and dry fuel such as charcoal produce very little CH_4 and H_2. It is therefore convenient to express the gas quality in terms of the conversion ratio $CO/(CO + CO_2)$. A conversion ratio

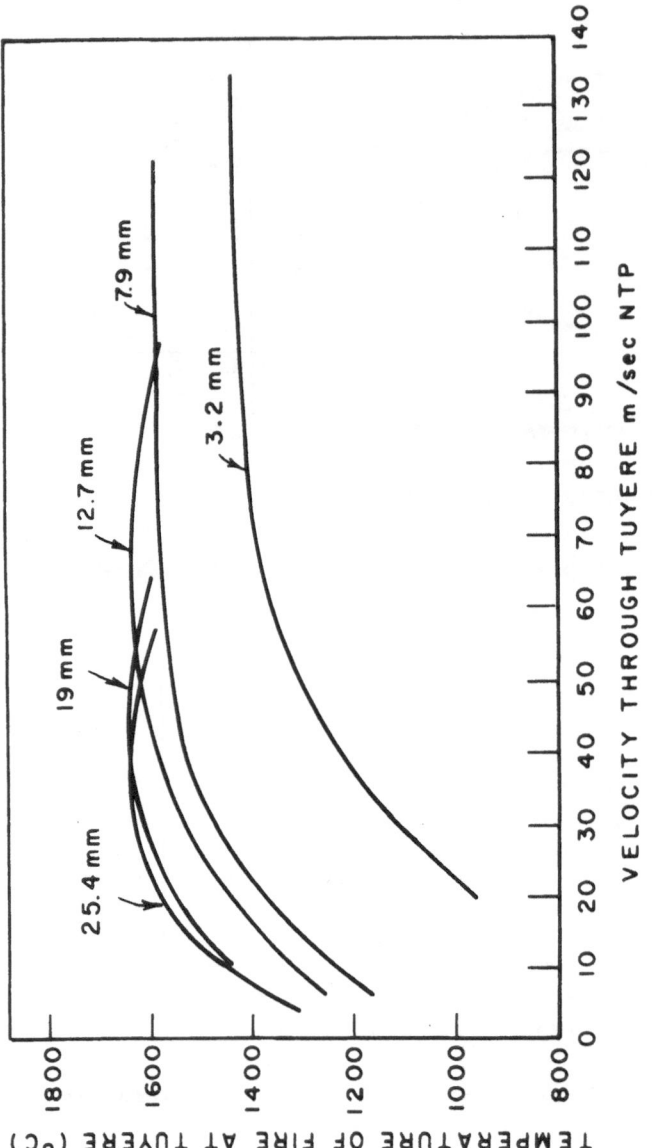

Figure 66. Fire Zone Temperature Versus Air Blast Velocity at Tuyere (16).

72

of 0.9 which corresponds to about 30% CO and 2% - 3% CO_2 represents a good quality gas in terms of the heating value. However, when talking about the quality of the gas one should keep in mind the final use of the producer gas. In most cases it is even more important to achieve a specific and constant amount of hydrogen in the gas. Internal combustion engines require a certain minimal amount of hydrogen in the gas necessary to achieve an appropriate flame speed during combustion in the engine cylinders. On the other hand too much hydrogen increases the chance of knocking and large fluctations in the hydrogen content lead to unsteady running conditions, since the advancement of the ignition or pilot oil injection depends on the hydrogen content of the gas.

Figure 67 shows the effect of the tuyere diameter and the air blast rate on the conversion ratio for charcoal. With regard to these experiments two important conclusions can be drawn: At air rates higher than 30 Nm^3/h, the tuyere diameter does not influence the conversion ratio and higher air blast velocities produce a better gas.

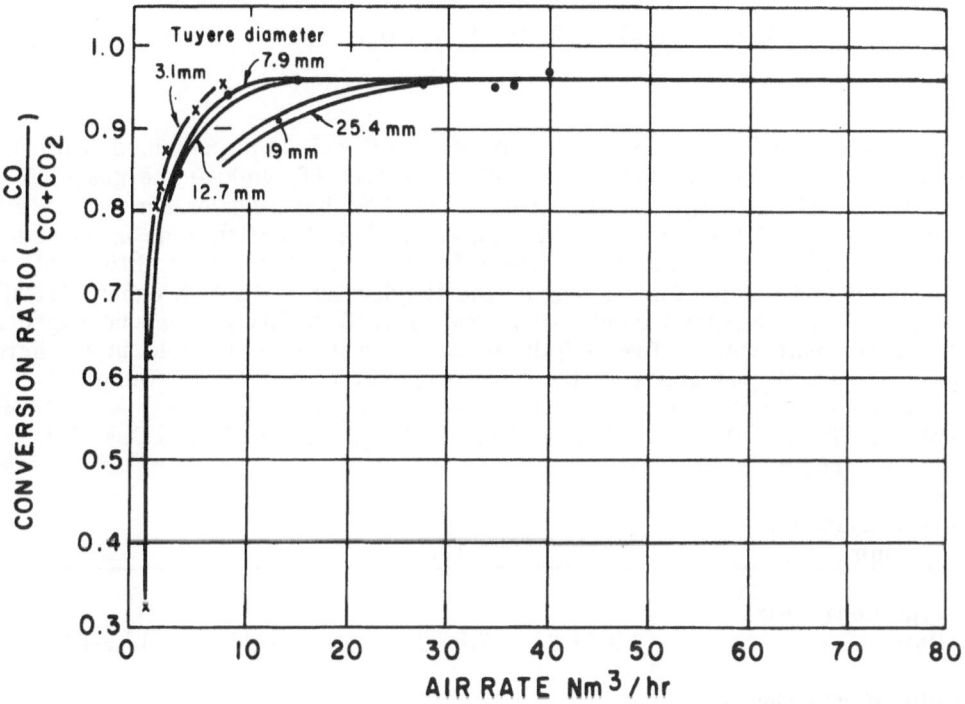

Figure 67.   Conversion Ratio Versus Air Rate for Various Tuyere Diameters (16).

In particular the last statement contradicts the usual belief and results of previous experiments that crossdraft gas producers generate a poor gas because of the short residence time of the gas and the small reduction zone. In this context, the reported data about the gas composition as a function of the air blast for a 3.2 mm tuyere (Table 2) is of interest.

Table 2. Gas Composition Versus Air Blast Velocity at Tuyere (16).

Air Blast Velocity m/s	Tuyere Fire Temperature °C	Gas Composition*				Conversion Ratio
		CO_2	CO	H_2	CH_4	
22.6	980	17.7	8.5	1.3	0.9	0.325
44.8	1300	9.1	20.3	4.2	1.1	0.693
72.3	1420	6.0	24.9	4.2	1.1	0.807
90.0	1400	5.6	27.5	4.3	1.2	0.832
115.0	1420	4.2	28.4	5.6	1.3	0.877
218.6	1520	2.6	30.1	6.5	1.3	0.922

$*N_2 = 100$ $(\% \ CO_2 + \% \ CO + \% \ H_2 + \% \ CH_4)$

The extraordinarily high air blast velocity of 218 m/s combined with high temperatures at the exit grate, as shown in Figure 68, yielded the gas with the highest chemical energy .or, in other words, the best conversion ratio. Such conditions are not favorable for the formation of a reduction zone as in updraft and downdraft gas producers to convert CO_2 into CO. In addition, all the reported data indicates that as long as the temperatures are high enough ($1500°C$ and higher) in the partial combustion zone, a good quality gas can be expected. The tested unit with a fire length of 33 cm was quite flexible in its actual power output by just changing the tuyere diameter as given in Table 3.

Table 3. Tuyere Diameter Versus Useful Range of Air Rate, Gas Rate and Engine Power (16).

Tuyere Diameter mm	7.9	12.7	19	25.4
Useful Range Air Rate m^3/h	6.3-11.9	7.6-27.2	12.4-47.6	13.9-64.6
Useful Range Gas Rate m^3/h	8.1-15.6	9.9-35.7	16.1-61.2	17.5-85
Useful Range, Engine Power hp	2.5-6	4-14	5-22	6-32

Finally, Table 4 shows the recommended minimum air velocity, air rate and gas rate for various tuyere diameters, to obtain a good conversion ratio of 0.9.

Table 4. Recommended Minimum Air Velocities, Gas and Air Rates, for Various Tuyere Diameters (16).

Tuyere Diameter mm	3.2	7.9	12.7	19	25.4
Air Blast Velocity m/s	146.0	35.0	17.0	12.0	7.5
Air Rate m^3/h	4.2	6.3	7.7	12.4	14.0
Gas Rate m^3/h	5.3	8.2	10.0	16.2	17.5

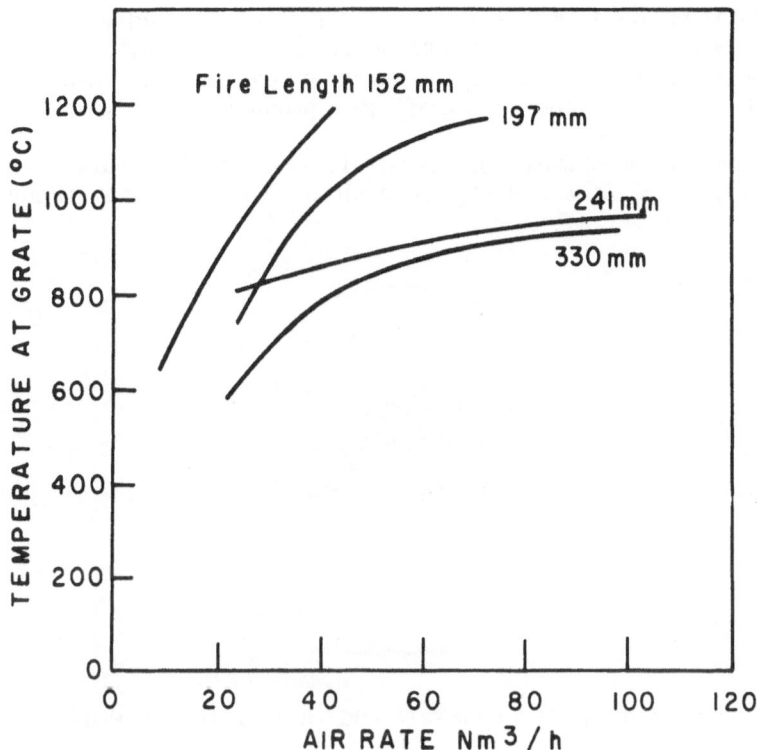

Figure 68. Exit Grate Temperature Versus Air Rate for Various Fire Lengths (16).

Historically, there has been a considerable controversy among engineers and researchers in gasification as to what extent the partial combustion zone and the reduction zone can be treated as two distinct zones existing independently. Although refined methods revealed some of the past mysteries on a microscale, this controversy still exists. The controversy centers around the question whether a considerable amount of CO could be generated directly through the reaction:

C + ½ O_2 = CO + 121,000 kJ/kg-mole at the carbon surface and the CO so produced may burn with excess oxygen in the void spaces of the bed. The arguments for the above reaction to occur and the CO to not burn is the high CO content of gas from crossdraft gas producers where there is not a distinct reduction zone, see Figure 9. In addition the reaction: C + O_2 = CO_2 + 393,800 kJ/kg-mole, if predominant at the tuyere, should yield a much higher temperature even if one accounts for the heat transfer and the incompleteness of the reaction. This question has been pursued by several authors and a summary of the results is given in Reference 5 and 19.

The gas exit of crossdraft gas producers looks much different than those of downdraft or updraft gas producers, where a funnel or a plain hole in the producer wall suffices in most cases. In general, it is important to design the exit grate in such a way that the considerable amount of coarse and fine particles entrained in the high speed gas stream are not carried out of the producer. This is accomplished by perforated grates and by taking off the gas vertical to the horizontal air blast (Figure 61). In addition, one could locate the gas exit port above or below the air blast level as done in the Bellay or Hamilton Motors Gas Producer. If carried out to the extreme, one ends up with a combination of updraft, downdraft and crossdraft gas producers as shown in Figure 69.

Although updraft, downdraft and crossdraft gas producers have been the types mostly built, there is a variety of gasifiers which do not really fit into

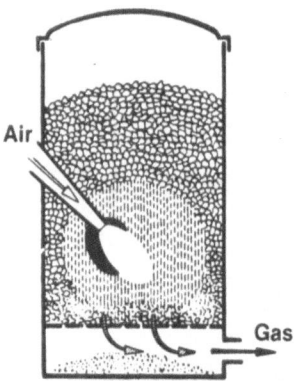

Figure 69. Combination of Downdraft and Crossdraft Gas Producer (25).

these categories. For instance, gasifiers with exit and inlet ports as shown in Figure 69. These units were built in an effort to combine the advantages of crossdraft with updraft or downdraft gas producers. Others like the Brush Koela Duo Draught plant could be operated either on updraft or some modified form of crossdraft. The crossdraft tuyere was usually used to quickly start up the plant and then operation switched to updraft which yielded a more regular gas composition and higher efficiency.

Another interesting design is shown in Figure 70. The Brandt Double Zone Producer works on the downdraft principle with the gas drawn off at the very top. The inner column was filled with high grade charcoal, whereas the outer annulus contained wood blocks. In a later design, the inside downdraft tuyeres were replaced by wall tuyeres at various heights in the lower part of the gas producer. The sole purpose of this unusual design is to get a very clean gas. This is achieved by passing all distillation gases through the partial combustion zone and then up through the entire charcoal column together with the products of the gasification process. This process was claimed to obtain a totally tar free gas.

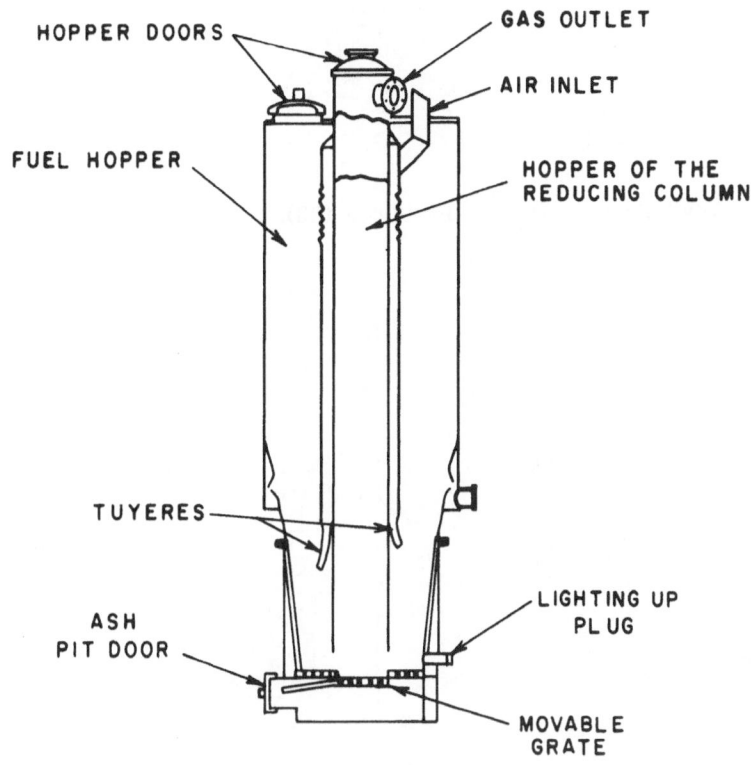

Figure 70. Brandt Double Zone Gas Producer (10).

Small portable or stationary gas producers have been built in all kinds of shapes. Figure 71 shows an early Volvo design in form of an eggshaped crossdraft gas producer for passenger cars. The unit was on a trailer which was hitched to the car and fueled with charcoal.

Gas producers can be built out of clay bricks as shown in Figure 72. This design by Groeneveld, et al. is being tested in Tanzania (12).

A downdraft gas producer applying the same middle tuyere principle has been built out of an oil barrel with a fire lining made of concrete and clay, Figure 73. The unit was used to drive a 4 cylinder Willis jeep engine connected to a 5 kW generator. The only cleaning equipment between the gas producer and engine was a cyclone. The gas producer fuel was charcoal.

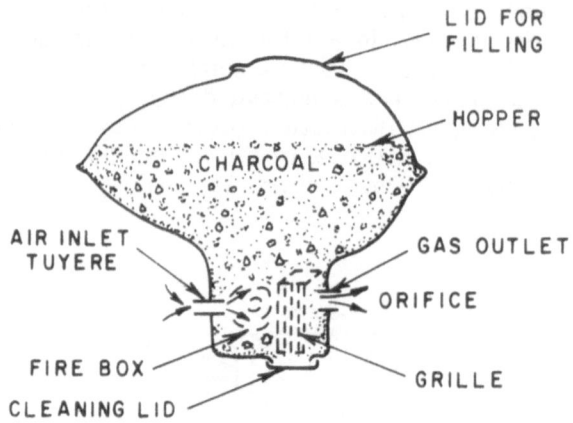

Figure 71. Volvo Crossdraft Gas Producer (28).

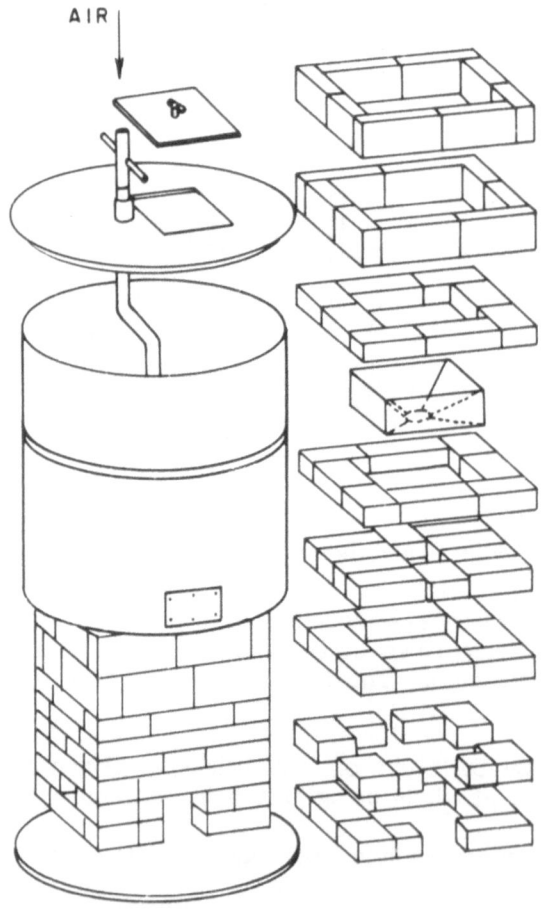

Figure 72. Gasifier Appropriate for Developing Countries (12).

78

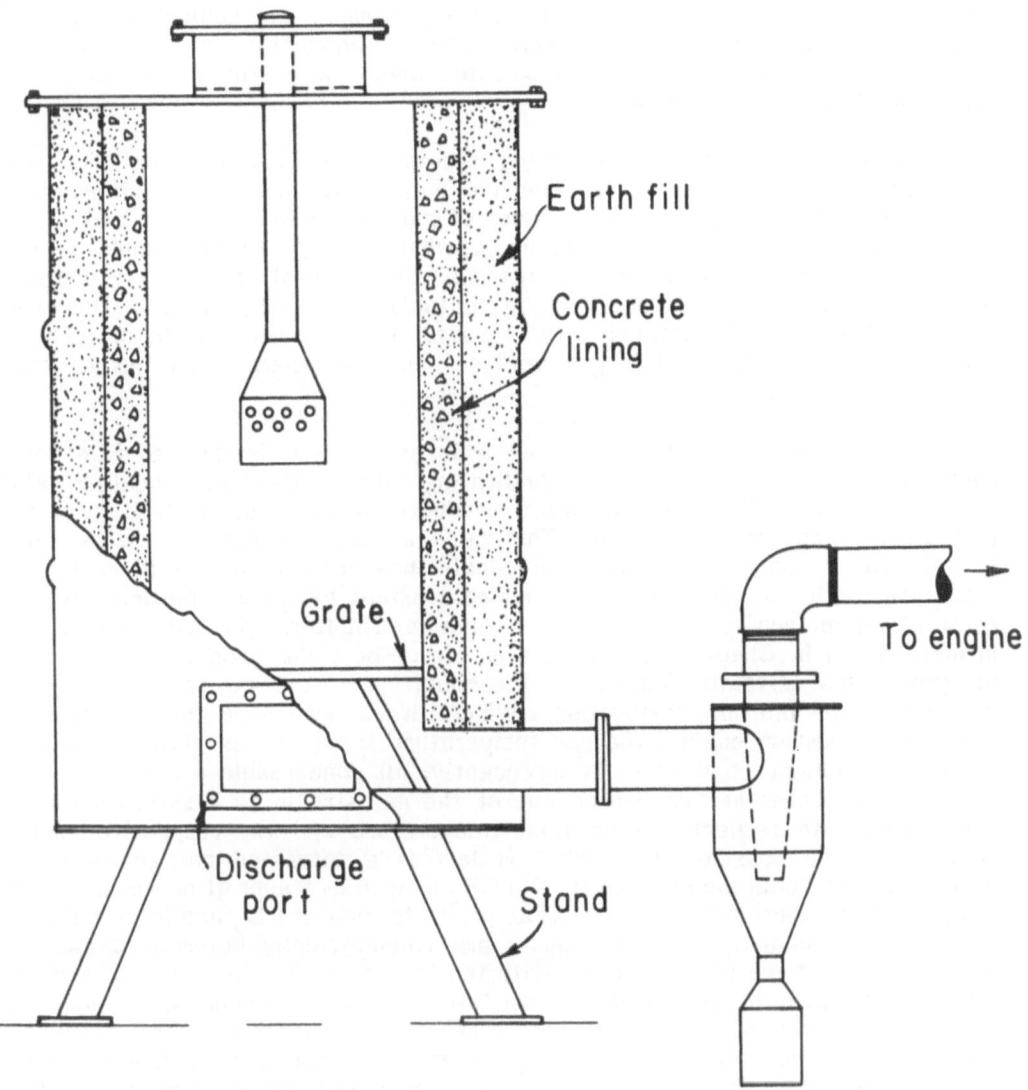

Figure 73. Design of a Simple Downdraft Gas Producer (6).

The history of gas producers reveals amazing designs and applications as well as hundreds of patents for gas producers during the booming years of gasification. References (10, 22, 24, 25) give a selected overview of the major designs of small, medium and large gasifiers.

When discussing an energy conversion system such as a gas producer, the efficiency of the unit will be the decisive factor with the present situation of tight fuel supplies. Reported efficiencies of gas producers should be taken with caution as long as it is not stated under what assumptions the numbers have been derived. The definition of the thermal efficiency of an energy conversion device is simply the ratio: (Useful energy output)/(total energy input). However, because of this simple definition there seems to be a wide range of opinions about the useful effect of a gas producer and the energy that has actually entered the system.

The energy contained in one kg of fuel is in most cases determined in a bomb calorimeter which measures the higher heating value. (The higher heating value includes the heat released by the water produced in the total combustion of the fuel when it condenses to liquid). This value is certainly not the energy going into the gas producer. The energy available in a gas producer is given by the net heating value of the fuel derived under constant pressure conditions and not constant volume conditions as given in a bomb calorimeter. This will be explained in more detail in Chapter V. The useful effect of a gas producer is a matter of opinion and certainly depends on the condition of the gas before it is used in a burner or internal combustion engine. When used as a fuel to drive an internal combustion engine, the gas temperature should be as close as possible to normal ambient conditions. Consequently, all condensable products such as tar, oils and water are condensed out of the gas stream to a saturation level determined with respect to the final temperature and the respective partial pressure of the constituents. What is left is a saturated gas consisting of combustible products such as CO, H_2 and CH_4 as well as traces of non-condensable higher hydrocarbons and non-combustible products such as CO_2 and large amounts of N_2. The quantities of these gases are volumetrically determined and the available heat of the gas calculated with the help of well established data about the heating value of the various constitutents. When fueling an engine, the thermal efficiency of a gas producer is around 70% under the most favorable conditions and can drop sharply to any lower level depending on how and with what fuel the plant is operated. The losses accounted for are due to radiation and convection from the producer body, solid carbon discharged with the ash, condensed products of distillation such as tars and oils and the considerable energy needed to evaporate the fuel moisture and heat it up to the prevailing equilibrium temperatures. In general these losses can be controlled to a certain extent through constructive measures such as preheating the air blast with the sensible heat of the outgoing gas stream and insulating the entire producer and parts of the piping system. In extreme cases such as portable units in subfreezing weather or high moisture content of the fuel, this is not only recommended but a necessity to obtain the required temperatures for the generation of a high quality gas. On the other hand, the allowable amount of sensible heat in the raw gas depends solely on the design of the purification system which, in most cases, operates within a narrow temperature range.

The useful energy from a gas producer when the gas is used to fire a burner is significantly higher. In practice the gas should enter the burner at a temperature as close as possible to the exit temperature of the gas producer. This will leave the tars and oils in vapor form, suitable for combustion in the burner. They will not be condensed out of the gas stream if the primary combustion air is above the lowest vapor dew point and are not considered a loss. In addition, almost no sensible heat is lost. This so-called hot gas efficiency can be as high as 95% under the most favorable conditions. Both the cold and hot gas efficiency of a gas producer are relatively high if compared to other energy conversion devices such as steam plants. In particular for small scale units, the advantage lies clearly with the gas producer system.

In a broad sense biomass gas producers use solar energy as a fuel and fall therefore within the category of new sources of energy called "appropriate energy" now being suggested as substitutes for oil. The discussion about how efficient appropriate energy generating devices are is fundamentally based on the net energy question: How efficient is an energy generating system in recovering the energy from non-renewable resources that have been used to build it? Although this question does not apply to gas producers as well as it does to other energy conversion devices, such as solar cells which in some cases can not recover the energy that has been used to build them, it is worthwhile to contemplate the position of a gas producer-engine system in this broader context (4).

The dimensioning of a gas producer is sometimes a difficult task, in particular when the unit is used for unsteady conditions or fired with fuels whose thermal behavior is not well known. The major part of the gasifier is used as a fuel storage space and its size depends on the bulk density of the feed material and the desired period for refueling the plant.

For instance, the average available net energy from one kg of wood with 10% moisture equals 16.4 MJ. This is considerably lower than the higher heating value of 20 MJ/kg. When the gas is used to drive a spark ignition engine, a cold gas efficiency of 70% and an engine efficiency of 18% may be reasonable. Consequently 1.3 kg of feed material (wood with 10% moisture) must be gasified each hour to run a 1 hp engine. In this context it is interesting to compare this calculated number with approximate fuel consumption per hp-hour as given by the manufacturer of various past systems. The actual fuel consumption depends heavily on where and how the vehicle is driven.

The load on a gasifier is most commonly expressed in terms of the specific gasification rate, the amount of dry fuel in kg that can be gasified per square meter of the "grate area" in one hour (kg/m^2-h). This definition can not readily be applied to crossdraft gas producers because the partial combustion zone can expand in all three dimensions very easily. The reader is referred to References (2) and (16) for sizing a crossdraft gas producer. In downdraft gas producers the "grate area" refers to the narrowest section of the throat. In updraft gas producers the section of the grate within the fire lining should be used as the relevant grate area. Gas producers, depending on the mode of running (up, down

Trade Name	Fuel	Heating Value (base not specified) MJ/kg	Consumption kg/hp-hour
Malbay	Charcoal	—	0.53
	Low Temperature Coke	29.5	0.56
	Anthracite	32.4	0.46
Wisco	Charcoal	—	0.40
	Low Temperature Coke	33.7	0.45
Imbert	Air dry wood	—	0.8-1.0
Humboltz Deutz	Anthracite	32.6	—
Gohin Poulence	Low temperature coke	—	0.47
Koela	Charcoal	32.2	0.45
	Low temperature Coke	30.7	0.45-0.49
	Anthracite	34.5	0.45-0.49
Swedish WW II model	Wood at 20% moisture	14.7	1 Avg.
Swedish Model (1957-63)	Birch wood 12% moisture	—	0.75-1.3

or crossdraft), can work only within certain limits of their specific gasification rate. For instance, a downdraft gas producer generates a highly tar laden gas when operated below a certain specific gasification rate. In addition, the CO and H_2 fractions in the producer gas are greatly favored by high temperatures and will therefore decrease at lower gasification rates and reach a point where the gas is not any longer suitable for combustion. Consequently a minimum specific gasification rate is required to maintain temperatures high enough for efficient tar cracking and good gas quality. On the other hand, too high a specific gasification rate leads to an excessive amount of unburned carbon in the ash and, in general, decreases the efficiency and increases the pressure drop and the temperature to a point where either the gas producer or the cleaning equipment are suspectible to damage. This latter case was well known to manufacturers of portable units that were usually sold with one or two sets of spare tuyeres, grates and even throats because overheating the plant was quite common on long uphill drives with truck engines. To what extent the allowable specific gasification rate varies with the fuel and whether the dependence is significant enough to shift the range established for coal and charcoal is difficult to answer. At the throat area in a gas producer all feedstock is present in a highly carbonized form and the allowable highest specific gasification rate depends heavily on physical and chemical properties of the fuel such as surface area and ash content. Intuitively, a highly reactive porous wood charcoal exposes a much larger and easily accessible surface to the reactant oxygen than densified coke. Charcoal can therefore be gasified faster per unit grate area than coke, provided the throat area can handle the high temperatures involved.

Extensive tests during the Second World War and the 1957 to 1963 period in Sweden established recommended dimensions and ranges of operation for downdraft gas producers. The numbers in Table 5 are derived from the experimental bench tests and road trails conducted in Sweden over several years. The gas producer tested is shown in Figure 74. The dimensions of the firebox, tuyeres, throat and grate as well as placement relative to each other are given in Figure 75.

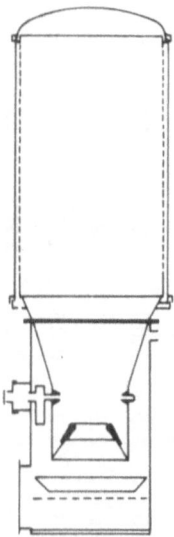

Figure 74. Swedish Downdraft Gas Producer (20).

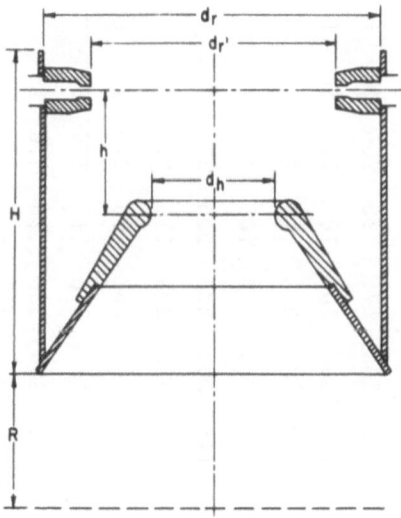

Figure 75. Dimensions for Downdraft Gas Producer with Wall Tuyeres (20).

Table 5. Dimensions for Swedish Downdraft Gas Producers (20).

d_r/d_h	d_h mm	d_r mm	d'_r mm	h mm	H mm	R mm	A no.	d_m mm	$\frac{A_m \times 100}{A_h}$	$\frac{d_r}{d_h}$	$\frac{h}{d_h}$	max. Nm^3/h	min. Nm^3/h	kg/h	v_m m/s
268/60	60	268	150	80	256	100	5	7.5	7.8	4.5	1.33	30	4	14	22.4
268/80	80	268	176	95	256	100	5	9	6.4	3.3	1.19	44	5	21	23.0
268/100	100	268	202	100	256	100	5	10.5	5.5	2.7	1.00	63	8	30	24.2
268/120	120	268	216	110	256	100	5	12	5.0	2.2	0.92	90	12	42	26.0
300/100	100	300	208	100	275	115	5	10.5	5.5	3.0	1.00	77	10	36	29.4
300/115	115	300	228	105	275	115	5	11.5	5.0	2.6	0.92	95	12	45	30.3
300/130	130	300	248	110	275	115	5	12.5	4.6	2.3	0.85	115	15	55	31.5
300/150	150	300	258	120	275	115	5	14	4.4	2.0	0.80	140	18	67	30.0
400/130	130	400	258	110	370	155	7	10.5	4.6	3.1	0.85	120	17	57	32.6
400/150	150	400	258	120	370	155	7	12	4.5	2.7	0.80	150	21	71	32.6
400/175	175	400	308	130	370	155	7	13.5	4.2	2.3	0.74	190	26	90	31.4
400/200	200	400	318	145	370	153	7	16	3.9	2.0	0.73	230	33	110	31.2

Column groups: Range of gas output (max. Nm^3/h, min. Nm^3/h); Maximum wood consumption (kg/h); Air blast velocity (v_m m/s).

Variables not given in Figure 75 are defined as follows for Table 5:

d_m = inner diameter of the tuyere

A_m = sum of cross sectional areas of the air jet openings in the tuyeres

A_h = cross sectional area of the throat

A = number of tuyeres.

On the average one cubic meter of producer gas before it is mixed with air in the carburetor contains the energy equivalent of 2 hp-hours. Assuming an engine efficiency of 0.20, the smallest model in Table 5 could provide satisfactory gas for an engine with a power range of 1.6 to 12 hp whereas the largest unit would perform well over a range from 14 to 92 hp.

For automobile, truck or bus applications, it is important to keep in mind that the gas producer should provide the engine with good gas under idling conditions as well as under full load. The turn-down ratio of a gas producer is expressed as the ratio:

$$\frac{\text{max. permissible specific gasification rate}}{\text{min. permissible specific gasification rate}}$$

The maximum permissible gasification rate for downdraft units is a well established number and given as 1 Nm^3/cm^2-h in gas output for wood. A minimal permissible specific gasification rate of 0.3 Nm^3/cm^2-h has been found suitable for Imbert type gas producers. However, the lower limit depends heavily on the shape of the throat and how well the plant is insulated as well as the number of tuyeres. Obviously, five or more tuyeres give a much better oxygen distribution and therefore a more homogenous firebed. In fact three tuyeres have been shown to be insufficient even under normal loads. A turn-down ratio of 4 to 6 for most gas producers seems to be sufficient for operation of an automobile because the ratio of highest to smallest number of crankshaft revolutions, which is directly correlated to the gas production, rarely exceeds 6 in normal operation. It is interesting to compare the limits of the specific gas production rate with the more practical quantity, the specific fuel consumption rate. For instance the 300/150 model with a throat area of 176.71 cm^2 can gasify at most 67 kg of wood per hour. This corresponds to 3,791 kg/m^2-h. The lower limit equals 509 kg/m^2-h, based on the minimal gas production rate of 18 Nm^3/h. One can see that these specific gasification rates are much higher than what is usually found for updraft gas producers. Fuel consumption below these established limits for the Swedish downdraft gas producer design does not mean a lower gas heating value. To the contrary the heating value of the gas will stay the same or become even higher, since CH_4 production will increase at lower temperatures. However, the gas will become unsuitable for internal combustion engines.

The UCD laboratory gas producer shown in Figure 76 is a modified form of the Swedish design, except for the castable throat that was replaced by a simple choke plate which is easier to build and not as susceptible to thermal stress. With this unit the specific gasification rates for crop residues were within the range of 225 kg/m^2-h to 5020 kg/m^2-h. Despite the great range in specific gasification rates, heating values of 6 to 8 MJ/m^3 were obtained. The gas was never used to run an internal combustion engine over a prolonged period since priority was given to evaluating the gasification characteristics of crop residues, in particular the tendency for slagging. Downdraft gas producers, from what is known so far about their performance, seem to operate best at a medium specific gasification rate. Units with more than 300 hp capacity, which corresponds to a fuel (wood) consumption of 250-300 kg/h, seem to be difficult to operate with small size or high ash fuels (12).

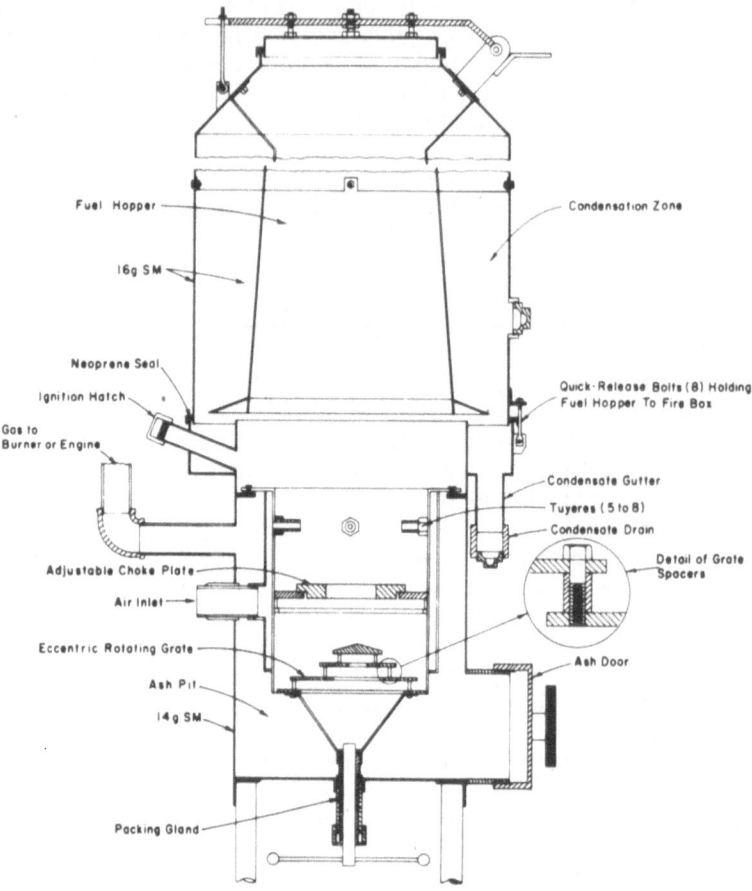

Figure 76. UCD Laboratory Gas Producer (11).

Specific gasification rates for updraft gasifiers are within the range 100 kg/m^2-h to 300 kg/m^2-h. The higher rates apply to slagging gasifiers whereas the lower rates prevent the ash from slagging in most low ash fuels. There is little concern about a lower limit in updraft gas producers because the tar production of an updraft gasifier takes place in the upper parts and the distillation products are drawn off without passing through an incandescent carbon bed. The lower limit of the specific gasification rate is therefore determined by the amount the temperature in the partial combustion zone drops below the limit where the gas generated becomes too wet and too difficult to ignite or sustain a flame.

It is not possible to compare the various gas producers and fuels and decide which combination would give an optimal performance. Heating values of producer gas are usually given on a dry basis. This value tells very little about the actual condition of the gas and its usefulness as a fuel for internal combustion engines. For instance, the gas might be too wet to ignite or loaded with tar and still have a high heating value, which may lead to premature conclusions about the suitability of the gas.

86

Figure 77. Rotating Eccentric Grate (11).

Comparison of updraft, downdraft and crossdraft gas producers is usually based on theoretical calculations of efficiencies and gas composition assuming equilibrium conditions and seldom on comparable experimental data. Using theoretical calculations for comparison, the updraft gas produer is most efficient and the crossdraft gas producer is least efficient. This result is mostly due to the higher gas exit temperature in downdraft and crossdraft gas producers. Neither theoretical calculations nor the existing experimental data are consistently specific on what base the data were derived. For instance, the common practice to report gas heating values per unit volume does not allow comparing data unless it is clearly stated what temperature and pressure the volume refers to. Gas heating values are usually calculated with the help of enthalpy tables. Since all these tables are in reference to some base, most commonly 0, 15 or $25^{\circ}C$ and 1 atm, and the physical state of the water can be either steam or liquid vapor, it is rather moot to compare heating values of producer gas. It is not uncommon to come across heating values from 4 MJ/m^3 to 6 MJ/m^3 for the same gas, depending on how the value was obtained.

Quite frequently the gas heating values for crossdraft and downdraft producers are less than the heating values of updraft units. For instance, an older report of the U.S. Bureau of Mines gives the average heating value of downdraft gas at 4.85 MJ/Nm^3 compared to 5.95 MJ/Nm^3 for updraft. Others have compared gas composition and heating value of various systems and some examples of their findings are given in Tables 6 and 7. Discussion about the heating content of the raw gas obtained from various systems are inconclusive and giving preference to an updraft gas producer based on the better heating value and efficiency is not always possible. In fact the heating value of the raw gas should receive the least consideration since differences in heating values of magnitude 20% have little effect on the power output of the engine and are in the range of what can be gained through improved piping connections or intake manifold.

Table 6. Gas Quality Versus Downdraft and Updraft Gas Producer for Charcoal and Anthracite Fuels (10).

	CO	H_2	CH_4	CO_2	O_2	N_2
Charcoal						
Updraft	30.8	12	0	3.6	0.4	53.5
Downdraft	23	14	0.9	7.0	0.2	54.9
Anthracite						
Updraft	29.3	6.8	1.6	1.4	0.3	60.6
Downdraft	22	12.0	1.1	6.0	1.0	57.9

Correlations between fuel, gas producer and gas composition are impossible to make since the process depends on too many variables which vary with the design of the gas producer. Reported gas composition data represents only the condition over a very small time interval. Tests conducted with the UCD laboratory gas producer and the UCD Civil Engineering gasifier revealed a rather unstable gas composition even over a short time period (see Figure 16). Tables 8-13 list various fuels gasified in various gas producers with and without steam injection to show that generalized correlations would be misleading.

Table 7. Gas Quality Versus Updraft and Crossdraft Gas Producers for Four Different Coal Fuels (17).

No. 1 — Updraft producer dry gasification

No. 2 — Updraft producer wet gasification

No. 3 — Crossdraft producer dry gasification

No. 4 — Crossdraft producer dry gasification

Fuel Analysis, Percent Weight

	No. 1.	No. 2.	No. 3.	No. 4.
Moisture	2.1	5.0	2.67	4
Ash	7.9	3.5	6.31	4
Volatile matter	9.6	5.5	6.25	6
Carbon	80.4	86.5	84.77	86.2
Sulphur	—	—	—	0.5-0.8
Water to coal ratio	15-40 % weight			
Water decomposed	80-85 % weight			

Gas Analysis, Percent Volume

	No. 1	No. 2	No. 3	No. 4
CO_2	0.6	1.6	1.7	1.0
CO	24.4	29.0	29.3	30.5
$H_2 + CH_4$	13.1	15.0	9.2	8.0
O_2	0.6	0.5	—	0.5
N_2	61.3	48.5	59.8	59.0
Heating value MJ/Nm^3	5.3	5.9	5.2	5.2
Percent combustible gas	37.5	44.0	38.5	38.5

Table 8. Gas Composition Versus Fuel. University of California, Davis Laboratory Downdraft Gas Producer (15).

Hogged Wood Manufacturing Residue

	% by weight
% fuel moisture content (wet basis)	10.8

Gas Composition

	% by volume
CO_2 in dry gas	6.6
CO in dry gas	29.0
H_2 in dry gas	13.6
O_2 in dry gas	0
CH_4 in dry gas	6.3
C_2H_6 in dry gas	0.3

Cracked Walnut Shell

	% by weight
% fuel moisture content (wet basis)	8.0

Gas Composition

	% by volume
CO_2 in dry gas	8.7
CO in dry gas	20.1
H_2 in dry gas	18.4
O_2 in dry gas	0
CH_4 in dry gas	4.845
C_2H_6 in dry gas	0.255

Prune Pits

	% by weight
% fuel moisture content (wet basis)	8.24

Gas Composition

	% by volume
CO_2 in dry gas	9.7
CO in dry gas	23.9
H_2 in dry gas	16.3
O_2 in dry gas	0
CH_4 in dry gas	8.17
C_2H_6 in dry gas	0.43

Corn Cobs

	% by weight
% fuel moisture content (wet basis)	11.0

Gas Composition

	% by volume
CO_2 in dry gas	10.2
CO in dry gas	21.7
H_2 in dry gas	16.9
O_2 in dry gas	0
CH_4 in dry gas	4.465
C_2H_6 in dry gas	0.235

75% Barley Straw, 25% Corn Stover (Cubed)

	% by weight
% fuel moisture content (wet basis)	6.9

Gas Composition

	% by volume
CO_2 in dry gas	10.9
CO in dry gas	20.9
H_2 in dry gas	13.4
O_2 in dry gas	0
CH_4 in dry gas	4.94
C_2H_6 in dry gas	0.26

Chipped Municipal Tree Prunings

	% by weight
% fuel moisture content (wet basis)	17.29

	% by volume

Gas Composition

CO_2 in dry gas	13.7
CO in dry gas	18.8
H_2 in dry gas	16.4
O_2 in dry gas	0
CH_4 in dry gas	4.75
C_2H_6 in dry gas	0.25

**1/4" Pellets: 75% Walnut Shell, 15% Rice Straw,
10% Saw Dust**

	% by weight
% fuel moisture content (wet basis)	7.1

Gas Composition

	% by volume
CO_2 in dry gas	8.4
CO in dry gas	26.1
H_2 in dry gas	12.4
O_2 in dry gas	0
CH_4 in dry gas	7.79
C_2H_6 in dry gas	0.41

Experiments in Finland with peat as a fuel in a downdraft gas producer yielded the gas composition shown in Table 9. The experiments also indicated that peat as moist as 50% and with a high fraction of fine material (30%) could still be gasified in this downdraft gasifier.

Table 9.   Gas Composition for Peat in Modified Imbert Gas Producer (7).

	% by volume
H_2	10.7 - 13.9
CO	11.0 - 21.3
CO_2	8.8 - 21.8
CH_4	0.5 - 1.0

Ebelmen's liquid slag gas producer fueled with charcoal and operated with dry air blast yielded the following gas composition:

Table 10. Gas Composition Obtained From One of the Earliest Updraft Gas Producers (22).

	% by volume
CO_2	0.5
CO	33.3
H_2	2.8
CH_4	None
N_2	63.4

An automotive crossdraft gas producer (6 hp) fueled with charcoal at 14.5% moisture content yielded the following gas composition:

Table 11. Gas Composition from a Wishart-H.S.G. Gas Producer Fueled with Charcoal (1).

	% by volume	
gas	dry blast	wet blast 9g H_2O per minute
CO_2	1.4	1.6
O_2	0	0.2
CO	31.9	33
CH_4	0.6	0.4
H_2	4.1	8.3
N_2	62.0	56.4

The Heller stationary updraft gas producer was one of the simplest gas producers ever built. Fuel gasified was coal with 29.3% moisture and 5.93% ash. With a dry air blast the following range in gas composition was obtained:

Table 12. Gas Composition of a Large Updraft Gas Producer (22).

	% by volume
CO_2	4.6 -7.8
CO	24.3 -29.4
CH_4	1.5 -3.6
H_2	16.8 -19.8
N_2	44.4 -48.3

Gas composition is shown in the following table for the Gohin-Poulenc Gas Producer fueled with semi-coke of 2.5% ash content and a dry air blast. The gas heating value deteriorates with distance traveled which occurs because of insufficient moisture inside the gas producer after a long journey without refueling.

Table 13. Gas Composition of an Automotive Crossdraft Gas Producer.

Change of Gas Composition during Use (3).

After a journey of	% by volume				
	CO_2	CO	H_2	CH_4	N_2
10 kilometers	1.7	24.9	15.9	2.1	55.4
54 kilometers	2.8	26.4	11.8	2.3	56.5
198 kilometers	2.7	27.8	7.4	1.8	60.2
218 kilometers	2.4	29.1	7.1	0.8	60.8
246 kilometers	3.1	30.5	4.1	0.5	61.8
250 kilometers	2.7	29.9	3.2	0.3	63.8

When comparing gas compositions the main interest lies in the amount of CO and H_2. Both gases have about the same heating value, Table 1, and are the major products of gasification. Methane production is low after the plant has been brought up to normal running temperatures. However, even small quantities of methane can contribute significantly to the heating value of the gas. This is shown in the nomogram below. The lower heating value of the producer gas can be determined provided the volume percentage of H_2, CO, CH_4 and higher hydrocarbons in the gas are known.

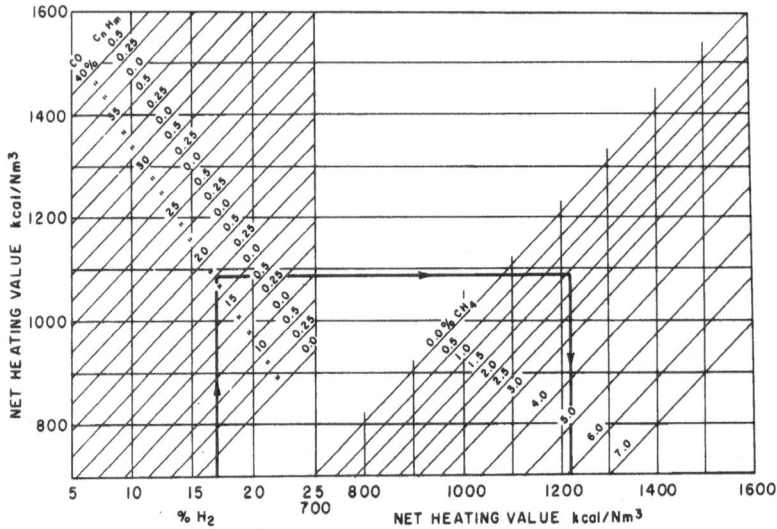

Nomogram for Lower Heating Value of Producer Gas (24). The example refers to a gas with 17% H_2, 20% CO, 0.25% C_nH_m and 1.5% CH_4.

The initial development of small portable gas producers was mostly based on experience with large updraft coal gasifiers. In such plants it was quite common to inject steam with the air blast into the partial combustion zone which leads to increased hydrogen production by decomposing water at high temperatures, Figure 17. Whether this practice is of any use in small portable or stationary gas producers is highly questionable.

All tests done with biomass fuels at UCD indicate that there is enough moisture in the fuel to generate a sufficient amount of hydrogen. These tests were conducted under mostly dry, hot climatic conditions with fuel moisture contents as low as 5%. There does not seem to be a downdraft gas producer designed for steam or water injection. To the contrary, too wet gas and too much moisture in the fuel were the most reported difficulties with downdraft gas producers. One has to keep in mind that the moisture content of air dried biomass fuels will rarely be below 15% in non-arid zones with high air humidity all year around. There have been a few small portable updraft and crossdraft gas producers on the market during the 1940-1950 period with steam or water injection. In case the fuel is extremely dry, steam injection will certainly have a beneficial effect on the heating value of the gas. However, the gain in heating value may not justify the additional complications for such a system. The amount of steam injected must follow the load on the gas producer, otherwise the partial combustion zone will be cooled down too much when idling the engine. In future research with high ash fuels having a tendency for severe slagging in any gas producer, steam or water injection may be the only practical solution to keep the temperature in the partial combustion zone below the ash fusion point. This method, however, will require sophisticated temperature sensing and a steam injection device capable of partially smoothing out the fluctations in gas composition that are unavoidable with alternate cooling and heating of the partial combustion zone.

Reference Chapter IV

1. Anonymous, Gas Producers, Automobile Engineer, November 5, 1942, pp. 433-464.

2. Anonymous, Gas Producer Tests, Automobile Engineer, v 31, n 417, 1941, pp. 418-420.

3. Anonymous, Producer Gas for Road Vehicles, Engineer, v 163, n 4248, 1937, pp. 682-684.

4. Baumol, W. J. and S. A. Blackman, Unprofitable Energy is Squandered Energy, Challenge, v 23, n 3, 1980, pp. 28-35.

5. Caram, H.S. and N.R. Amundson, Diffusion and Reaction in a Stagnant Boundary Layer About a Carbon Particle, Part 1, Ind. Eng. Chem. Fundam. v 16, n 2, 1977, pp. 171-181.

6. Cruz, I. E., A Status Report on Studies Conducted by the University of Philippines College of Engineering on Alternative Fuels for Internal Combustion Engines, University of Philippines, Mechanical Engineering, Diliman, Quezon City, 1980.

7. Ekman, E. and D. Asplund, A Review of Research of Peat Gasification in Finland, Technical Research Centre of Finland, Fuel and Lubricant Research Laboratory, 02150, Espoo 15 Finland, 1972.

8. Edmister, W. C. et al., Thermodynamics of Gasification of Coal with Oxygen and Steam, Transaction of the American Society of Mechanical Engineers, v 74, July 1952, pp. 621-636.

9. Foster-Wheeler Co., Air Gasification, Retrofit '79 Proceedings, The Solar Energy Research Institute, SERI/TP-49-183, Seattle, Wash., February, 1979.

10. Goldman, B. and N. C. Jones, The Modern Portable Gas Producer, Institute of Fuel, London, v 12, n 63, 1939, pp. 103-140.

11. Goss, J. R., An Investigation of the Down-Draft Gasification Characteristics of Agricultural and Forestry Residues: Interim Report, California Energy Commission, p500-79-0017, November, 1979.

12. Groeneveld, M. J. and K. R. Westerterp, Social and Economical Aspects of the Introduction of Gasification Technology in the Rural Areas in Developing Countries, American Chemical Society, Symposium on Thermal Conversion to Solid Wastes and Biomass, Washington, D. C., September, 1979.

13. Gumz, W., **Gas Producers and Blast Furnaces**, John Wiley and Sons, N.Y., 1950.

14. Harter-Seberick R., Motor Gas-Producers and Their Fuels, Their State of Development with Special Reference to the Use of Fossil Fuels, Fuel, London, v 16, n 1, 1937.

15. Jenkins, B.M., Downdraft Gasification Characteristics of Major California Residue-Derived Fuels, Ph.D. Thesis in Engineering, Univ. of California, Davis, 1980.

16. Kaye, E. and A. Burstall, Gas Producers for Motor Vehicles, Institution of Engineers, Australia Journal, v 14, n 4, 1942, pp. 81-93.

17. Lowe, R., Gas Producer as Applied to Transport Purposes, Journal of the Junior Institute of Engineers, June 1940, pp. 231-253.

18. Lustig, L., New Gas Producer for Dual Fuel Engines, Diesel Progress, v 13, n 5, 1947.

19. Mon, E. and N.R. Amundson, Difussion and Reaction in a Stagnant Boundary Layer About a Carbon Particle, Part 2: An Extension, Ind. Eng. Chem. Fundam., v 17, n 4, 1978, pp. 313-321.

20. Nördstrom, Olle, Redogörelse för Riksnämndens för ekonomisk Försvarsberedskap forsknings- och försöksverksamhet på gengasområdet vid Statens maskinprovninger 1957-1963, (from) Overstyrelsen for ekonomisk forsvarsberedskap, Sweden, January, 1962.

21. Parker, H. W. and L. H. Holmes, Alternative Energy Sources for Agricultural Applications Including Gasification of Fibrous Residues, Texas Energy Advisory Council, Energy Development Fund, 1979.

22. Rambush, N. E., **Modern Gas Producers**, Van Nostrand Co., New York, 1923.

23. Savage, P. R., Slagging Gasifier Aims for SNG Market, Chemical Engineering, September 12, 1977, pp. 108-109.

24. Schläpfer P. and J. Tobler, **Theoretische and Praktische Untersuchungen über den Betrieb von Motorfahrzeugen mit Holzgas**, Schweizerische Gesellschaft für das Studium der Motorbrenstoff, Bern, Switzerland, 1937.

25. Skov, N. A., and M. L. Papworth, **The Pegasus Unit,** Pegasus Publisher. Inc., Olympia, Washington, 1975.

26. Takeda, S., Development of the Gas Engine, The Bulletin of the Faculty of Agriculture, Mie University, Tsu, Japan, n 58, 1979, pp. 137-141.

27. Takeda, S. and J. Sakai, Research on Gas Engine Driven by Agricultural Waste, The Bulletin of the Faculty of Agriculture, Mie University, Tsu, Japan, n 53, 1976, pp. 187-203.

28. Taylor, G., Gas Generators Capture European Interest, Automotive Industries, v 82, n 1, 1940, pp. 22-25.

29. Vigil, S. A. and G. Tchobanoglous, Thermal Gasification of Densified Sewage Sludge and Solid Waste, Water Pollution Control Federation Conference, Las Vegas, Nevada, October, 1980.

30. Williams, R. O. and B. Horsfield, Generation of Low-BTU Gas from Agricultural Residues, Experiments with a Laboratory-Scale Gas Producer, Proceedings of Cornell Agricultural Waste Management Conference, New York State College of Agriculture and Life, 1977.

31. Woods, M. W., Producer Gas Vehicles, Institution of Engineers, Australia, v 10, n 3, 1938, pp. 89-96.

CHAPTER V: FUEL

During 140 years of commercial gasification almost every possible lignocellulosic or carbonaceous fuel has been more or less successfully gasified. However, the development work was done with the most common fuels such as coal, wood and charcoal. The normal approach was to build a gasifier and then search for a fuel that could be gasified in the unit. This practice has led to a misleading classification of fuels into suitable and unsuitable for gasification. There are fuels which have a long history of gasification such as coal and wood. From gasification of both these fuels three typical modes of gas producers evolved: updraft, downdraft and crossdraft. However the increasing use of producer gas for internal combustion engines made it necessary to obtain producer gas that was clean and cool. It was recognized that less obvious fuel properties such as surface, size distribution and shape have an important role in gasification as well as moisture, volatile matter and carbon content.

The most common classification of fuels is with regard to their gasification suitability in updraft, downdraft and crossdraft plants. Fuels with a high ash content and low ash melting point are troublesome when gasified in a downdraft or crossdraft gas producer. Fuels with the tendency to generate a considerable amount of tar when carbonized are less suitable for updraft gasification. Such a classification should serve only as a rough guideline. They have led in many cases to false expectations. The key to a successful design of a gasifier is to understand the properties and thermal behavior of the fuel as fed to the gasifier.

An attempt to classify potential fuels for gasification according to their parameters which have the greatest influence on gasification follows:

1. Energy content of fuel
2. Fuel moisture content
3. Size and form in which the fuel is gasified
4. Size distribution of the fuel
5. Bulk density of the fuel
6. Volatile matter content of the fuel
7. Ash content and ash chemical composition
8. Ultimate analysis of the fuel.

Energy content of fuel: The energy content of solid fuels is, in most cases, obtained in an adiabatic, constant volume bomb calorimeter. The values obtained are the higher heating values which include the heat of condensation from the water formed in the combustion of the fuel. The fuel heating value is also reported on a moisture and ash free basis, or on a moisture free basis only. In all cases these data do not represent the amount of energy available to the gasifier. The chemical process in a gasifier is most suitably described by a constant pressure process. In addition, much energy is needed to vaporize water and this energy is usually not recovered. Therefore, the energy that can be extracted from the fuel is less than most reported heating value data.

In order to avoid serious errors in the dimensioning of a gasifier and its economic assessment, the net heating value of the fuel should be assumed to be the

100

energy available to the gasifier per kg fuel as fed to the plant. The higher heating values for 19 major crop residues as obtained from bomb calorimeter tests can be estimated by the formula (21):

Heating value in kJ per kg oven dry matter = -8419.7 + 479.3 C + 667.6 H + 58.8 0 - 1207.7 S where C, H, 0 and S are carbon, hydrogen, oxygen and sulfur, respectively. This correlation has been obtained through a multiple regression analysis and is quite accurate for the crop residues listed in Table 15.

The computation of the net heating value is presented for cellulose ($C_6H_{10}O_5$). Cellulose contains 44.4% carbon, 6.2% hydrogen and 49.4% oxygen by weight. Assuming all the hydrogen in the fuel reacts with oxygen to form water, 0.558 kg of water are formed when combusting one kg of cellulose. In an air blown gas producer, one cannot assume that all the hydrogen reacts with oxygen to form water. However, the loss due to evaporation of the water is considerable and amounts to 1,365 kJ per kg of dry cellulose. Combined with the loss due to reaction at constant pressure, the net heating value of oven dry cellulose is 18.4 MJ/kg compared to 19.9 MJ/kg as obtained from bomb calorimeter tests. In all practical cases the fuel is fed into the gasifier with a certain moisture content, which is defined as the water driven of by heating at 105 $^{\circ}$C leaving oven dry cellulose as the final product. Figure 78 shows the considerable loss in fuel energy with moisture content in the case of cellulose. Consequently, the net heating value should be used when assessing the energy a potential biomass fuel can supply to the gasifier. Higher heating values of a select group of fossil and biomass fuels are given in Tables 14 and 15.

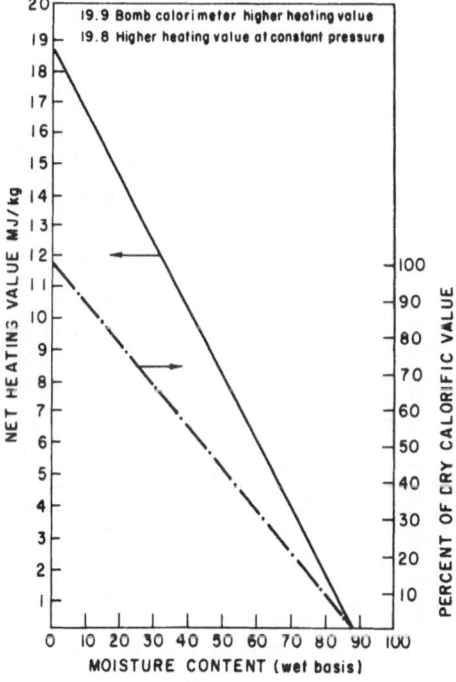

Figure 78. Net Heating Value of Celluose as a Function of Moisture Content (21).

Table 14. Moisture Content and Heating Values of Fossil Fuels

Fuel	Moisture Content % Wet Weight	Average Higher Heating Value MJ/kg Dry Basis	Reference
Coal, air dried			
Lignite	45	19.7	20
Subbituminous C	30	22.1	20
Subbituminous B	25	25.6	29
Subbituminous A	17	30.2	20
High Volatile C bituminous	17	30.2	20
High Volatile B bituminous	10	32.6	20
High Volatile A bituminous	4	33.7	20
Medium Volatile bituminous	5	34.8	20
Low Volatile bituminous	4	36	20
Semi Anthracite	3	34.9	20
Anthracite	3	33.7	20
Meta Anthracite	5	31.4	20
Bituminous Coal Char	—	28.1	27
Peat (Finland), average	40-70	22.5	13
Milled Peat, 40%-50% moisture, dry basis	—	7.5-12*	13
Sod Peat 30%-40% moisture, dry basis	—	11-14*	13
Peat briquettes 10%-15% moisture, dry basis	—	17-18.5*	13
Peat pellets 10%-20% moisture, dry basis	—	16.8-18.9*	13
30%-40% moisture, dry basis	—	12.6-14.7*	13
Gasoline	—	43.6	24
Diesel Oil	—	45	24

*Wet basis, Net Heating Value.

Table 15. Moisture Content and Heating Values of Non Fossil Fuels

Fuel	Moisture Content % Wet Weight	Average Higher Heating Value MJ/kg Dry Basis	Reference
Alfalfa seed straw, air dried	8	18.4	21
Almond shell, air dried	7	19.4	21
Barley straw	8-20	17.3	21
Bean straw	8-20	16.8	21
Beef cattle manure	—	14.6	24
Coffee hulls	70	28.8	4
Corn cobs	8-20	18.9	21
Corn stalks	8-20	18.3	21
Cotton gin trash	20	16.4	21
Cotton stalks	25-45	15.8	24
Flax straw, collected off ground	—	20	24
Furfural Residue	50	20	4
Olive pits, air dried	10	21.4	21
Peanut husks, air dried	—	19.7	4
Peach pits, air dried	11	23	21
Prune pits, air dried	8	23.3	21
Rice hulls	—	15	21
Sunflower hulls, oil type	—	20	24
Sunflower stalks, grown in greenhouse	—	21	24
Screened composted sewage sludge, 22% inorganic	—	9.9	24
Sewage sludge and wood chips, composted, 14% inorganic	—	15.2	24
Safflower straw cubes	9	19.5	21
Walnut shell (cracked)	7-10	21.1	21
Walnut shell (6 mm pellet)	7-10	20.4	21
Walnut hull	25-45	—	18
Wheat straw with 50% corn stalks	8-20	16.9	21
Wheat straw, collected behind a combine	—	18.9	4

Table 15 continued

Fuel	Moisture Content % Wet Weight	Average Higher Heating Value MJ/kg Dry Basis	Reference
Wood average	—	20	12
Pine bark	40–60	21	4
Pine, freshly felled	40	19.9	4
Fir, freshly felled	37	11.4*	4
Fir, seasoned	15–20	14.9*	4
Fir, kiln dried	8	17.8*	4
Beech, freshly felled	40	19	12
Birch, freshly felled	31	19	12
Oak, freshly felled	35	18.3	12
Wood Charcoal — mixed forest wood,			
Keyna native burned	—	31.3	17
Yarura wood British Guiana	—	30.1	17
English mixed hard wood, stationary retort	—	32.2	17
Japanese hard wood	—	31.9	33
Japanese Palm nut	—	32	33
Wood charcoal, average	2–10	29	17

*Wet basis, Net heating value

Fuel moisture content: In most cases there is very little choice of the fuel moisture content which may be desired for ease of operation, efficiency, optimal gas yield and heating value of the raw gas. The moisture content of most biomass fuels is determined by the type of fuel, its origin, and treatment before it is used as a fuel for gasification. Moisture in biomass can be fundamentally subdivided into three categories:

1. Inherent moisture is the moisture a fuel can hold when it is in equilibrium with ambient atmosphere at 96-97 percent relative humidity. Inherent moisture is held in capillary openings in the biomass.
2. Surface moisture is the moisture which occurs on the surface and is in excess of inherent moisture.
3. Decomposition moisture is the moisture formed from organic compounds of the fuel as they are decomposed by heating. Generally temperatures in the range of 200 $^{\circ}$C to 225 $^{\circ}$C are required, which is well above the temperatures required for expelling surface and inherent moisture.

The moisture content of fuels cited in the literature usually refers to inherent moisture plus surface moisture. Tables 14 and 15 list the average moisture content of the most common fuels under various conditions. Values in Tables 14 and 15 should only serve as an indication of the wide range of fuels with various moisture contents have been gasified in the past. These numbers are certainly not representative, since location and processing methods influence the moisture content of a fuel strongly. For instance, in most humid zones the air dried biomass fuel will seldom have a moisture content below 20%. Whereas, moisture contents below 10% are not rare in arid zones for air dried biomass fuels.

It is desirable to use fuel with low moisture content, because the loss due to evaporation of the fuel moisture is considerable and in most practical cases never recovered. Any fuel moisture will be heated up and evaporated from the heat supplied to the gas producer from partial combustion of the fuel. For the case of 25 $^{\circ}$C fuel temperature and 300 $^{\circ}$C raw gas exit temperature, 2875 kJ per kg moisture must be supplied by the fuel to heat and evaporate the moisture. This heat will not be recovered in most practical cases. The losses associated with the evaporation of the fuel moisture are given in Figure 79. The reader is cautioned that this heat loss represents only the heat of evaporation of inherent and surface moisture, not the heat loss caused by the decomposition moisture.

The theoretical limit of 88% moisture content for cellulose at which the fuel combustion is not any longer self-sustaining is indicated in Figure 78. In practice the moisture content at which fuel combustion can be sustained is much lower. For example, the point of highest moisture content for lignocelluosic material such as wood and crop residues at which combustion remains self-sustaining is about 70% for fuel with a higher heating value of 18.6 MJ/kg on a dry basis (22). Fuels with moisture contents as high as 50% have been gasified in downdraft gasifiers (13,21). The economics of gasifying fuel with such a high moisture content is questionable. Igniting the fuel becomes increasingly more difficult and the gas quality and yield are very poor.

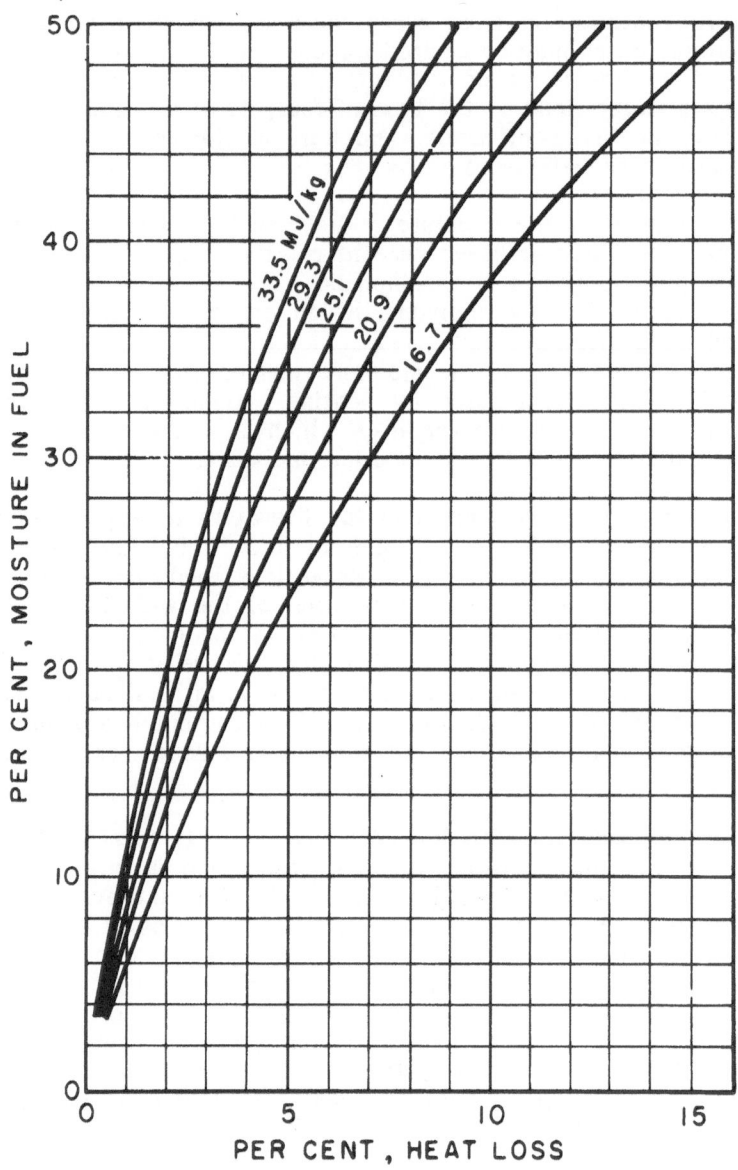

Figure 79. Moisture Content Versus Heat Loss for Fuels with Various
Higher Heating Values (27).

Schläpfer and Tobler have computed the gas composition and other properties
of the raw gas for various moisture contents (Figures 80 and 81). Although
these figures do not resemble the exact composition and properties of the raw
gas obtained from gasification of wood and crop residues with various moisture
contents, they show clearly the general trend of decreasing gas yield, heating
value and power output of the internal combustion engine with increasing moisture
content of the fuel. Experiments carried out at the University of California,
Davis, confirm this. The results are given in Figure 82.

The assumptions made are for usual running conditions of a portable wood gas producer and are given as follows:

Utlimate analysis of wood: 50% C, 6% H, 44% 0

Heating value: 18,834 kJ/kg dry basis

The watershift reaction:

$$H_2O + CO = CO_2 + H_2O + 41,854 \text{ kJ is in equilibrium at } 700 .^{o}C$$

Loss through convection and radiation: 15% of net heating value of the fuel

Exit gas temperature: 350 ^{o}C

CH_4 production: 0.040 Nm^3/kg dry wood.

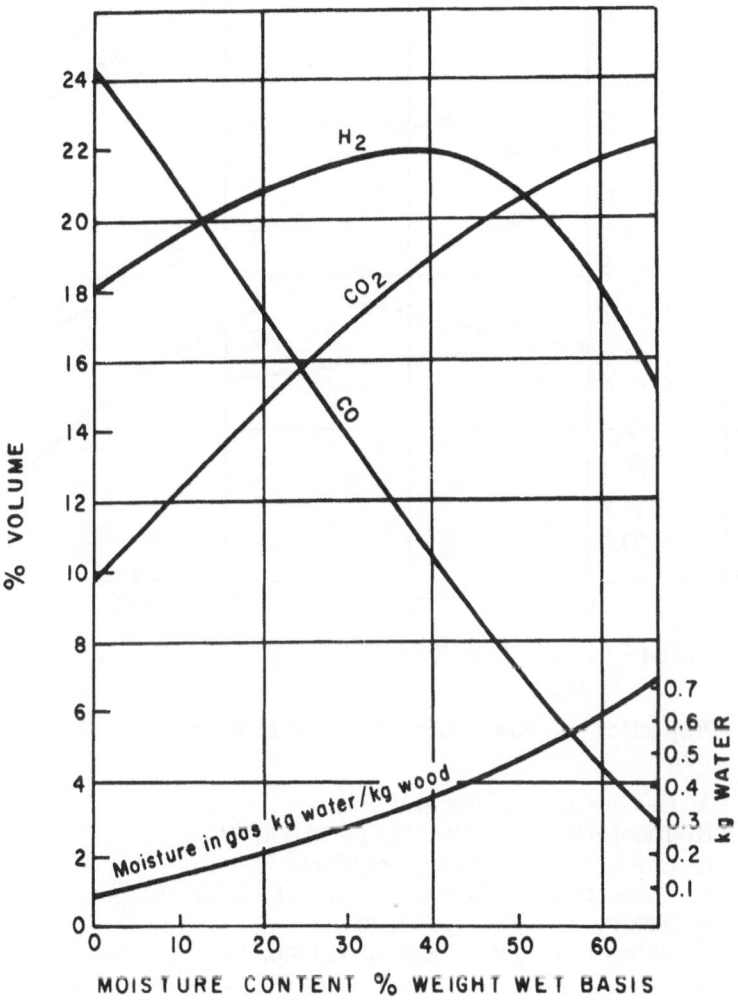

Figure 80.   Gas Composition as a Function of Moisture Content (30).

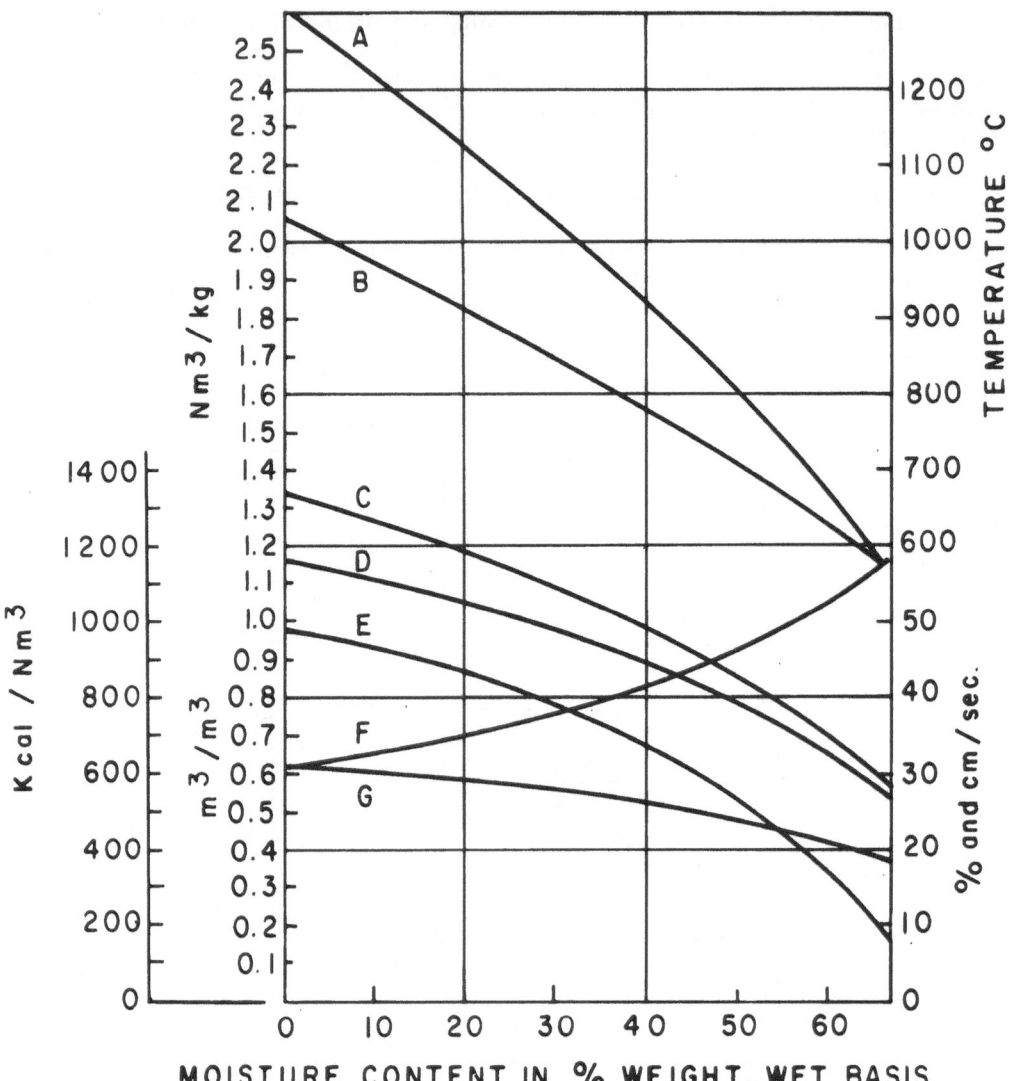

Figure 81. Properties of Wood Gas as a Function of Moisture Content (30).

A Gas yield, Nm^3/kg wet basis
B Theoretical reaction temperature, $^{\circ}$C
C Lower heating value, kcal/Nm^3
D Theoretical air/gas ratio, m^3 of air/m^3 of gas
E Maximum flame speed, cm/sec
F Theoretical power loss in engine, %
G Heating value of gas-air mixture at 0 $^{\circ}$C and 1 atm, kcal/Nm^3

Of importance is also the hydroscopic behavior of the fuel. For instance all biomass fuels will adjust their moisture content according to the relative humidity

108

of the air. In case of charcoal which is notorious for its ability to store water, the moisture content will quickly go up with the humidity of the surrounding air as shown in Figure 83.

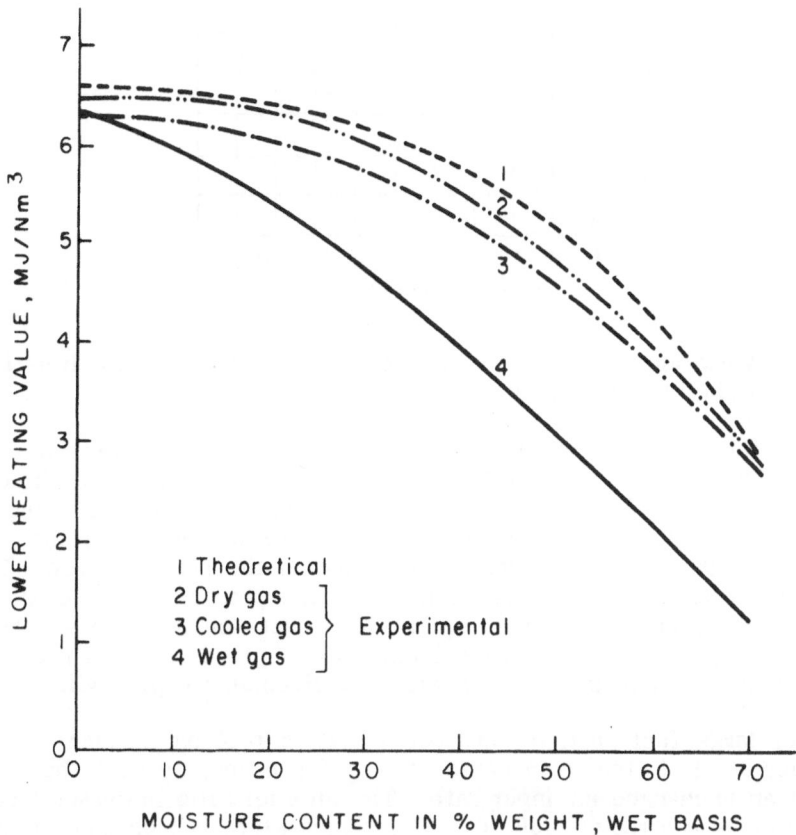

Figure 82.   Heating Value of the Gas as a Function of the Moisture Content of the Fuel on a Wet Basis (29).

The vast majority of reports and government regulations conclude that a moisture content below 15% by weight is desirable for trouble free, economical operation of a plant. In rare cases, the fuel may be even too dry and cause overheating of the reactor vessel, but such a situation is corrected with a wet air blast.

Fuel size: The fuel size influences the pressure drop across the gasifier and therefore the power that must be supplied to draw the air and gas through the plant. In the case of engine operation, the natural suction of the intake manifold has to overcome the pressure drop across the entire system. In theory it would be desirable to offer the incoming air as much fuel surface as possible to obtain a favorable gasification rate and high physical reaction speed. In the case of fine, mulled fuel the gasification could also be completed within a smaller fuel column. However, practical experience in portable gas producers has shown that there are certain limits to this.   Experiments with the small laboratory gas

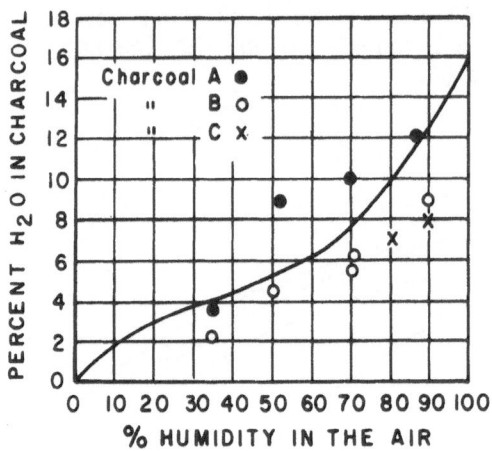

Figure 83. Variation in Moisture Content of Charcoal with Relative Humidity of the Air (2).

producer at the University of California, Davis, have shown pressure drops across the gas producer from 4 cm H_2O in the case of gasification of densified cotton stalk cubes (3 cm x 3 cm x 5 cm) to 45 cm H_2O for cracked walnut shells. The air input rates of 20 m^3/h and 28 m^3/h were too close together in order to contribute much to the observed difference in the pressure drop. The fuel size has, therefore, a considerable impact on the pressure drop across the gas producer and oxygen penetration depth. More data on how the air input rate influences the pressure drop across a plant for a given fuel and pressure drop observed in commercial portable systems are given in Chapter VII.

Bridging of large fuel particles has often been a problem in small, stationary gas producers. It is the main cause of slag formation, because the fuel stops flowing at an unchanged air input rate. The air-fuel ratio increases locally and the temperature can reach 2000 °C. This temperature is high enough to induce slagging in all fuels. It has been established and experimentally confirmed (21), that the ratio of fuel size (largest dimension) to smallest cross section —usually the throat or choke plate of a gasifier — should be at least 6.8 in order to avoid bridging.

The vast experience with portable units (20-100 hp) fed mostly with wood, wood charcoal and various coals have established recommended fuel sizes. The recommended fuel size is related to the gas producer or source whenever possible in Table 16.

In general, undersized particles increase the pressure drop through the fuel bed, oversized particles cause bridging and incomplete carbonization because of the short fuel residence time in the carbonization zone and too much void space.

Because larger pieces require a longer time for complete gasification than smaller pieces, the depth of the fuel bed is related to the fuel grading. Table 17 lists rough guidelines that have been established for medium and large-size gas producers.

110

Table 16. Recommended Fuel Size for Small Gas Producers (20-100 hp).

Plant	Fuel	Size mm	Reference
Malbay	low temperature coke, anthracite	10-25	17
Wisco	charcoal, peat coke	20-40	17
	low temperature coke	15	17
Gohin Poulence	charcoal	15-22	17
	anthracite	5-15	
Brandt	wood	80x40x40	17
Koela	charcoal	10-20	17
	low temperature coke	10-15	17
	anthracite	5-10	17
UCD Laboratory Gas Producer	wood	20-40 cubes	21
	hard durable cubes of corn stalks, alfalfa and cereal straw	30x30x50	21
	hard durable rice hull pellets	larger 10	21
	hammermilled corn cobs	40	21
	fruit pits	15-30	21
Imbert	wood, birch	60-80 length 50-60 diameter	17
	oak	20x40x60	17
Swedish Gas Producers 1939-45	sawed and split fire wood,		
	thick blocks	8 cm x 25 cm^2	2
	thin blocks	6 cm x 20 cm^2	2
	cylinders	8-9 cm, 25-75 dia.	2
	sticks	6 cm, 25-50 dia.	2
	charcoal, coarse grade	10-60	2
	charcoal, fine grade	10-30 no more than 10%, may be of 10-20	2
British Government Regulations	fine grade wood	length 20-50 largest cross section, 25 cm^2	
	coarse grade wood	length 30-80, largest cross section, 30 cm^2	

Table 17. Fuel Size and Depth of Fuel Bed (27).

Fuel	Grading mm	Smallest Economical Depth cm
Anthracite	10-20 beans	30-60
"	25-40 nuts	75-90
Coke	20-30 cubes	75
"	30-50 cubes	115
"	50-75 cubes	180
Coal	15-20 nuts	55
"	As mined	145-200
Wood	Large blocks	150-210
"	Sawdust and shavings	120-150

Grading: The size distribution of the fuel should be as small as possible. Trouble free, reliable gasification is best accomplished through a fuel bed of uniform size. If the size range is too large the air blast and gas are forced through an uneven fuel bed caused by separation of the fine and coarse particles. Hot spots and cold spots which lead to channeling and clinker formation are the final result. Undersize and oversize not exceeding $\pm$ 10% was the general rule adapted during the 1930-1950 period (17). Successful gasification of peat containing 30% fine material has been reported (13). Apart from the effect of the grading upon the fuel bed depth, the quantity of fine particles will influence the specific gasification rate. A high fraction of fine particles decreases the specific gasification rate as shown in Table 18.

Table 18. Fine Particle Fraction and Specific Gasification Rate (27).

Fuel		Grading mm	Specific gasification rates kg/m^2-h
Bituminous coal,	Washed nuts	25-50	126
	Rough slack	under 40 with 20% under 6	106
	Rough slack	under 20 with 50% under 6	87
Coke Nuts		20-40	145
Coke		under 20 with 50% under 6	72

Figure 84 illustrates in a simple sketch, the appearance of a potential fuel for gasification. In particular, charcoal, peat and brown coal briquettes have the tendency to crumble in the fuel hopper and are only suitable for gasification if the fraction of fine parts can be controlled within limits.

Fuel Form: The form in which the fuel will be gasified has some economical impact on the system. For instance, a 30 hp engine with an overall efficiency of 15% requires 500 MJ/h of cold, clean gas. Assuming a gasification efficiency of 70% for the gas producer-purification system and a heating value of 18 MJ/kg

112

BROWN COAL

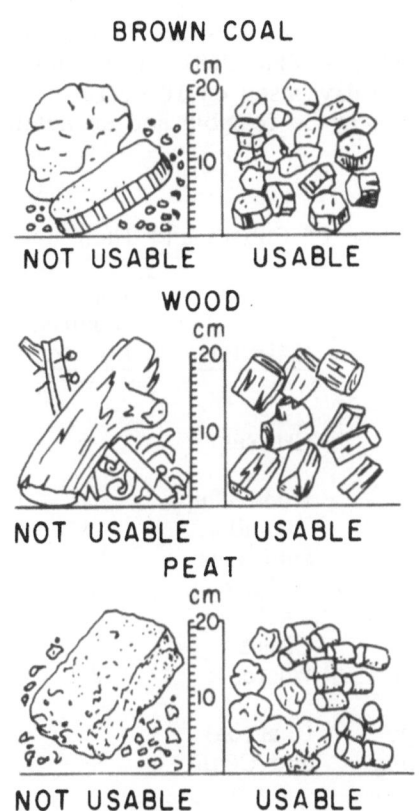

Figure 84. Desired Grading Of Gas Producer Fuel (2).

for the feed material, 28 kg/h must be supplied to the gasifier. For this small scale a batch feed operation is appropriate. On the average, the hopper size for an automotive gas producer should handle 56-112 kg, the amount necessary to run the plant for 2-4 hours without refilling. From the calculation, it is obvious that the bulk density which depends on the size of the fuel plays an important role. For instance, gasifiers have been successfully operated with loose shredded cereal straw and rice hulls. However, this requires a continuous feed or a suitable large container above the gasifer from which the feed can be fed into the gasifier in short intervals and a large ash container.

Densifying biomass has been a development in the U.S. for the past 25 years. Cubers and pelletizers densify biomas and all kinds of municipal and industrial waste into "energy cubes." The energy cubes are in most cases delivered in cylindrical or cubic form and have a high density of 600 to 1300 kg per m^3. Their specific volumetric energy content in MJ/m^3 is consequently much higher than the raw material they are made from. The uniform size is very desirable as already pointed out. In the case of biomass fuel from saw mill residues and logging activities which generate a fuel that tends to form a packed, highly resistent bed in a gasifier, densification may be the only way to make these residues available for fixed bed gasification.

The densifiers manufactured in the U.S. are in most cases large (at least 5 t/h output) and expensive ($80,000-$200,000). For small-scale units which require about 200-500 kg of fuel a day based on 10 hours of operation, the development of a hand-operated cuber may be feasible and economical under certain circumstances.

Table 19. A Selection of Cubers and Their Specific Properties (24).

Name	Method	Materials	Energy required kWH/kg	Briquet Density kg/m³
John Deere Cuber	Ring & Die	Grasses	5	769-881
Prestolog	Die	Wood Chips and Sunflower Hulls	31	1200
Citrus Pulp Pelleter	Screw Extrusion	Citrus Wastes	2.7	640
American Roll Type	Roll Type Briquetting	Charcoal and Coal	—	1200
Japanese Roll Type	Roller Extrusion Type	Sawdust	—	993
NDSU Test Apparatus	Press	Organic Wastes	—	1200

A worldwide survey of densification systems is given in Reference (34).

Bulk density: As far as the storage capacity of the charging hopper is concerned the bulk density of the fuel is significant. The volume occupied by a stored fuel depends not only on the specific density of the single fuel particles and the moisture content but also upon the grading and whether the fuel is piled loosely or compacted. The storage capacity of a biomass gas producer will be only one fifth of that of a comparable coal gasifier when the biomass is not in densified form. Table 20 lists the average bulk density of the most common fuels used in gasification.

In general, biomass fuel used for gasifiers occupies about 20-75% of the container volume. Fruit pits occupy about 65% of the container volume. The bulk density has a considerable impact on the gas quality because it influences the fuel residence time in the fire box, the fuel velocity, the fuel bed density and the gas flow rate.

Table 20. Bulk Density of Various Fuels.

Fuel	Grading	Bulk Density kg/m^3	Reference
Saw dust	loose	177	27
Saw dust	briquets 100 mm long 75 mm diameter	555	27
Peat	dust	350-440	13
	briquets 45x65x60 mm	550-620	13
	hand cut	180-400	13
Charcoal (10% moisture)	beech	210-230	2
	birch	180-200	2
	softwood blocks	150-170	2
	softwood slabs	130-150	2
	mixed 60% hard/40% soft	170-190	2
Wood	sizes as in Table 16		
	hardwood	330	2
	softwood	250	2
	mixed 50/50	290	2
Straw	loose	80	—
	bales	320	—
Alfalfa seed straw	cube 30x30x50 mm, 7% moisture	298	21
Barley straw	cube 30x30x50 mm, 7% moisture	300	21
Bean straw	cube 30x30x50 mm, 7% moisture	440	21
Corn cobs	11% moisture	304	21
Corn stalks	cube 30x30x50 mm	391	21
Cotton gin trash	23% moisture	343	21
Peach pits	11% moisture	474	21
Olive pits	10% moisture	567	21
Prune pits	8% moisture	514	21
Rice hulls	cube 30x30x50 mm	679	21
Safflower straw	cube 30x30x50 mm	203	21
Walnut shells	cracked	336	21
	8 mm pellets	599	21
Wood, blocks	17% moisture	256	21
chips	10% moisture	167	21
Coal	anthracite	830-900	27
	bituminous	770-930	27
Coke	hard	380-530	27
	soft	360-470	27
Brown coal	air dry lumps	650-780	27

The fuel residence time determines to what extent the partial combustion and reduction reactions take place. This is given by the degree to which the equilibrium state is reached at a given temperature. Too short a residence time causes incomplete conversion of CO_2 into CO, poor gas quality and too much unburned carbon in the ash. Too long a residence time may increase slag formation (21). Figure 85 relates the fuel residence time to the fuel consumption rate for various bulk densities. This data was obtained with the University of California, Davis, Laboratory gas producer.

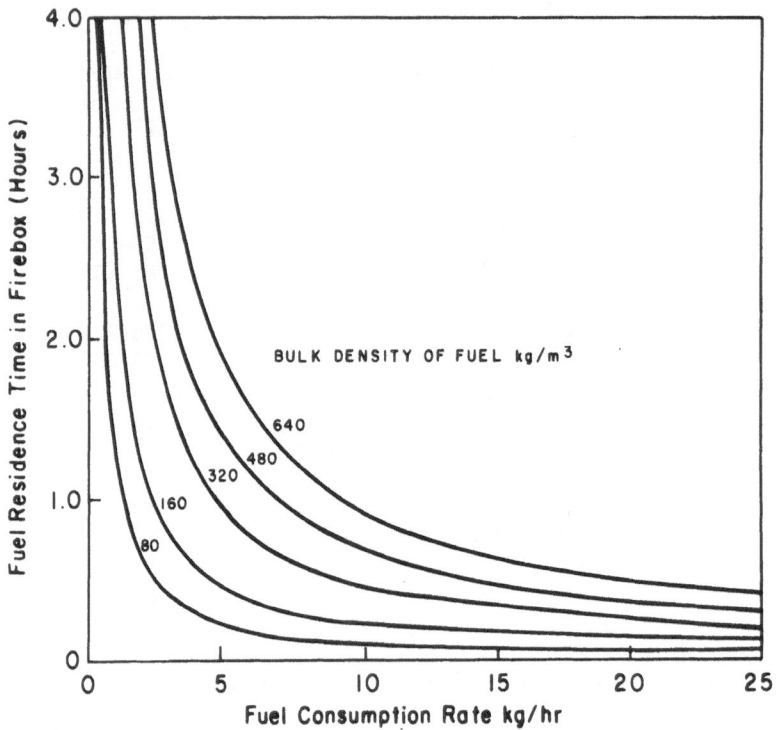

Figure 85.   Fuel Consumption Rate Versus Fuel Residence Time for Various Bulk
Densities (21).

Volatile matter: Volatile matter, fixed carbon, moisture and ash are the products
obtained from a proximate analysis of solid fuels. In this analysis the moisture,
fixed carbon, ash and volatile matter are determined by specific procedures.
The amount of fixed carbon in a fuel is defined by difference as follows:

FC (% weight) = 100 - (% moisture + % ash + % volatile matter)

The proximate analysis provides information on the combustion characteristics
of the fuel.   Table 21 lists the volatile matter of common fuels used in
gasification on a dry weight basis.

The volatile matter plus the inherent and chemically bound water in the fuel
are given up in the distillation zone at moderate temperatures of 100-500 ^{o}C
and form a vapor consisting of water, tar, oils and gases. On first glance it
seems obvious that fuels high in volatiles have greater problems from tars and
oils that condense at about 120-150 ^{o}C and, as mentioned in Chapter VI, must
be removed before the gas is used in an internal combustion engine.   However,
how much tar and vapor leaves the gas producer depends mostly on the design
of the plant.   It has been reported that a high volatile fuel such as peat can
be gasified with no tar in the raw gas (13).   Successful gasification of high
volatile fuel into mostly tar-free gas can be accomplished by careful control of
the firebox temperature and the physical properties of the fuel.

116

Table 21. Volatile Matter of Fuels for Gasification.

Fuel	Volatile Matter % Weight	Reference
Crop residues	63–80	21
Wood	72–78	21
Peat	70	13
Coal		
lignite	40	20
subbituminous A,B,C	45	20
high volatile bituminous	40–45	20
low volatile bituminous	20–30	20
semianthracite	8	20
anthracite	5	20
meta anthracite	1–3	20
Charcoal	3–30 and over, depends strongly on manufacture	17

The common practice of using anthracite, coke and charcoal in portable gas producers during the 1930-1950 period avoided the tar problem. In portable units used to drive internal combustion engines, the continuous change in the output from the gas producer favored the tar generation, even in downdraft gas producers, because the gasifier was never in an equilibrium state at constant temperature. It was therefore necessary to use specially prepared fuel that has little volatile ash such as anthracite, coke or high-quality charcoal with volatile matter below 5%.

Ash: The mineral content in the fuel that remains in oxidized form after complete combustion is usually called ash. In practice the ash produced in a gasifier also contains incompletely burned fuel in the form of char. The ash content of a fuel and the ash composition have a major impact on the trouble-free operation of a gasifier. It is obvious that a high ash content of the feed lowers the amount of energy available from the gas producer and more space must be provided where the ash can be discharged. If conditions in the firebox are conducive to melting of the ash, then the degree of slagging will, of course, be more severe for the higher ash content fuels. For instance, cotton gin trash produces about 20% ash whereas wood chips only 0.1%. In the case of cotton gin trash, 2,000 grams of mineral matter need to be passed through the generator each hour, whereas wood chips would yield only 10 grams. This calculation has been based on the fuel consumption for a 7-10 hp engine and shows clearly that the ash content is the major limiting factor for a successful operation of a gas producer. In the case of wood chips only 10 grams of ash an hour could possibly fuse together and form clinkers which inhibit fuel flow and finally stop operation altogether.

It is well known that the mineral content in the fuel has a catalytic effect on the reaction in the oxidation zone and can increase the reactivity of the fuel.

Figures 86 and 87 present tests conducted at Battelle Laboratories in Columbus, Ohio. It was shown that the treatment of wood with a 1.5% ash slurry had a considerable effect on the H_2/CO mole ratio and the reactivity. When wood was heated with its own ash in form of a slurry sprayed on it, the reactivity and H_2/CO ratio increased two fold (14).

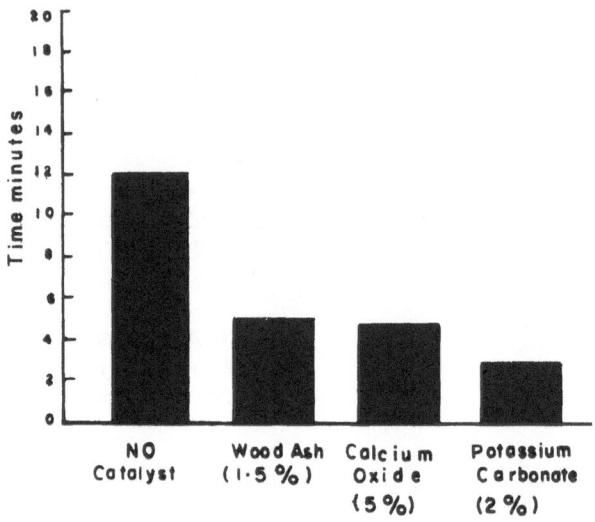

Figure 86. Time Required to Convert 95% of the Wood at 750 $^{\circ}$C (14).

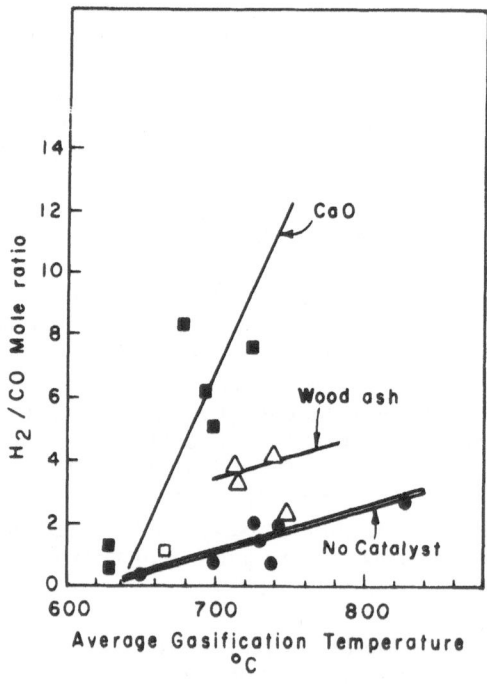

Figure 87. Effect of Gasification Temperature and Catalyst on Product Gas and H_2/CO Ratio (14).

The average ash content of major fuels for gasification are listed in Table 22. The numbers are derived on a dry fuel basis obtained from a proximate analysis of the fuel.

The common belief that all wood is low in ash is incorrect. There are several tropical species which have an ash content that exceeds those of coal, such as Strychnos Ignatii 7.5-8.3%, and Picrasma Excelsa 7.8% (35). Several attempts have been made to differentiate between the ash content of softwood and hardwood, or between the ash content of sapwood and heartwood. No generalization has been found.

Table 22. Ash Content of Major Fuels (20,21,24).

Fuel	% weight ash, dry basis	Fuel	% weight ash, dry basis
Alfalfa seed straw, cubed	6.0	Municipal tree prunings	3.0
Almond shell	4.8	Olive pits	3.2
Barley straw mix	10.3	Peach pits	0.9
Bean straw	10.2	Peanut husks	1.5
Charcoal	2-5	Peat (average)	1.6
Coffee hulls	1.3	Douglas fir wood blocks	0.2
Coal	5-17	Prune pits	0.5
Corn cobs	1.5	Refuse derived fuel	10.4
Corn stalks	6.4	Rice hulls	16-23
Cotton gin trash	17.6	Safflower straw	6.0
Cubed cotton stalks	17.2	1/4" pelleted walnut shell mix	5.8
Pelleted rice hulls	14.9	Walnut shell (cracked)	1.1
Furfural residue	12	Wheat straw and corn stalks	7.4
Hogged wood manufacturing residue	0.3	Whole log wood chips	0.1

The melting temperature of ash has been the topic of several papers and books (7,21,25,27,31). The individual melting point of the minerals gives some indication of how the mixture will behave under high temperatures. However, the ash minerals form an eutectic mixture which will start melting at the lowest possible melting point, dependent of the fractions of the individual species. The most common base to determine the composition of the ash of biomass and coal is the SiO_2-Al_2O_3-Fe_2O_3-TiO_2-CaO-MgO-Na_2O-K_2O-SO_3 system because the oxides of these minerals amount to at least 95% of all minerals found in the ash. More than 22 trace elements have been identified that are different from those listed above. Unfortunately, the variations in the coal and biomass ash are large and depend too much on location and history of the fuel in order to give a narrow range of the fractions found. Figures for American coal are given in Table 24. The U.S. Bureau of Mines gives the average analysis for SiO_2, Al_2O_3, and Fe_2O_3 of ash from coal as 45.7%, 26%, and 18.1%, respectively. These three constituents generally make up about 90% of the ash from bituminous coals.

If the temperature in the firebox rises above the melting point, the mixture will melt and the molten material will flow together and forms large clinkers, clinging to internal surfaces, tuyeres and grates. The fuel flow finally will be obstructed which will increase the air fuel ratio and the temperature. The gas then will become so poor that it cannot be combusted. In case the air-fuel ratio reaches the stochiometric value for combustion, serious damage to the plant may occur.

The complexity involved in the determination of a possible slagging temperature of ash based on its mineral components has been thoroughly examined by several authors. The results are not conclusive and only general guidelines of the slagging potential of a fuel can be given.

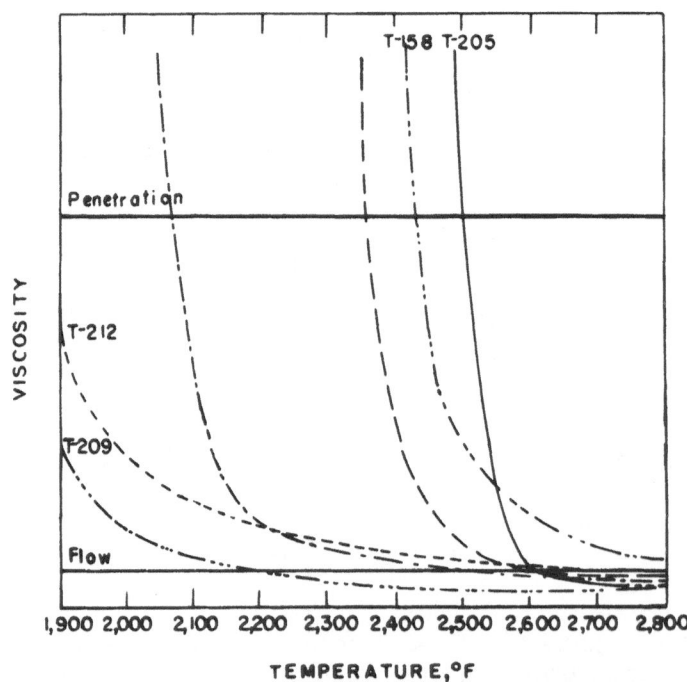

Figure 88. Viscosity of Various Coal Ashes Versus Temperature (7.).

In practice, the ash of a fuel to be gasified should be tested under laboratory conditions before any decisions are made as to what type of gasifier should be used. The fusion characteristics of ash depends greatly on the state of oxidation of the iron contained in it. In general, it could occur as Fe_2O_3, FeO and Fe. The degree of oxidation of the iron in slag has a marked effect on its viscosity between 50 and 100 poises. For comparison, water at 25 ^{o}C has a viscosity of 0.01 poise whereas light motor oil at the same temperature has a viscosity of 1 poise. In general, the viscosity of slags decreases rapidly at first and then more gradually as the flow temperature is approached. This is illustrated in Figure 88 which shows the viscosity of various ashes as a function of temperature. Flow is depicted to be where the slag could easily be tapped (7).

Penetration is the viscosity at which a rod could be poked into a slag without too much effort. Figure 88 illustrates that some ashes have a very narrow temperature range between softening and liquid state, while others show large differences between the penetration and flow temperature. A small change in composition may require a large change of temperature to produce softening, but the more complex the composition the less the effect some changes exert. The difficulties in controlling slagging can be overcome by two totally different types of operation of a gasifier.

1. Low temperature operation that keeps the temperature well below the flow temperature of the ash.

2. High temperature operation that keeps the temperature above the melting point of the ash and in addition fluxes are added to lower the flow temperature even more.

The first method is usually accomplished by steam or water injection or the natural moisture in the fuel. It has been suggested that slag formation can be controlled completely by saturating air with water vapor to a wet-bulb temperature of 50° to 55 ^{o}C for slag melting temperatures of approximately 1,200 ^{o}C (19). The latter method requires provisions for tapping the molten slag out of the oxidation zone. Either method has its advantages and disadvantages. Deciding what method should be used depends on the specific case.

Tests have been conducted to determine the influence of adding fluxes such as iron ore, feldspar, fluorspar, salt cake, limestone and dolomite to the fuel to obtain a desired flow temperature suitable for the specific application of the gas producer. Figure 89 shows the influence on Na_2O on the flow temperature of ash. It has been generally accepted that alkali salts usually given as Na_2SO_4 lower the flow temperature. This has been a serious problem in the gasification of certain German brown coals because the flow temperature of the ash was below 900 ^{o}C. Some crop residues contain a considerable amount of Na which will be oxidized to Na_2O and lower the flow temperature to a point where gasification below the melting point of the ash will not be practical.

Based on experimental data, the melting temperature of coal ash may be predicted within a SiO_2 - Fe_2O_3 - CaO - MgO system. Figure 90 describes such a nomogram which permits the determination of the viscosity of coal ash slag as a function of composition and temperature in the SiO_2 - Fe_2O_3 - CaO - MgO system.

Whereas for coal a SiO_2 - Fe_2O_3 - CaO - MgO system is sufficient to determine the slagging potential of its ash, this system is inadequate for biomass fuels. As listed in Table 24, the bulk of the minerals in biomass lies within the SiO_2 - K_2O - Na_2O - CaO system for most fuels tested.

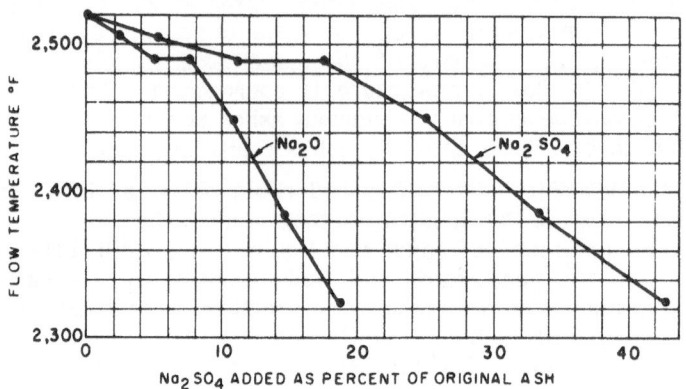

Figure 89.   Effect of Na_2SO_4 on the Flow Temperature of Ash (7).

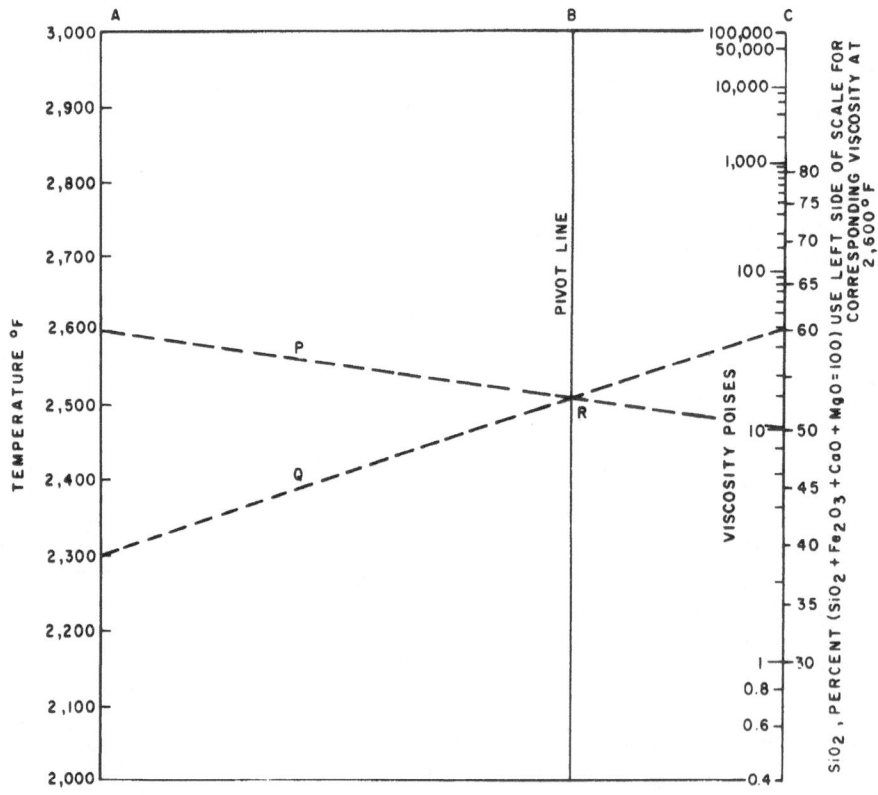

Figure 90.   Viscosity of Coal Ash Slag as a Function of Temperature and Ash Composition (7).

METHOD OF USING NOMOGRAM: Scale C shows relationship directly between percent SiO_2 and liquid viscosity at $2,600^\circ F$. To find viscosity at any other temperature: 1 Connect $2,600^\circ F$ point on scale A with desired composition or viscosity on scale C. 2 Note pivot point on line B. 3 Draw line through desired temperature on scale A through pivot point; intersection on scale C is liquid viscosity at new temperature. Example: At 50 percent SiO_2, liquid viscosity is 10 poises. Line P has pivot point at R, and line Q shows that liquid viscosity at $2,300^\circ F$ is 40 poises.

The flow temperatures and thermal behavior of two component and three component mixtures have been extensively studied and their phase diagrams are known (25). Figures 91 to 94 list those which are relevant for ash obtained from gasification of biomass.

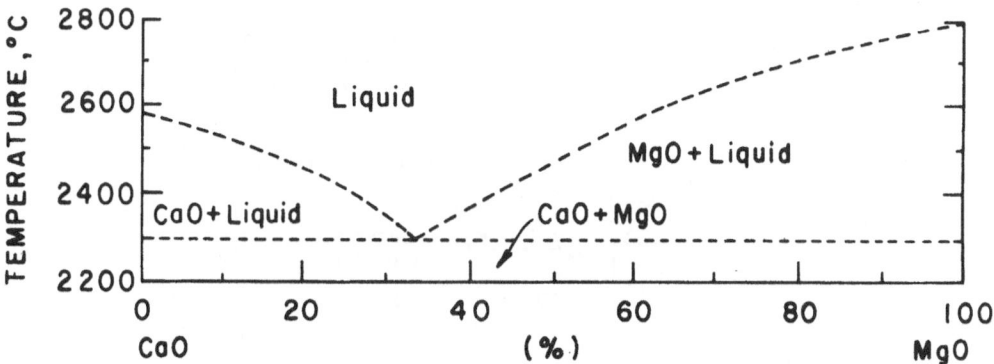

Figure 91. System CaO - MgO (25).

Although it would be desirable to have a four-dimensional outlay of the SiO_2 -K_2O - Na_2O - CaO system, which is not available because of the complexity involved, the figures identify clearly the components (such as K_2O, Na_2O) that have a fluxing influence and lower the melting point temperature of the ash.

It has also long been recognized that the most troublesome components of the ash are SiO_2 and the alkalies, Na_2O and K_2O. In many biomass fuels and coals, the SiO_2 content makes up 50% of the ash and can reach extreme values up to 97% in case of rice hulls. Na_2O and Ka_2O are also relatively high in some biomass fuels. The danger lies not only in their influence to lower the flow temperature but in their tendency to vaporize at temperatures easily obtained in a gas producer. This is particularly true if the alkalies are in the form of chlorides and sulfides. Consequently, a small amount of sulfur and chlorine in the fuel makes things even worse.

Although the silicon oxides have a fairly high melting point, it has been shown that considerable amounts of SiO_2 evaporate at $1550^\circ C$ despite the fact that the boiling point of SiO_2 lies much higher at $2230^\circ C$. SiO vapor then reacts with oxygen from an oxygen carrier in the gas stream such as water vapor and sometimes reaches the filter and engine in an extremely fine (0.1 micron) and highly abrasive, glassy state. Evaporation of silicon can be easily recognized

123

as a white coating inside the connecting pipes to the internal combustion engine. A similar reaction takes place in case the silicon can react with sulfur. The SiS and SiS$_2$ vapors react with oxygen and reach the engine and filter in form of very fine fly ash. All three products cannot be removed efficiently from the gas stream with conventional mechanical filters and are not water soluble.

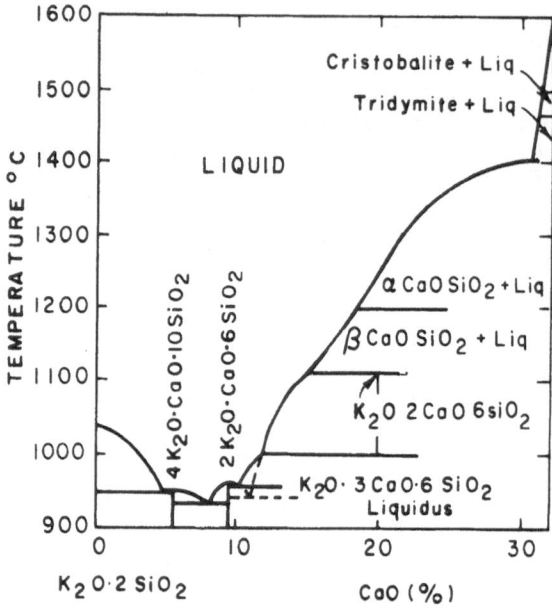

Figure 92. System K$_2$O - SiO$_2$ - CaO (25).

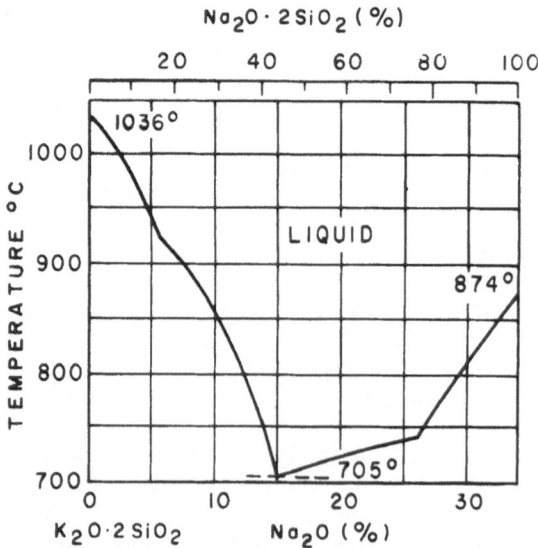

Figure 93. System K$_2$O • 2SiO$_2$ - Na$_2$O (25).

124

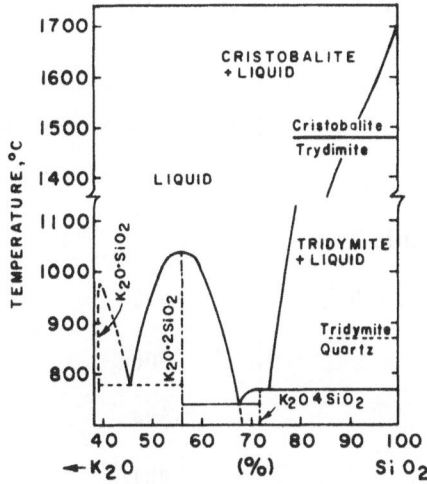

Figure 94. System K_2O - SiO_2 (25).

Tests with a portable gas producer have shown that the evaporation of SiO_2 was particularly high in dry gasification and surprisingly low with wet gasification (10). The flow temperatures of the most common constituents and their products in coal and biomass ash are listed in Table 23.

Table 23. Flow and Boiling Point Temperatures of Common Ash Constituents.

Mineral	Flow temperature ^{o}C	Boiling temperature ^{o}C
SiO_2	1460–1723	2230
CaO	2570	2850
Fe_2O_3	1560	—
Metallic Fe	1535	--
FeO	1420	—
MgO	2800	3600
Al_2O_3	2050	2210
$MgO \cdot Al_2O_3$	2135	—
$MgO \cdot Fe_2O_3$, forms above 700 ^{o}C	1750	—
$CaO \cdot Fe_2O_3$, forms above 600 ^{o}C	1250	--
$3 Al_2O_3 \cdot 2 SiO_2$	1930	—

125

Table 23 continued

Mineral	Flow temperature $^\circ$C	Boiling temperature $^\circ$C
$Al_2O_3 \cdot SiO_2$, converts into $3\ Al_2O_3 \cdot 2\ SiO_2$ above 1550 $^\circ$C	1930	—
NaCl	800	1465
Na_2SO_4	884	—
$Na_2S_2O_7$	401	—
NaS_2	920	—
KCl	790	1405 (1500)
K_2SO_4	1096 (588 transition)	—
$K_2S_2O_7$	larger 300	—
K_2S_5	206	—
$CaCl_2$	765	1600
$CaSO_4$	1450	—
$MgSO_4$	1127	—
$Fe_2(SO_4)_3$	480	—
FeS	1195	—
FeS_2	1171	—
SiS	—	940
SiS_2	1090	—
$Al_2(SO_4)_3$	770	—
Al_2S_3	1100	—
P_4S_{10}	290	514
P_4S_3	172	407

Table 24. Mineral Oxides in Coal and Biomass Ash (20,21,35).

Coal	% SiO$_2$	% Al$_2$O$_3$	% Fe$_2$O$_3$	% TiO$_2$	% CaO	% MgO	% Na$_2$O	% K$_2$O	% SO$_3$	% CL
Anthracite	48–68	25–44	2–10	1.0– 2	0.2– 4	0.2– 1	—	—	0.1– 1	—
Bituminous	7–68	4–39	2–44	0.5– 4	0.7–36	0.1– 4	0.2– 3	0.2– 4	0.1–32	—
Subbituminous	17–58	4–35	3–19	0.6– 2	2.2–52	0.5– 8	—	—	3.0–16	—
Lignite	6–40	4–26	1–34	0.0–08	12.4–52	2.8–14	0.2–28	0.1–1.3	8.3–32	—
Biomass										
Wheat straw	56.8	—	0.5	—	5.8	2.0	6.0	14.8	7.6	5.0
Corn stover	18.6	—	1.5	—	13.5	2.9	13.3	26.4	8.8	0.9
Rice straw	78.46	1.38	0.14	0.1	2.2	3.03	1.79	9.93	0.34	—
Residue derived fuel	31	27	4	6.0	6	1	7	6	—	—
Rice hulls	90–97	—	0.4	—	0.2–1.5	0.1–2	0–1.75	0.6–1.6	0.1–1.13	0.15–0.4
Wood	0.09–?	1–75	0.5–3.3	—	10–60	1.4–17	under 10	1.5–41	—	—

The list indicates the wide range of possible ash compositions for various coal and biomass fuels. Knowing the ash composition is especially important for high ash fuels, since any clinker formation will quickly obstruct the gas and fuel flow and stop operation. High ash fuels combined with low ash melting point are the most difficult to gasify, due to the poor gas quality one obtains at fire zone temperatures below 1,000 C.

127

A realistic picture of the slagging potential of biomass fuels can, of course, only be obtained through actual trials with a gas producer. Tests conducted at the University of California, Davis, with the small laboratory gas producer specifically for slagging resulted in the following classification:

Table 25. Slagging Behavior of Crop Residues and Wood (21).

Slagging Fuels	% Ash	Degree of Slagging	Non-Slagging Fuels	% Ash
Barley straw mix	10.3	Severe	Cubed alfalfa seed straw	6.0
Bean straw	10.2	Severe	Almond shell	4.8
Corn stalks	6.4	Moderate	Corn cobs	1.5
Cotton gin trash	17.6	Severe	Olive pits	3.2
Cubed cotton stalks	17.2	Severe	Peach pits	0.9
RDF pellets	10.4	Severe	Prune pits	0.5
Pelleted rice hulls	14.9	Severe	Walnut shell (cracked)	1.1
Safflower straw	6.0	Minor	Douglas Fir wood blocks	0.2
1/4" pelleted walnut shell mix	5.8	Moderate	Municipal tree prunings	3.0
Wheat straw and corn stalks	7.4	Severe	Hogged wood manufacturing residue	0.3
			Whole log wood chips	0.1

It was observed that independent of the chemical composition of the ash, slagging occurred with most fuels having an ash content of more than 5%. However, one has to keep in mind that no attempts were made to keep the temperature of the fire zone below the melting point of the ash. The official British Government regulations for portable gas producer units requiring no more than 4% ash in natural fuels and smaller than 5% in carbonized fuels reflects this general trend.

It cannot be emphasized enough that commercial gasification as practiced for the last 140 years has avoided problematic fuels, those high in ash or with a tendency for slagging, because of the difficulties involved in achieving reliable operation over a continuous period without too much attention to the gasifier.

Ultimate analysis: The ultimate analysis of coal and biomass fuels, although it does not reveal the suitability of a fuel for gasification, is the main tool for predicting gas compositions and temperature limits through a mass and energy balance of the gasification process. Existing data is usually given on a C-H-O or C-H-O-N basis. Since the nitrogen content of most fuels is below 3%, there is not much difference between the two systems. Fuels high in total carbon as given by the ultimate analysis tend to yield less tar in the raw gas because of the small fraction of volatiles. In order to avoid confusion, it is best to split the total carbon in the fuel into base carbon and volatile carbon. Base carbon represents the carbon that remains after devolatilization, whereas volatile carbon is defined as the difference between total carbon and base carbon. Base

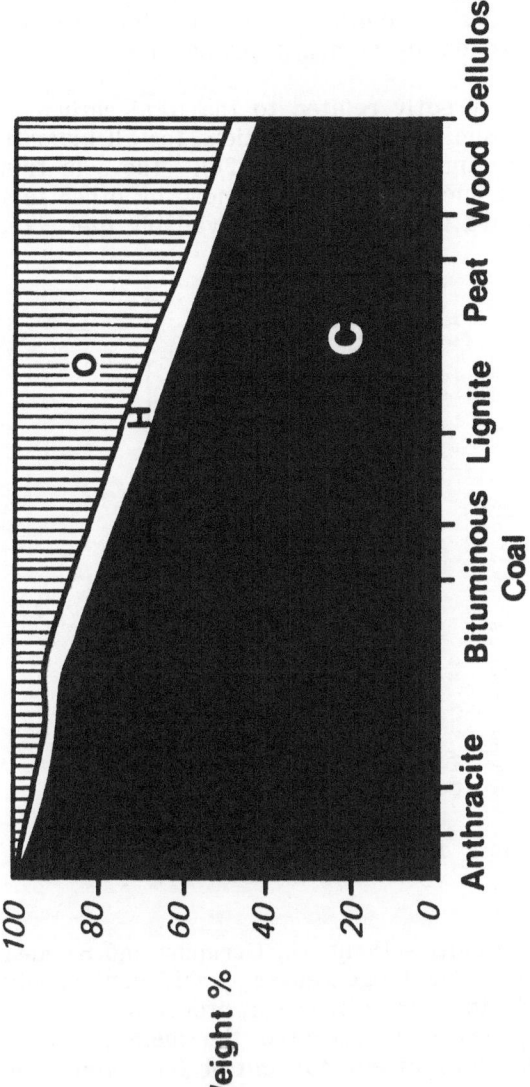

Figure 95.   Ultimate Analysis of Wood and Coal (32).

129

carbon does not equal the fixed carbon as given by the proximate analysis, because the fixed carbon fraction includes in addition to carbon other organic components which have not been evolved during standard devolatilization.

The higher heating value of the fuel is directly related to the total carbon in the fuel. It is of interest to notice the similar C-H-O fractions for all biomass fuels tested at the University of California, Davis (Figure 96). The selection of a biomass fuel for gasification is consequently highly influenced by other fuel properties such as ash content, ash chemical composition and available fuel size.

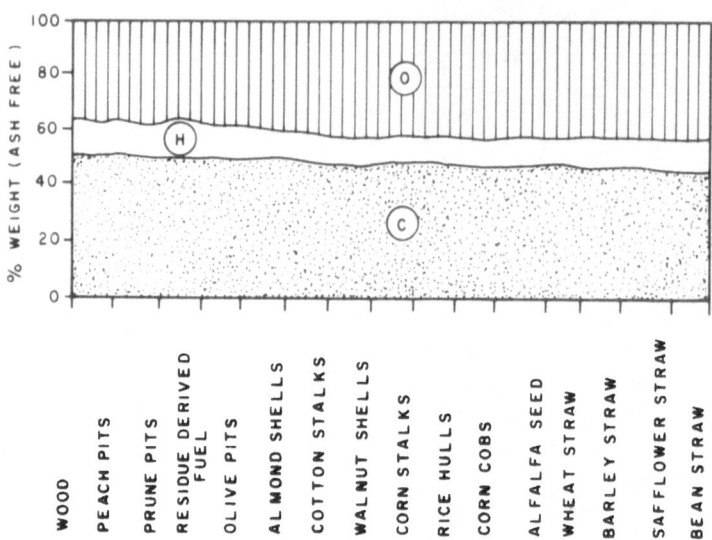

Figure 96. Ultimate Analysis of Biomass Fuel Tested at the University of California, Davis.

It is illustrative to recall the logistic difficulties Denmark, Germany and Sweden had before and during the Second World War to guarantee a sufficient supply of suitable fuels. It was soon recognized that the production, processing, storing and distribution of suitable wood and charcoal presented the main problem. Strict government regulations were put into effect to control fuel properties for gas producers. Nevertheless, many people lost their money or became deeply discouraged about gas producers because the available fuel did not meet the gas producers capability. Regulations became more strict, even governing with what kind of tool and in what direction the wood should be cut. Buying gas producer charcoal became almost as difficult as buying a precious stone, because the quality of the charcoal varied so much and could not be determined by its physical appearance. In addition, many gas producer manufacturers built and sold gas producers without being concerned how sensitive the unit was to changes in the fuel characteristics. The public demand for convenience and fast starting properties of the automobile led to designs such as the Kalle gas producer which became more and more sensitive to even the smallest changes in the fuel properties. The proper functioning of the unit was only guaranteed if a specially prepared fuel was used. Currently, the situation may not be much different if a large demand were to develop for gas producers. The logistic problems associated with the proper fuel supply will outweigh the technical problems in Third World Countries. In addition, there is serious doubt that current manufacturers of small gas producer units could successfully guarantee the frequently made claims that their units can be operated on all kinds of biomass. A selection of fuels that have been gasified at the University of California, Davis, is shown in Figures 97 to 104 together with some qualitative explanations about how physical properties of the fuel influence the gasification process. It should be mentioned that, although these fuels have been gasified in the UCD Laboratory Downdraft Gasifier, not all of them resulted in reliable continuous gasification.

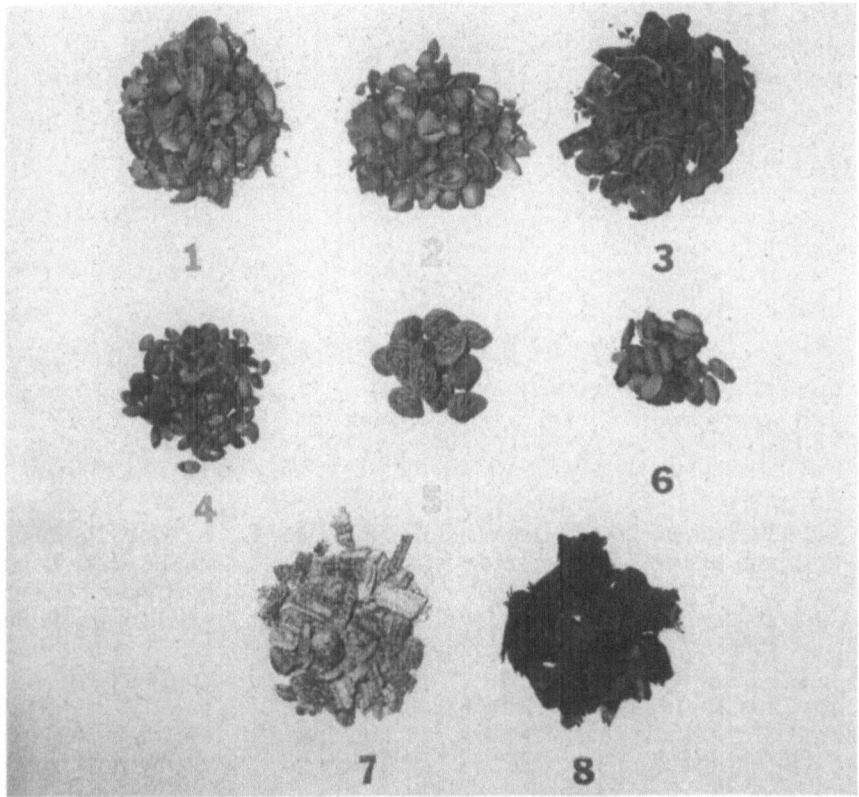

Figure 97. Various Undensified Crop Residue Gas Producer Fuels and Shredded Tires. All are excellent fuels as evidenced by continuous 6-h gasification tests at specific fuel rates of 48 to 81 kg/h.

1. Hard-Shell Almond Shell
2. Soft-Shell Almond Shell
3. Cracked Walnut Shell
4. Olive Pits
5. Peach Pits
6. Prune Pits
7. Broken Corn Cobs (average size 30 – 40 mm long, 20 mm diameter).
8. Shredded Tires (up to 25% by weight added to wood blocks).

Figure 98. Various Loose and Densified Gas Producer Fuels. Only the Eucalyptus
and lignite cubes were satisfactorily gasified in the UCD Laboratory
Downdraft Gas Producer. Cubes made with a standard John Deere
stationary hay cuber. Pellets made with a California Pellet Mill
tapered round hole ring die.

1. Loose Rice Hulls (average size 8x2 mm).
2. Pelleted Rice Hulls (average size, 20 mm long, 10 mm diameter).
3. Pelleted Sawdust (average size, 20 mm long, 6 mm diameter).
4. Mixture of 90 kg Newspaper
            10 kg Sewage Sludge (cubes 30x30x50 mm).
5. Refuse Derived Fuel (RDF) Pellets (average size, 20 mm long, 13 mm
    diameter).
6. Eucalyptus Young Growth (small branches complete with leaves).
7. Lignite Cubes (30x30x50 mm).
8. Mulled Walnut Shell (average size, chips 4x4 mm).
9. Mixture of: 75 kg Hammer Milled Walnut Shell
                15 kg Hammer Milled Rice Straw
                10 kg Sawdust
                Pellets 10 mm long, 5 mm diameter. The rice straw and
                sawdust served as natural binders to produce a pellet from
                walnut shells.

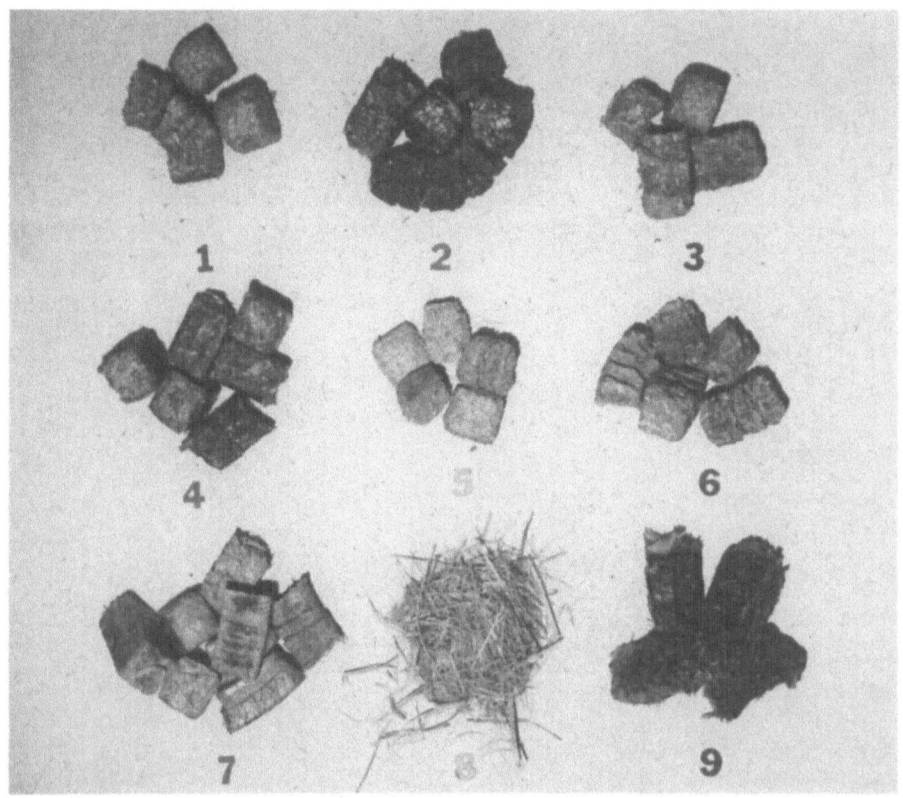

Figure 99. Cube densified agricultural residues made with a standard John Deere Stationary Cuber except the rice straw which was cubed with a special die in the Deere cuber. The average size of the cubes is 30x30x50 mm.

1. Alfalfa Seed Straw
2. Mixture of 75 kg Barley Straw
   25 kg Corn Fodder
   2.7 kg Binder (Orzan)
Corn fodder needed along with Orzan (50% by weight in water), a ligno-sulfanate from paper pulping liquor to make a stable cube from barley straw.
4. Corn Fodder
5. Safflower Straw
6. Mixture of 50 kg Wheat Straw
   50 kg Corn Fodder
Corn fodder used as a natural binder.
7. Rice Straw
8. Coarse Screen Hammer Milled Rice Straw
9. Clean Cotton Gin Trash

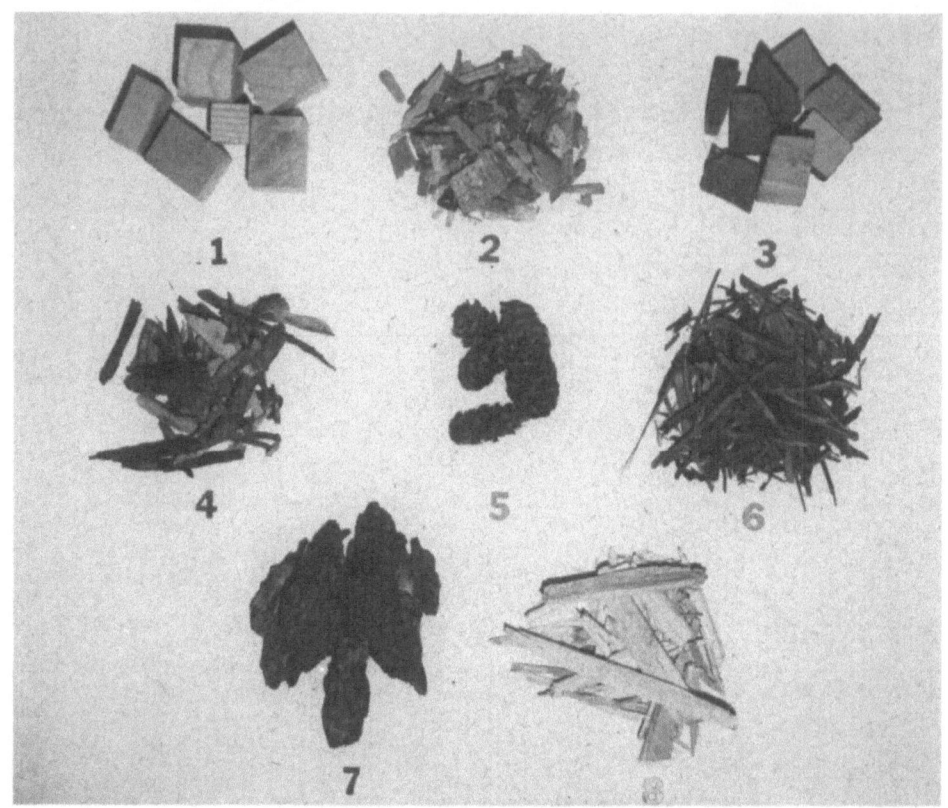

Figure 100.   Wood Fuels for Gasification.

1.   White Fir Wood Blocks (average size, 50x40x30 mm).
2.   Whole Log Chips (for paper pulping)
3.   Douglas Fir Wood Blocks (average size, 50x40x30 mm).
4.   Chipped Tree Prunings
5.   Douglas Fir Cones (average size, 60 mm long, 40 mm diameter).
6.   Prune Tree Chips
7.   Bark (must not be from skidded logs which have a large soil fraction in the bark).
8.   Hogged Wood Waste (kiln-dried trim and waste lumber from a wood products manufacturing plant).

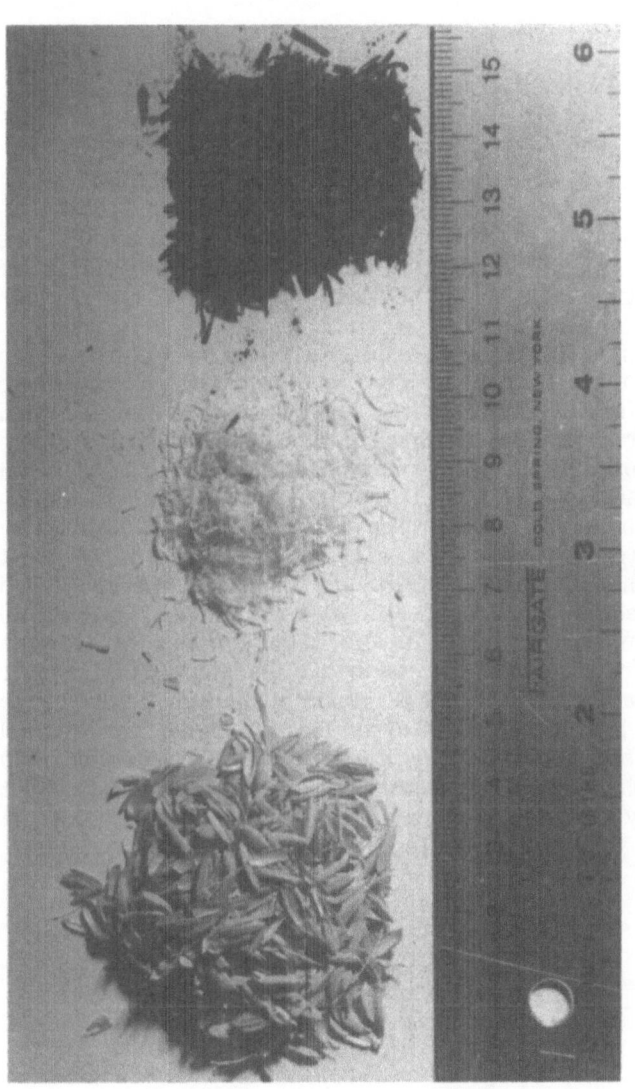

Figure 101. Rice Hulls in Three Different Stages.

The left pile shows natural rice hulls as received from rice mills in the Sacramento Valley. The right pile represents the hulls after gasification. The middle pile is the gasification residues totally stripped of carbon by heating in a muffle furnace at 1,250 °C. The pure white color is due to the 90 to 97% silicon dioxide contained in the ash. It can be seen that the size reduction of rice hulls during gasification is small and the physical shape and appearance of a single rice hull is not altered significantly due to the remaining silicon skeleton. This unusual behavior of rice hulls combined with their low bulk density of 95 kg/m^3 requires different gas producers capable of handling a large volumetric throughput of feed material with a short residence time.

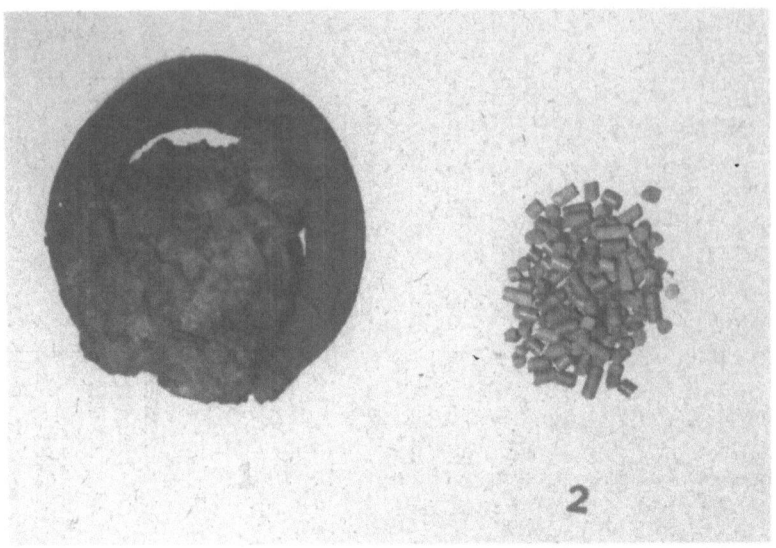

Figure 102.   Clinker Formation (1) on the Choke Plate in a Downdraft Gas
Producer Fueled with Pelletized Rice Hulls and a Sample (2) of
the Pelleted Fuel Used During the Test.

Pelletized rice hulls have a much higher bulk density (700 kg/m^3) compared to
loose rice hulls.   They can, therefore, be gasified in batch fed gas producers
with a considerably smaller fuel hopper.   On the other hand, the densification
of rice hulls does not necessarily improve the gasification characteristics of rice
hulls.   The total surface exposed to the air stream and the void space will be
signficantly reduced.   This leads to a decreased reactivity, sensitivity to locally
overheating the fuel bed and clinker formation.   Tests conducted at the University
of California, Davis, demonstrated that it is possible to increase the air blast
rate   four   fold   within   3 minutes   in   an   updraft   gas   producer   without   much
alternation of the gas produced from a loose rice hull bed.   The 2 m high 30
cm diameter rice hull column reacted very flexibly to any increase of oxygen
supply and expanded rapidly from an incandescent bed of height 30 cm to over
80 cm. It is not possible to say what caused the clinker formation at the choke
plate of the downdraft gas producer as shown in picture above.   The usual
explanation of temperatures above the melting point of the ash describe only
the secondary cause of this phenomena.   Temperatures above the melting point
of the ash globally or locally confined may have many causes.   Most of them
can be controlled as soon as they are identified.

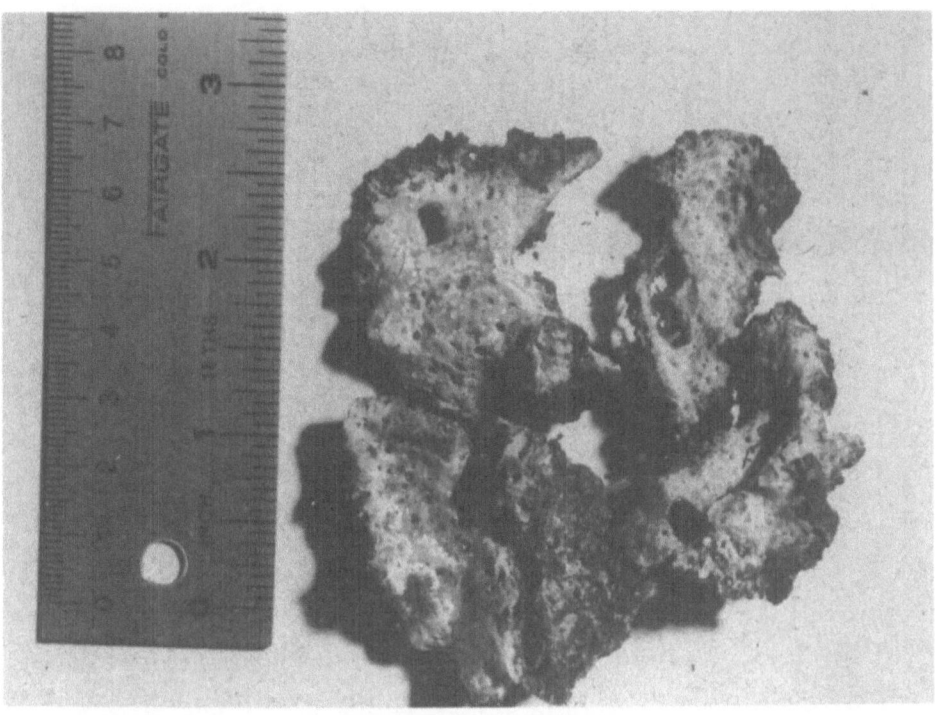

Figure 103. Two Stages of Rice Hulls Produced in an Updraft Gas Producer Blown at a Very High Air Rate.

This picture shows the two phases of thermal decomposition of loose rice hulls in a gas producer operated at a very high air rate but with little change in the combustible properties of the hot gas. The white coating of molten silicon dioxide is due to locally confined excess of oxygen supply to the loose rice hull bed. The characteristic caves obtained from this test may be explained as follows: The locally created oversupply of oxygen when blowing a gas producer too hard generates open flame conditions which cause melting of the considerable amount of ash (90 to 97% silicon). However, any further clinker formation is restricted by the protective layer of molten silicon that greatly restricts the oxygen transport to the lower carbon-rich layers. This leads to characteristic caves in the otherwise homogeneous fuel bed which are covered inside with molten silicon dioxide. It should be clearly pointed out that this type of clinker formation is not due to operating the plant at a temperature above the melting point of the ash. The clinker formation is also not due to bridging of the fuel which stops the downward flow of the feedstock at an unchanged air rate and consequently leads to open combustion at this point in the fuel bed. So far the literature about gasification has not distinguished between clinker formation due to physical insufficiencies of the fuel, chemical composition, plain operational mistakes, or operating a gas producer outside the range it was designed for.

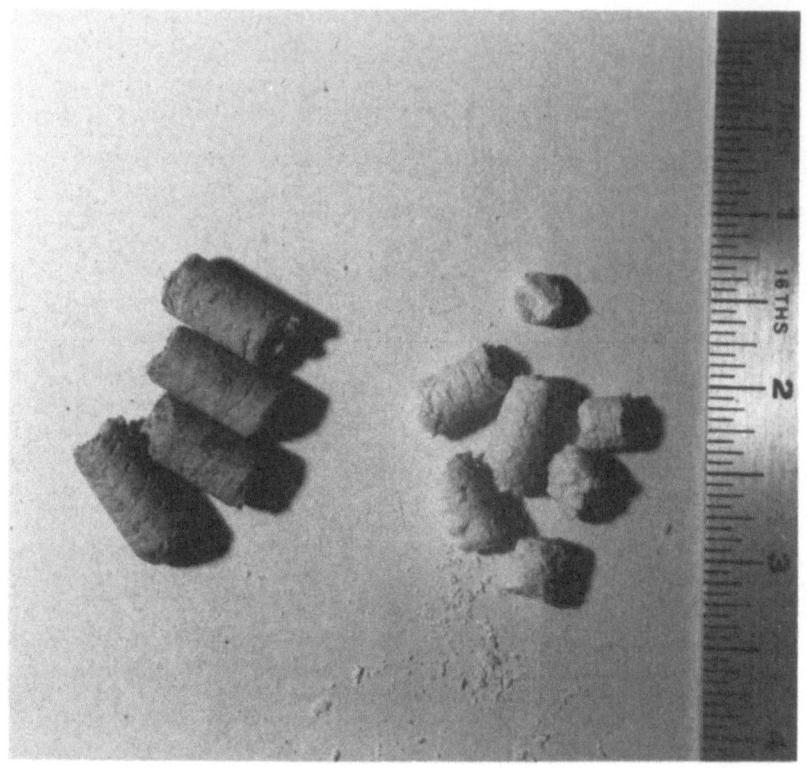

Figure 104. Size Reduction of Pelletized Rice Hulls Under Thermal Decomposition.

The minimal size reduction of pelletized rice hulls when stripped of all carbon
is shown in Figure 104. The pure white ash residues adjacent to the scale
represent about 16% of the uncharred rice pellets, by weight. As can be seen
from the figure, the size reduction is only about 10-20% in each direction. It
should be clearly emphasized that the decay of pelletized rice hulls strongly
depends on the time rate of heating the sample. The above residues were
obtained through fast heating up to 1,200 ^{o}C over a period of one hour, starting
at 500 ^{o}C. Slow heating of rice pellets showed a complete decay of the structure
of the pellets.

Reference Chapter V

1. Anonymous, Alternative Fuels for Motor Vehicles, Engineering, v 148, n 3847, 1939, pp. 387-388.

2. Anonymous, Generator Gas The Swedish Experience From 1939-1945, Solar Energy Research Institute, Golden, Colorado, SERI/SP 33-140, January, 1979.

3. Asplund, D., Peat as a Source of Energy in Finland, Peat as a Fuel, Finn Energy '79 Seminar, The State Fuel Centre, Jyvaskyla, Finland, January, 1979.

4. Bailie, R.C., Current Developments and Problems in Biomass Gasification, Sixth Annual Meeting, Biomass Energy Institute, Winnipeg, Manitoba, Canada, October, 1977.

5. Blackwood, J. D. and F. McGrory, The Carbon-Steam Reaction at High Pressure, Australian Journal of Chemistry, v 10, 1957, pp. 16-33.

6. Bulcraig, W. R., Components of Raw Producer Gas, Institute of Fuel Journal, v 34, n 246, 1961, pp. 280-283.

7. Corey, R. C., Measurement and Significance of the Flow Properties of Coal-Ash Slag, U.S. Department of Interior, Bureau of Mines, Bulletin 618, 1964.

8. Dobbs, R. M. and I. A. Gilmour, Combustion of Coal in a Fluidized Bed Boiler, Department of Chemical Engineering, University of Canterbury, New Zealand, April, 1976.

9. Dolch, P., Die Verflüchtigung von Kieselsäure und Silizium als Silizium-sulfid, Eine technologische Studie, Chem. Fabr., v 8, n 51, 52, 1935, pp. 512-514.

10. Dolch, P., Über die Verflüchtigung von Silizium und Kieselsäure durch Schwefel und ihre Bedeutung für die Praxis, Montan. Rundschau, v 27, n 1, 1935, pp. 3-4.

11. Dolch, P., Vergasung von Steinkohle im Fahrzeuggaserzeuger, Brennstoff, Chemie, v 17, n 4, 1936, pp. 67-69.

12. Dowson, J. E. and A. T. Larter, **Producer Gas**, Longmans Green and Co., London, 1907.

13. Ekman, E. and D. Asplund, A Review of Research of Peat Gasification in Finland, Technical Research Centre of Finland, Fuel and Lubricant Research Laboratory, Espoo, Finland.

14. Feldman, H.F., et al., Conversion of Forest Residues to a Clean Gas for Fuel or Synthesis, TAPPI Engineering Conference, New Orleans, Louisiana, November 26, 1979.

15. Funk, H. F., Treating Waste Material to Produce Usable Gases, United States Patent, N 3970524, July 1976.

16. Garret, D. E., Conversion of Biomass Materials into Gaseous Products, Thermochemical Coordination Meeting, Fuels from Biomass Program, Energy Research and Development Administration, Columbus, Ohio, April, 1978.

17. Goldman, B. and N.C. Jones, The Modern Portable Gas Producer, Institute of Fuel, London, v 12, n 63, 1939, pp 103-140.

18. Goss, J. R., An Investigation of the Down-Draft Gasification Characteristics of Agricultural and Forestry Residues: Interim Report, California Energy Commission, P500-79-0017, November, 1979.

19. Gumz, W., et al., **Schlackenkunde**, Springer-Verlag, Gottingen, 1958.

20. Hendrickson, T. A., **Synthetic Fuels Data Handbook**, Cameron Engineers Inc., Denver, Colorado, 1975.

21. Jenkins, B. M. Downdraft Gasification Characteristics of Major California Residue-Derived Fuels, Ph.D. Thesis, Engineering, University of California, Davis, 1980.

22. Jones, J. L. and S. B. Radding, Solid Wastes and Residues Conversion by Advanced Thermal Processes, American Chemical Society, Symposium Series, n 76, Washington, D.C., 1978.

23. Knutson, J. et al., Crop Residues in California — Some Factors Affecting Utilization, University of California, Division of Agricultural Sciences, Leaflet 2872, 1978.

24. La Rue, J. and G. Pratt, Problems of Compacting Straw, Sixth Annual Conference, Biomass Energy Institute, Winnipeg, Manitoba, Canada, October 13, 1977.

25. Levin, Ernest M. and Carl R. Robbins, **Phase Diagrams for Ceramists**, The American Ceramist Society, Columbus, Ohio, 1964.

26. Perry, Robert H. and Cecil Chilton, **Chemical Engineers Handbook**, McGraw-Hill Company, New York, 1973.

27. Rambush, N. E., **Modern Gas Producers**, Van Nostrand Company, New York, 1923.

28. Reisner, W. and M.V. Eisenhauf Rothe, Bins and Bunkers for Handling Bulk Materials, Trans. Tech Publications, 1971.

29. Redding, G. J., The Effect of Fuel Moisture Content on the Quality of Gas Produced from the Gasification of Crop and Forest Residues, Master's Thesis, Agricultural Engineering, University of California, Davis, 1979.

30. Schläpfer, P. and J. Tobler, **Theoretische und Praktische Untersuchungen über den Betrieb von Motorfahrzeugen mit Holzgas,** Schweizerische Gesellschaft für das Studium der Motorbrenstoffe, Bern, Switzerland, 1937.

31. Selvig, W.A. and F.H. Gibson, Analysis of Ash from United States Coals, U.S. Department of Interior, Bureau of Mines, Bulletin 567, 1956.

32. Skov, N. A., and M. L. Papworth, **The Pegasus Unit,** Pegasus Publisher Inc., Olympia, Washington, 1975.

33. Takeda, S., Development of Gas Engine, The Bulletin of the Faculty of Agriculture, Mie University, Tsu, Japan, N 58, 1979, pp. 137-141.

34. Tatom, J. W. Survey of International Biomass Densification Equipment, Solar Energy Research Institute, Golden, Colorado, Contract AD-8-1187-1, September, 1978.

35. Wise, L.E., **Wood Chemistry,** Reinhold Publishing Corporation, New York, 1944.

CHAPTER VI: CONDITIONING OF PRODUCER GAS

The gas leaves the producer as a mixture of N_2 (nitrogen), H_2 (hydrogen), CO_2 (carbon dioxide), CO (carbon monoxide), CH_4 (methane), small amounts of C_2H_2 (acetylene), C_2H_4 (ethylene), C_2H_6 (ethane), tar vapor, mineral vapor, water vapor, dust (mostly carbon and ash), sulfur and nitrogen compounds. The only constituents which are combustible are H_2, CO, CH_4, C_2H_2, C_2H_6, C_2H_4 and the tar vapor. All the others, including tar, are corrosive, produce pollutants or may seriously interfere with the operation of burners or internal combustion engines. It is therefore essential to clean the gas to a certain extent. The degree of purification of the gas depends on the use of the producer gas.

A. Sulfur compounds in the producer gas (24,41): They are undesirable because their condensates are corrosive and are pollutants in the exhaust gases. The sulfur compounds occurring in the crude gas are H_2S (hydrogen sulfide), CS_2 (carbon disulfide), COS (carbonyl sulfide), SO_2 (sulfur dioxide), S_2 (sulfur as a gas) and traces of C_4H_4S (thiophene), $CH_3 \cdot C_4H_3S$ (methyl-thiophene), C_2H_2 HS (acetyl-mercaptan), $CH_3 \cdot HS$ (methanethiol), $CH_3 \cdot S \cdot CH_3$ (methyl-disulfide). The bulk of the sulfur contained in fuel gasified in a gas producer will exist in the raw gas as hydrogen sulfide (H_2S) (94% to 97%). Carbon disulfide (CS_2) and sulfur as gas occur only in "dry" gasification at high temperatures and no moisture in the air blast. This is unlikely to happen in practice. The SO_2 generated in the partial combustion zone will be converted into H_2S and COS. The completeness of this conversion depends on the temperature of the fuel bed and the moisture content of the air blast. Any SO_2 in the raw gas results from insufficient steam addition to the air blast or is a product of the distillation zone.

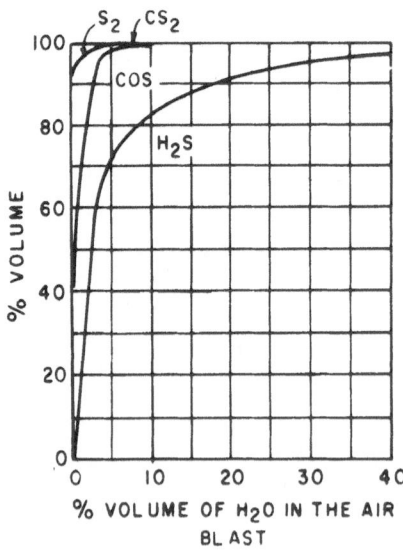

Figure 105. Gaseous Sulfur and Gaseous Sulfur Compounds in Raw Producer Gas as a Function of Water Vapor in the Air Blast (24).

The distribution of the sulfur compounds as a function of the water content in the air blast is shown in Figure 105.

A typical analysis of the fuel, gas and sulfur compound composition of an anthracite coal gasifier is given in Table 26. The specific gasification rate was 150 kg/m^2-h.

Table 26. Typical Analysis for a Gas Producer Fueled With Anthracite Coal (9).

Approximate Analysis of the Fuel (% weight, dry basis)		Gas Composition (% volume, dry basis)	
H_2O	5.2	CO_2	3.8
Volatile Matter	33.1	O_2	0.3
Fixed Carbon	53.7	CO	29.3
Ash	8.0	CH_4	3.3
		H_2	7.9
		N_2	55.4

Ultimate Analysis of the Fuel (% weight, dry basis)		Distribution of Sulfur in the Fuel (% weight, dry basis)	
C	84.0	pyrite sulfur	56
H_2	5.1	sulphate sulfur	43
S	1.1	organic sulfur	1
O_2+N_2	9.8		

Distribution of Sulphur Compounds in the Products of Gasification (% weight).

Sulfur in the char	7.5
Sulfur in the soot and dust	5.6
Sulfur in the tar	0.9
Sulfur in the gas: as H_2S	66.3
as SO_2	13.3
as COS	6.4

143

The amount of sulfur compounds in the raw gas depends primarily on the sulfur content of the fuel as given in the ultimate analysis. The sulfur content of biomass fuels has been investigated by several authors during the past 60 years. Their results are given in Table 27.

Table 27. Sulfur Content of Biomass Fuels

Biomass	% sulfur, dry weight basis		References
Alfalfa Seed Straw		0.3	5
Almond Shells	less than	0.02	5
Barley Straw		0.14	5
Coffee Hulls		0.2	37
Corn Cobs		0.001-0.007	5,39
Corn Fodder		0.15	5,37
Corn Stalks		0.05	5
Oat Straw		0.23	37
Cotton Gin Trash		0.26-0.31	5
Flax Straw, Pelleted	less than	0.01	5
Furfural Residue		0.4	8
Olive Pits		0.02	5
Peach Pits		0.04	5
Peanut Husk		0.1	8
Peat (Finnish)		0.05-0.2	19
Peat, General		1.5-2.0	41
Rice Hulls		0.16	5
Rice Straw		0.10	5
Walnut Shells		0.03-.09	5
Wheat Straw		0.17	37
Wood, Chipped		0.08	5
Wood, General		0.02	5,30
Wood, Pine Bark		0.1	8
Wood, Green Fir		0.06	8
Wood, Kiln Dried		1.0	8
Wood, Air Dried		0.08	8

Not all of the data in Table 27 were derived on an elemental basis for H, C, N, O, S and ash. The list shows clearly that biomass residues contain very

144

little sulfur. Consequently, the generation of H_2S is of little importance in the gasification of biomass, as long as the sulfur content does not exceed 0.5%. However, the introduction of steam or water to the air blast increases the H_2S content considerably as shown in Figure 105. At normal loads the total sulfur in the gas is about 1 g/m^3 for a crossdraft unit mounted on a truck fueled with anthracite coal (29).

The sulfur content of coal fuels such as anthracite, bituminous and lignite, are in general much higher than biomass-fuels. The sulfur content of coal also depends strongly on the origin of the coal. Some typical values are given in Table 28.

Table 28.   Sulfur Content of Coal (19,41)

Source	% sulfur, dry weight basis
Welsh anthracite	0.5
German anthracite	1.7
Bituminous coal	1-2 (generally)
Scotch bituminous coal	4
German bituminous coal	3
American coal	0.13-6.8
Lignite	1.5-2 (generally)
Spanish lignite	6
Finnish peat	0.05-0.2

B. Nitrogen compounds in the producer gas: Under the working conditions of a gas producer at atmospheric pressure and moderate temperatures between 1000 and 2000 $^{\circ}$C, NH_3 (ammonia) and HCN (hydrocyanic acid) can be found in the raw gas. Several authors (24,41) agree that the bulk of the generation of NH_3 and HCN comes from the nitrogen found in the fuel and that the amount of NH_3 and HCN formed by the reaction with the nitrogen in the air-blast is insignificant. Consequently the higher the nitrogen content of the fuel, the higher is the amount of NH_3 and HCN in the raw gas.

Table 29 indicates that it is safe to assume a nitrogen content of less than 2% for gas producer fuels. Figure 106 shows the amount of NH_3 and HCN obtained when anthracite with a nitrogen content of 1.1% is gasified. The activated anthracite was treated with a 1% solution of sodium silicate. The experiment was performed in a small gasifier with a hot gas output of 62.3 Nm^3/h. The effect of the injection of preheated air and up to 0.45 kg of steam per kg of fuel consumed on the yields of NH_3 and HCN was negligible.

Portable wood gas generators usually yield 2-6 grams of NH_3 in each liter of water condensed out of the gas stream (43).

Table 29. Nitrogen Content of Gas Producer Fuels.

Biomass Fuels	% nitrogen dry weight basis	References
Barley, Straw	0.59	5
Corn Cobs	0.16-0.56	5
Corn Fodder	0.94	5
Cotton Gin Trash	1.34-2.09	5
Corn, Stalks	1.28	5
Flax Straw, Pelleted	1.1	5
Oat Straw	0.66	37
Olive Pits	0.36	5
Peach Pits	1.74	5
Peat	0.5-3.0	19
Prune Pits	0.32	5
Rice Hulls, Pelleted	0.57	5
Safflower, Straw	0.62	5
Walnut Shells	0.260-0.4	5
Wood (General)	0.009-2.0	5,41

Coal Fuels		
Anthracite	less than 1.5	41
German and English bituminous coal	0.5-1.9	41
American coal	0.5-2	41
Brown coal and lignites	0.5-2	41

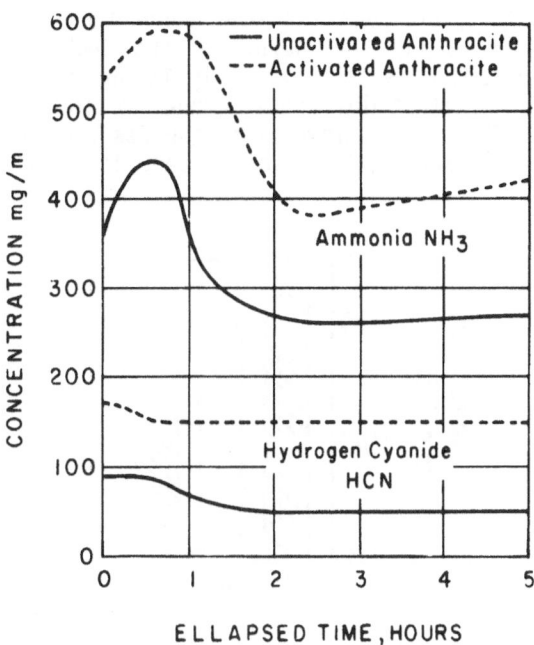

Figure 106. Ammonia and Hydrogen Cyanide Distribution in Producer Gas Made
With Preheated Air and Steam (31).

C. Dust in the producer gas: It is of greatest importance that the gas delivered
to an engine be free from dust of an abrasive nature and that it should contain
the absolute minimum of corrosive constituents. The full development of power,
freedom from excessive cylinder wear and conservation of lubrication oil depends
on these conditions. The amount of dust entrained in the raw gas depends on
many factors such as: type of gas producer, type of fuel and the specific
gasification rate and the temperature of the partial combustion zone. The type
of dust carried by the gas seems to play an even more important role than the
quantity. Dust loads in the raw gas from mobile units which operate with a high
specific gasification rate were found to contain approximately 50% of the total
solid particulates in the raw gas in the form of ash which had been fused into
a hard granular material of very abrasive nature, resembling quartz sand (52).
This material was insoluble in dilute acids. On the other hand, the dust from
stationary plants (operated at a much lower specific gasification rate) was 90%
soluble in water and the remainder in weak acid. The abrasive nature of the
dust is determined by the temperature in the partial combustion zone and the
chemical composition of the ash. The SiO_2 (silicon oxide) and Fe_2O_3 (iron oxide)
contents are especially important because of their abrasive nature. In addition,
the gas generated by mobile units was found to contain an appreciable quantity
of soot (carbon dust). The soot in the raw gas, which deposits in the connecting
pipes and coolers, is partly caused by the reaction $2CO \rightarrow CO_2 + C$ that takes
place to a certain extent after the gas leaves the gas producer.

Figure 107 shows the variation of the dust concentration in producer gas with
the method of operating an automotive crossdraft gas producer.

147

It will be observed that, although the dust concentrations were of the same order with the dry air blast, 180 mg/m³ of the dust was below 5 microns diameter, whereas with the wet air blast only 14 mg/m³ was below 5 microns diameter. This is an important difference because particles below 5 micron are considerably more difficult to remove from the gas stream. In addition, wet scrubbers are not very effective in removing particles of this small size.

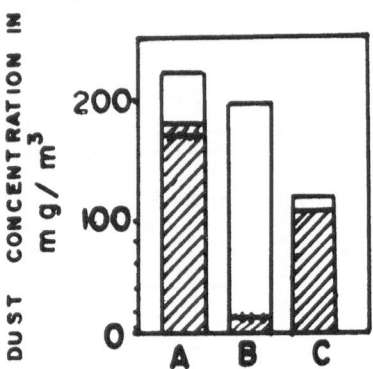

Figure 107. Variation of Dust Concentration in Producer Gas With the Method of Operating the Producer (32).

The shaded area represents the concentration of particles smaller than 5 microns in diameter. The dotted line represents the dust concentration at the outlet of the wet scrubber. A. Anthracite, dry air blast. B. Anthracite, wet air blast. C. Activated anthracite, wet air blast.

The amount of dust that can be tolerated in the gas entering an internal combustion engine has been the subject of numerous papers (29,32,52) and extensive testing during the 1930-1950 period. It was found that with a dust concentration up to 10 mg/Nm³ the engine wear was the same order as that obtained with gasoline, but beyond that amount it increased rapidly, being up to five times as great when the concentration reached 50 mg/Nm³.

The dust content also depends strongly on the specific gasification rate at which the gasifier is operated. The lower average of 200-600 mg/Nm³ may be well above 2000 mg/Nm³ when the gasifier is run on overload. The rating of the gasifier also has an effect on the size distribution of the dust. At normal load only a small quantity, about 4%, of the dust reaching the filters consists of coarse particles, mainly partly burned fuel (66 to 1000 microns). The bulk of the remainder is extremely fine, being only 3 to .3 microns in diameter. At high specific gasification rates, which implies a higher gas stream velocity, the proportion of coarse material is greatly increased and may total about 25%. Other authors (35,43) report somewhat higher numbers, 1-3 g/Nm³, under normal load conditions in wood gas generators. This amount of dust in the raw gas corresponds to about 2-6 grams of dust per kg wood gasified.

Figure 108 shows the gas velocity in the annular space inside an Imbert downdraft gasifier as a function of the load on the gas producer.

148

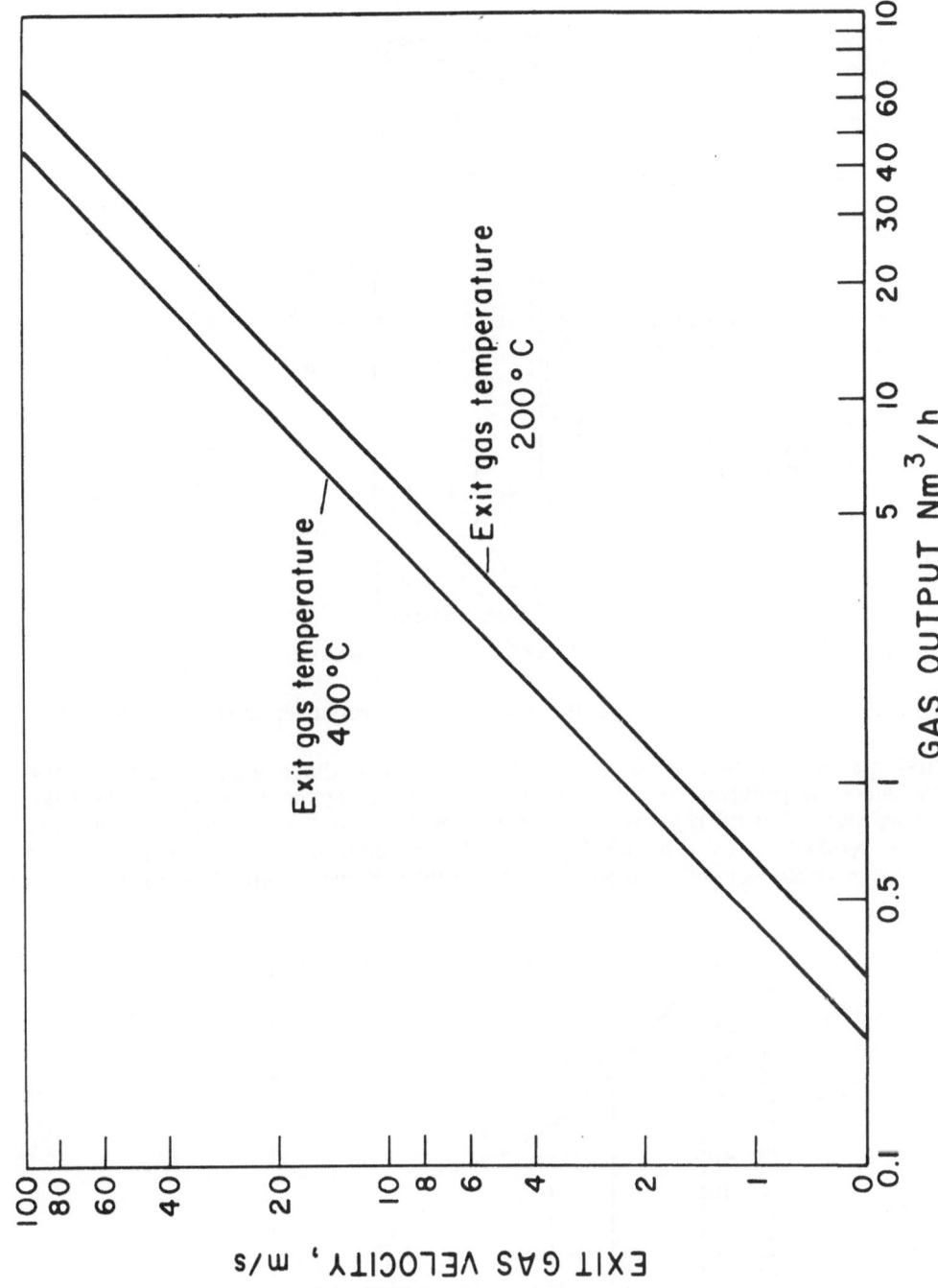

Figure 108. Hot Gas Velocity in the Annular Space of a Downdraft Gas Producer (see A, Figure 109) (35).

149

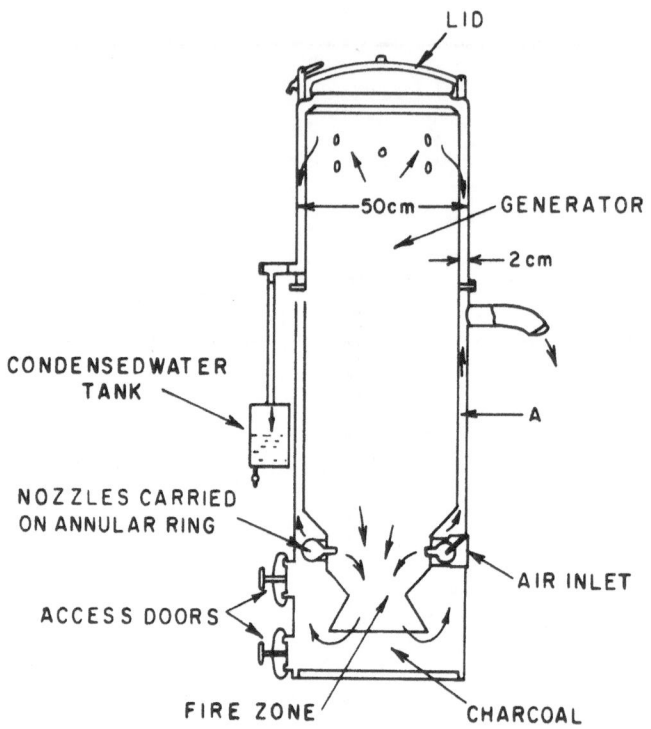

Figure 109. A Cross Section of the Imbert Gas Producer, Circa WW II (21).

On first glance it seems to be attractive to separate the coarse material in the gas producer by gravitational settling in the annular space (see A, Figure 109). However, tests showed that only 2% of the dust carried out could be separated at a gas production rate of 60 Nm^3/h. The amount of dust separated at the lowest permissible rate of 5-10 Nm^3/h was slightly higher, being 10% (35). Other

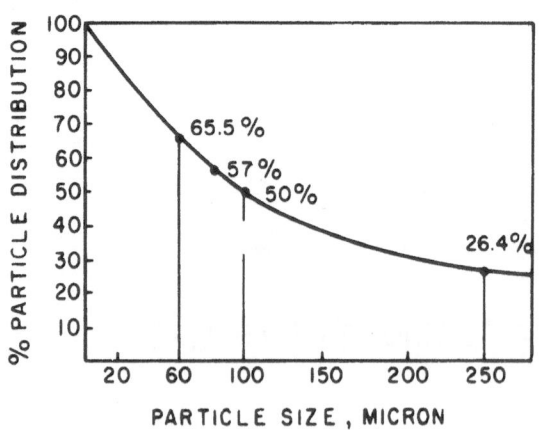

Figure 110. Particle Size Distribution (6).

test reports with the Imbert type gas producer give the particle size distribution of the dust in the raw gas as shown in Figure 110 and Table 30.

Table 30. Particle Size Distribution (6).

Particle size, micron	%
Over 1000 (1 mm screen)	1.7
1000 - 250	24.7
250 - 102	23.7
102 - 75	7.1
75 - 60	8.3
Under 60	30.3
Losses	4.2
	100.0

The same source reports the dust concentration in the raw gas for wood gas generators of various makes as a function of the gas production rate. A gas output of 100 Nm^3/h was considered the maximum for most automotive gas producers (Figure 111).

Figure 112 shows the dependence of the dust concentration in the raw gas on the type of fuel used. For this type of gas producer with a maximum gas output of 80 Nm^3/h the amount of dust in the raw gas was considerably lower when wood was used as a fuel instead of charcoal.

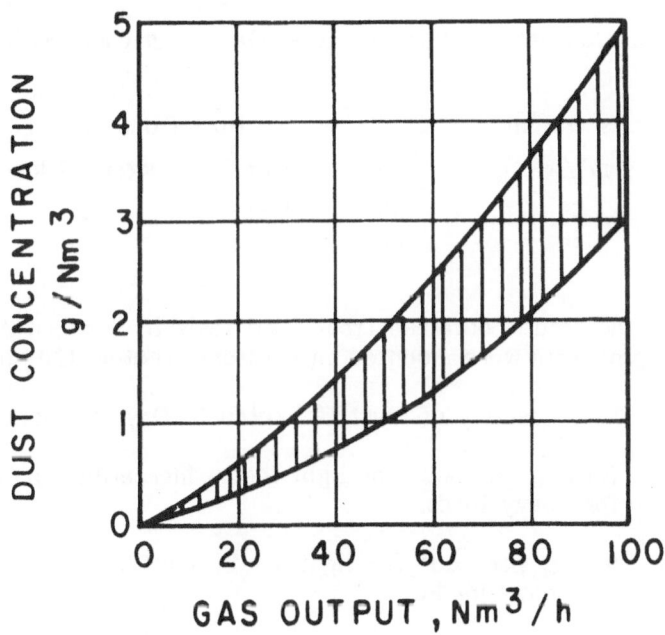

Figure 111. Dust Concentration in the Raw Gas for Commercial Automotive Gas Producers (6).

151

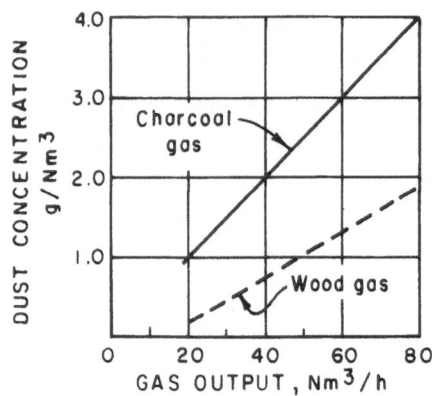

Figure 112.   Dust Concentration as a Function of the Fuel (6).

Road tests conducted in Sweden with modified Imbert downdraft gas producers mounted on tractors and trucks yielded about 2.1-3 grams of dust per Nm^3. The dust content was obtained under normal driving conditions (47-58 km/h) and a gas consumption between 40 Nm^3/h and 68 Nm^3/h.

The dust concentration in raw gas obtained from crossdraft gas producers is in general high, due to the high gas velocity, and varies considerably with the specific gasification rate. Table 31 lists values obtained in road tests at different speeds.

Table 31. Dust Concentration in the Raw Gas for a Crossdraft Gas Producer (1).

Gas Output m^3/hr	Speed km/h	Dust Content g/Nm^3	Dust Content kg/1000 km
59	56	0.65	0.7
85	80	6	6.3

Table 32 gives the results obtained from field tests in Western Australia.   The downdraft gas producers were mounted on kerosene tractors (20 hp at 1050 rpm).

Table 32.   Quantity of Dust Collected in Dry Cleaners (11).

Powell Plant: 0.13 kg per hour for light loads, increasing to 0.6 kg per hour for heavy loads.

Herbert Plant: 0.3 kg per hour for light loads, increasing to 0.9 kg per hour for heavy loads.

The dust carried over by the raw gas is considerable with fuels containing an excessive quantity of fine particles or the tendency to break up in the gas producer, such as lignites, brown coal, peat and soft charcoal. Since the dust-carrying capacity of a gas varies with the sixth power of the velocity, it is obvious that the best means to prevent dust from leaving the gas producer is to keep the exit temperature as low as possible. This can be done by an empty space above the fuel bed in an updraft gas producer, where the velocity of the gas stream will decrease and the dust can settle down.

It is also often overlooked that the air blast drawn into a suction gas producer can contain a considerable amount of dust as shown in Table 33.

Table 33. Dust Content of Air (6).

Air in	mg/m^3
Rural areas and suburbs	0.5 – 1
Cities	2
Industrial centers	4
Streets with heavy traffic	20
Dusty highways, excavation and graveling work, farm work with tractors, etc. .	over 200

In summary, the dust content in the raw gas varies considerably from as little as 0.2 g/Nm^3 to over 6 g/Nm^3. Its composition and chemical nature also gives rise to large changes from highly water soluble, soft ash to insoluble, very abrasive sintered material which is highly damaging to any internal combustion engine. The bulk of the dust in the raw gas consists of ash carried out by the high gas velocity in the fuel bed. Consequently a chemical analysis of the ash gives a good indication of how abrasive the dust actually is.

D. Moisture in the raw gas: Moisture leaving the gas producer in the form of steam as part of the raw gas, has in general several sources:

1. Moisture in the air in the form of water vapor.

2. Moisture injected into the combustion air in the form of water vapor or steam.

3. Moisture in the fuel in the form of:

    a) Inherent moisture held in the capillary openings.
    b) Surface moisture which occurs on fuel surface and is in excess of the inherent moisture.
    c) Decomposition moisture as released from organic compounds in the 200-225 $^{\circ}$C range.

4. The water generated by chemical reactions within the H-C-O system.

The amount of water leaving the gas producer as steam depends on the exit temperature, the moisture input, and the chemical processes in the gas producer.

The amount can be predicted by thermodynamical considerations and calculations as listed in Table 34. The results are based on a moisture content of 0%, 16.7% and 28.6% of the wood fuel, an equilibrium temperature of the watershift reaction $CO + H_2O = CO_2 + H_2 + 41,200$ kJ/kg-mole at 700 $^{\circ}$C, a loss through radiation and convection of 15%, a heating value of the dry wood of 18.8 MJ/kg and an ultimate chemical composition of the wood of 50% C, 6% H, and 44% O on a H-C-O basis.

Table 34. Moisture Content of the Raw Gas (43).

Exit Temperature of the raw gas $^{\circ}$C	Moisture Content of the wood	Water in the raw gas kg/kg wood
0	0.0	0.063
150	0.0	0.075
350	0.0	0.093
500	0.0	0.109
0	16.7	0.140
150	16.7	0.155
350	16.7	0.178
500	16.7	0.199
0	28.6	0.217
150	28.6	0.235
350	28.6	0.262
500	28.6	0.286

The assumption of a higher equilibrium temperature of 900 $^{\circ}$C increases the amount of water in the raw gas about 25%, everything else being constant.

The black condensate generated in most gas producers is on the average a mixture of 80-95% water and 5-20% tars and oil. Consequently, the gasification of wet fuel or steam injection in updraft gas producers increases the amount of condensates drastically. This is by no means an indication that tar generation has been greatly increased but merely a sign of too much moisture escaping the decomposition process. In updraft gas producers not much can be done to curtail the amount of moisture in the gas as outlined in Chapter III and V. But, high moisture fuels also do not influence the updraft gasification process very much, because the fuel moisture can not be decomposed in the partial combustion zone. In downdraft gas producers any moisture released by the fuel will have to pass through the throat area and should be decomposed. However, this highly endothermic reaction cools down the partial combustion zone and therefore generates favorable conditions for an increased amount of uncracked tar and moisture in the raw gas.

Extensive tests with gasification of biomass done at the University of California, Davis gave the following results:

Table 35. Amount of Water Condensed Out of the Raw Gas of a Downdraft Gas Producer (30,42).

Fuel	Moisture Content % Weight Wet Basis	Gas Temperature at Exit of Producer $^{\circ}$C	Water Condensed Out of the Hot Gas kg/kg Wet Fuel
Cubed Alfalfa Straw	7.9	288	0.165
75% Barley Straw* 25% Corn Stover	6.9	231	0.092
Cubed Bean Straw*	13.0	329	0.202
Corn Cobs	11.0	327	0.225
Cubed Corn Stalks*	11.9	355	0.368
Cubed Cotton Gin Trash*	23.5	260	0.477
Cubed Cotton Stalks*	20.6	260	0.475
Rice Hulls, Pelleted*	8.6	214	0.182
50% Wheat Straw* 50% Corn Stover	15.0	311	0.55
Wood Blocks, Douglas Fir	5.4	315	0.40
Chipped Municipal Tree Prunings	17.3	221	0.35
Hogged Wood, Manufacturing Residues	10.8	260	0.37
Whole Log Wood Chips	18.0	unknown	0.37
Whole Log Wood Chips	32.0	287	0.67
Whole Log Wood Chips	51.7	326	1.04

*These fuels are not acceptable for sustained downdraft gas producer operation because of slag formation in the partial combustion zone.

The fuel was gasified in a downdraft gasifier, with a hearth zone diameter of 30.5 cm and a capacity of 360 MJ/h of clean cold gas, corresponding to the energy requirement to drive a 30 hp engine. The raw gas was cooled to ambient temperature of 30-35 ^{o}C and left the condenser in saturated condition.

For dry gasification the moisture content of the raw gas can be predicted from the fuel moisture content by the equation:

Gas Moisture Content (% by volume) = 3.6686 + 0.59216 x Fuel Moisture Content, where the fuel moisture content is given in % by weight, wet basis (30,42).

Figure 113 shows the condensed water collected as a function of the fuel consumption rate. Tests were performed with the UCD Laboratory Gas Producer on 11 major California crop residues as listed in Table 29. In general it can be said that about 0.2 to 0.7 kg of water per kg of wet fuel gasified must be removed from the raw gas before it is used as fuel for an internal combustion engine.

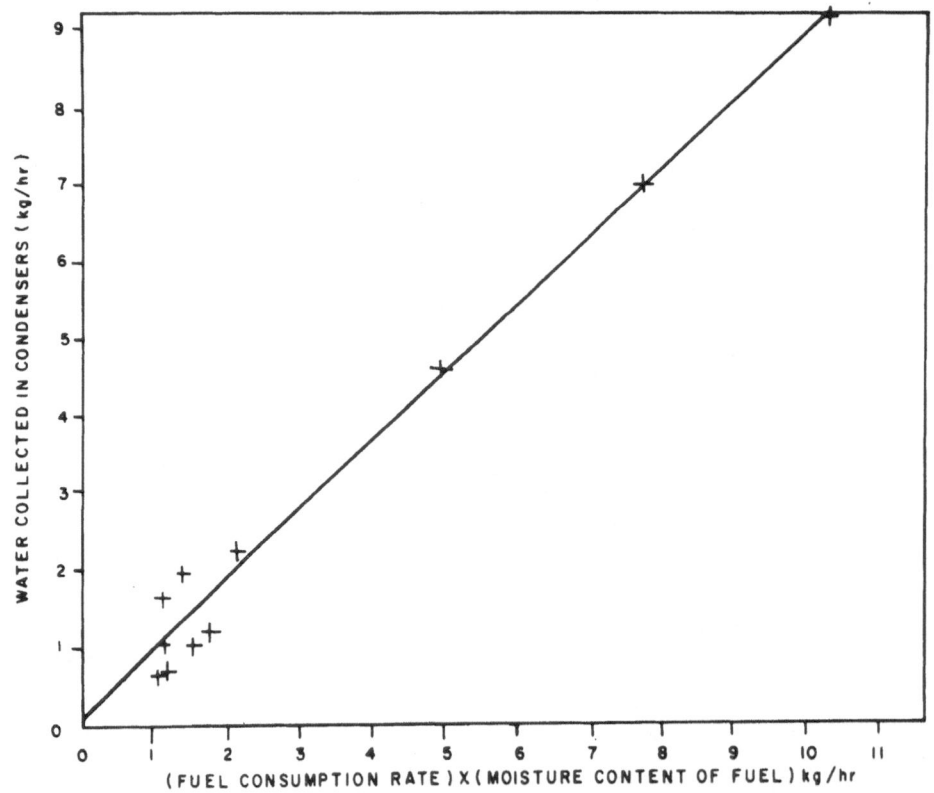

Figure 113. Linear Regression Model to Predict Amount of Water Collected in Condensers from the Fuel Consumption Rate Multiplied by the Fuel Moisture Content (Wet Weight Basis) (30,42).

156

E. Temperature of the raw gas leaving the producer: The exit temperature of the raw gas has a considerable impact on the choice of the cleaning equipment and its arrangement. For instance the exit gas temperature determines the dimension of the condenser and the choice of the filter media for dry filtration of the gas. The temperature itself depends on so many variables which in turn are not independent of each other, that a quantitative analysis is not given. However, much can be said about how different designs and modes of operation as well as choice of fuel influences the exit gas temperature. Actual measured temperatures for different types of gas produces are given in Table 36.

Table 36. Gas Temperature at Outlet for Various Gas Producers

	Temperature at Producer Outlet in $^{\circ}$C	Fuel Used	Reference
Malbay updraft	180-220	charcoal	21
Malbay updraft	150-160	low temperature coke	21
Malbay updraft	160-175	anthracite	21
Wisco updraft	400	charcoal	21
Humboldt Deutz updraft	280-300	anthracite, charcoal	21
Koela	180-230	anthracite, charcoal	21
Mie University updraft gas producer	20-80	wood charcoal, palm nut charcoal, anthracite	50
Modified Imbert downdraft (Swedish design)	200-580 depending on specific gasification rate	peat briquettes, wood chips, charcoal	35
Gohin Poulence crossdraft	400-500	charcoal, low temperature coke, anthracite	21
University of Kentucky updraft gasifier	160-380 depending on depth of fuel bed and fuel	corncobs	39
Purdue University modified Imbert type	275	corncobs	40
UC-Davis Laboratory Downdraft Gas Producer 360,000 kJ/h	220-360	crop residues (see Table 29)	30
UC-Davis Civil Engineering downdraft gas producer 600,000 kJ/h	300	paper cubes, solid waste cubes, wood chips	56
UC-Davis Downdraft Pilot Plant 5,000 MJ/h	240-454 depending on moisture content of the fuel	wood chips	

157

The exit temperature changes significantly with the specific gasification rate as demonstrated in Figure 114. The test units were two modified Imbert downdraft gas producers (Swedish design) with a hearth zone diameter of 26 cm and throat diameters of 12.4 cm and 8.7 cm, respectively.

In general updraft gas producers have a lower exit gas temperature than downdraft and crossdraft gas producers because the upward moving gas releases its heat to the downward moving fuel. In most cases an exit temperature of 150-600 ^{0}C can be expected for small-sized gas producer plants (400 MJ/h of cold clean gas, equivalent to the energy required to drive a 30 hp engine).

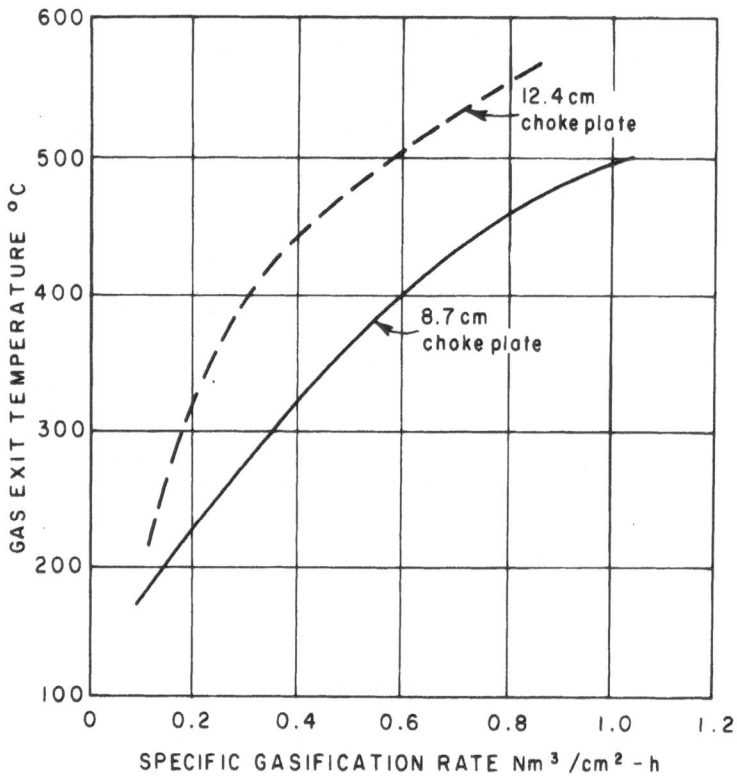

Figure 114. Exit Gas Temperature Versus Specific Gasification Rate (35).

Portable crossdraft gas producers, due to the very short chemical reaction zone of 15-25 cm and the insignificant reduction and distillation zone the gas passes through, generate unusually high exit temperatures as given in Figure 115.

The height of the fuel bed also influences the exit temperature as shown in Figure 116. This data was obtained from a corn cob-fueled, updraft gasifier of 0.155 m^{2} cross-sectional area and a specific gasification rate of 256 kg/h-m^{2}.

158

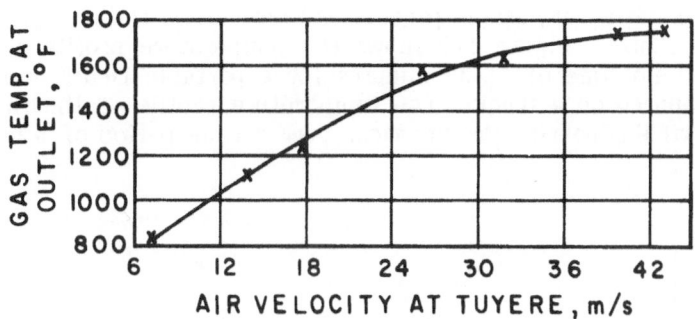

Figure 115. Gas Exit Temperature as a Function of Air Velocity Through the Tuyere for British Emergency Gas Producer (1).

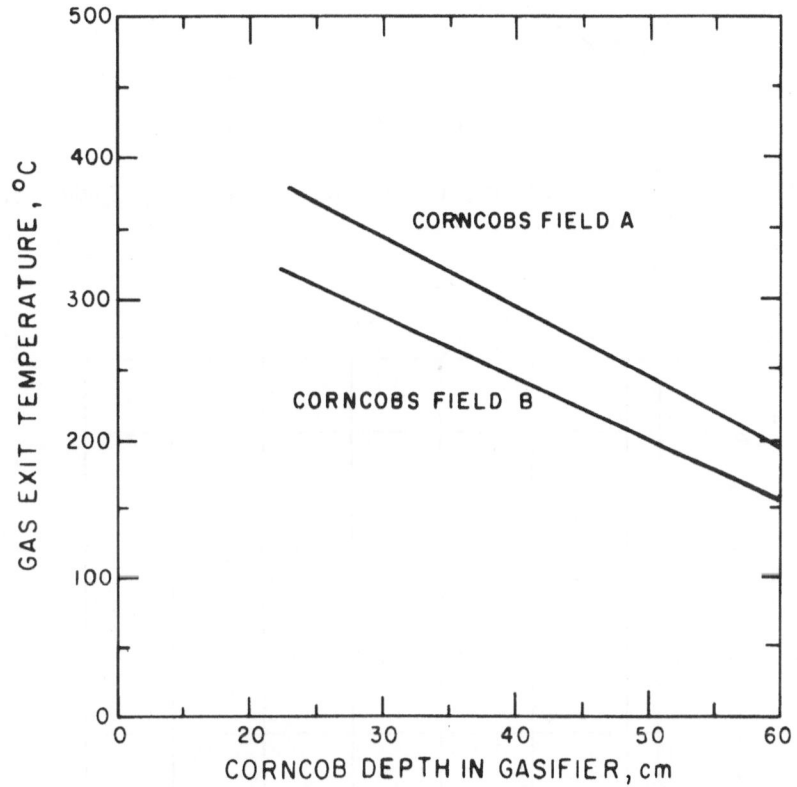

Figure 116. Temperature of the Gas Exiting the Gasification Chamber as a Function of Corncob Depth in the Gasifier (39).

159

Gasification is not a very stable process and changes in the exit gas temperature occur within limits. Figure 118 shows the temperature profile of the raw gas during a start-up time of 16-32 minutes for a portable Deutz and Kromag gas generator mounted on a truck. The temperature profile of the raw gas of the UC-Davis Civil Engineering gas producer over a time period of 5 hours is shown in Figure 117.

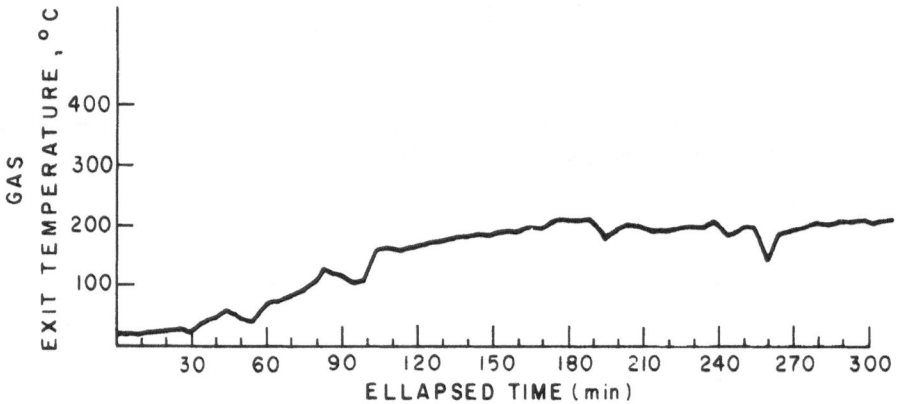

Figure 117. Gas Exit Temperature Versus Time for a 40 hp Sludge Waste Gas Producer (56).

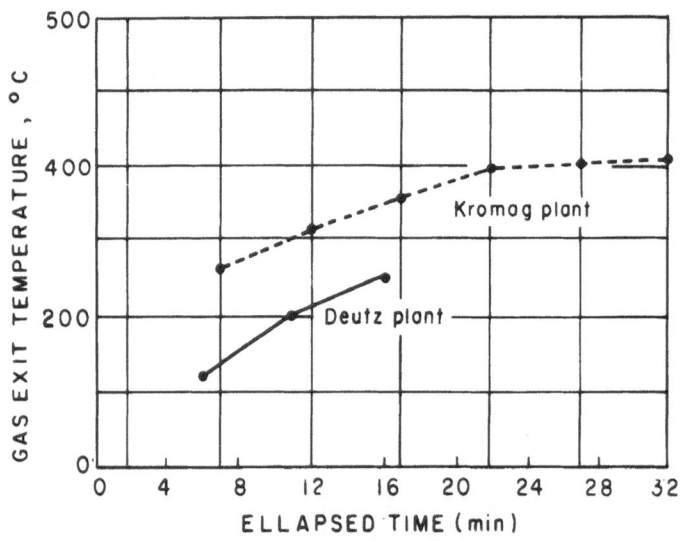

Figure 118. Exit Gas Temperature Versus Start Up Time for Two Automotive Gas Producers (43).

160

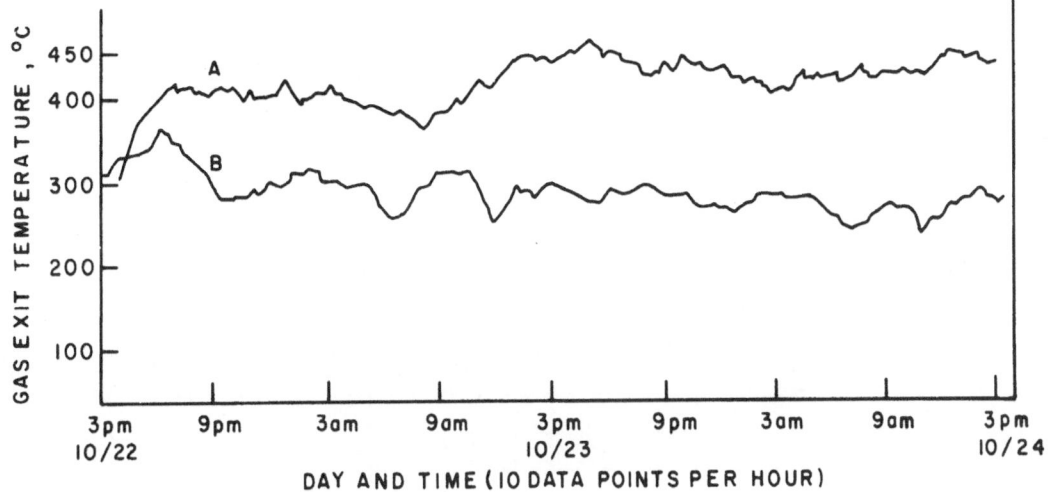

Figure 119. Influence of the Fuel Moisture Content on the Gas Exit Temperature
Over a 48-hour Period (23).

The moisture content of the fuel also influences the exit temperature of the
gas. Figure 119 shows the temperature profile of a continuous run over 48
hours with the UC-Davis Pilot Plant. Curve A represents the temperature
profile for wood chips with a moisture content of 11% whereas Curve B describes
the temperature profile for wood chips with a moisture content of 25% (23).

In dry gasification a low exit temperature is desirable for two reasons. First,
all the sensible heat in the hot gas is lost to the cooler when the gas is used
to drive an internal combustion engine which requires cool gas. Secondly, a
low temperature gives some indication of how well the reduction of CO_2 into
CO has taken place in the reduction zone. The governing endothermic reaction
is CO_2 + C = 2 CO - 172,600 kJ (at 25 $^{\circ}$C, 1 atm). This reaction consumes
heat as can be seen from the equation. Figure 120 shows the considerable
amount of sensible heat that is lost when cooling the exit gas. The gas mixture
is split up into its moisture, CO_2 and CH_4 components for simplicity.

The exit temperature of an updraft gas producer is in general low, because the
gas has to pass through the entire fuel column above the partial combustion
zone. Updraft gas producers are therefore very efficient and produce a gas
with a high heating value if it is combusted in a furnace immediately upon
leaving the gas producer. Crossdraft gas producers are at the opposite end of
the scale with high gas exit temperature and low heating value of the raw gas.

161

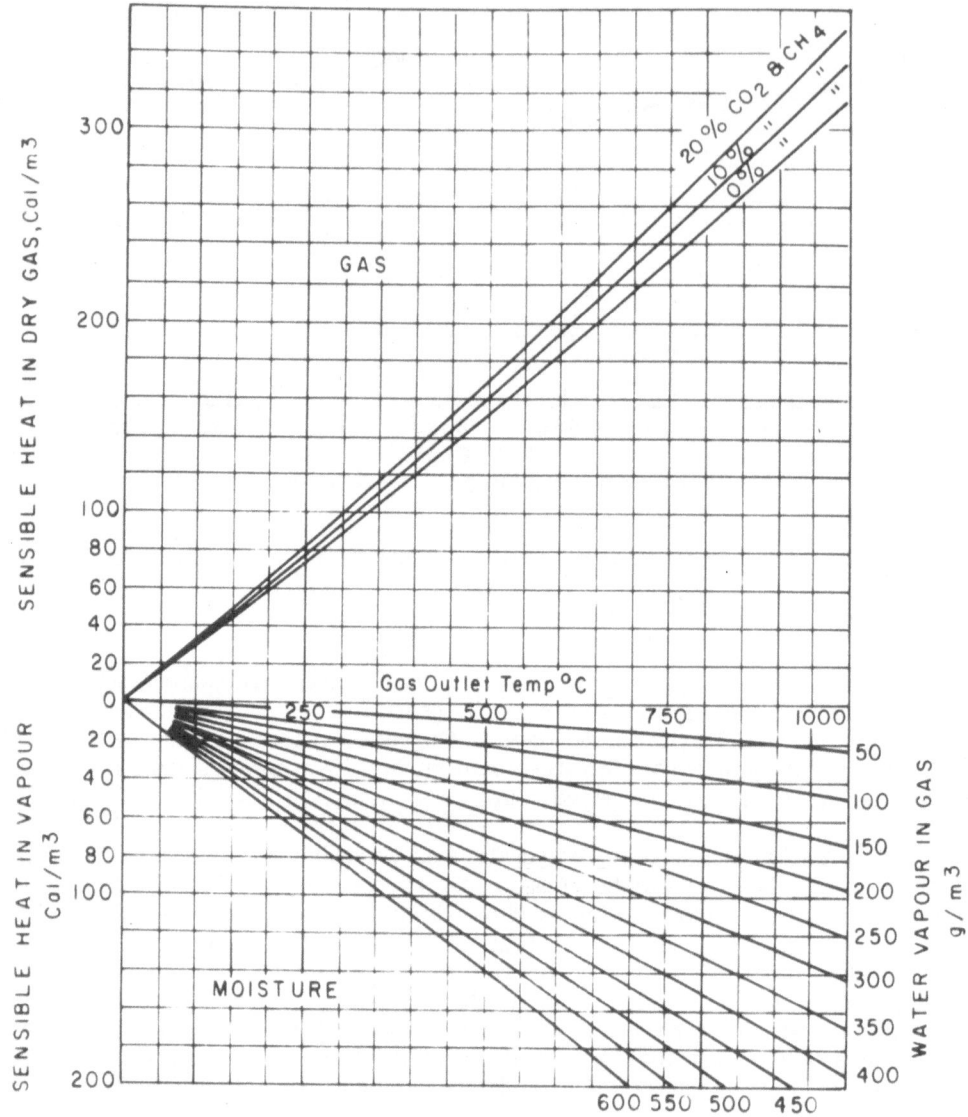

Figure 120. Sensible Heat in the Exit Gas (41).

F. Distillation products in the raw gas: The very nature of gasification generates a gas that is actually a mixture of three gas streams obtained from the partial combustion zone, reduction zone and distillation zone. The extent of these

162

zones depends on the type of gas producer. The distillation products are the least understood and consists mainly of tar, light and heavy oils, noncondensable gases and water vapor. In order to understand the mechanism of distillation in the thermal decomposition of biomass and coal, one needs to rely on experiments carried out under laboratory conditions. The results obtained do not represent the situation in a gas producer with regard to the quantitative yield of the various distillation products. However, the mechanism of thermal decomposition is the same in both cases and must be well understood before any decision can be made on how to reduce the tar in the raw gas by cracking inside the gasifier.

The removal of tar from the producer gas is one of the more difficult problems in gas cleaning. The difficulties that can arise in using tar laden gas range from inconvenient and expensive cleaning equipment to serious failure of the entire system.. Tar laden gas in engines will quickly gum the valves and necessitate a stoppage of the producer and engine. It can be so severe that the entire engine has to be pulled apart and be cleaned. Tar is a very complex substance and is one of the products of destructive distillation of biomass and coal. The exact composition will depend on a large number of factors, the most important of which is the temperature at which it is formed. Because tar removal outside the gas producer is troublesome, expensive and insufficient with medium technology devices, every effort must be undertaken to generate a gas as tar free as possible if the gas is used in an internal combustion engine.

As mentioned, tar is one of the products of thermal decomposition of solid fuels. The tar yield is therefore related to the amount of volatile matter in a fuel. However, it is not true that a fuel with a high percentage of volatile matter necessarily results in more tar in the raw gas than a low volatile matter fuel. The percentage of volatile matter in a fuel and the tar yield depends on the method applied in the laboratory and the reported results are difficult to compare. Tests conducted at the Fuel Research Station, London showed no correlation of potential tar and volatile matter of different kinds of treated and untreated coals (Table 37 and 38). In this context potential tar is defined as the maximum yield of tar from coal. The method applied was to heat the test fuel to a temperature of 600 $^{\circ}$C, keep the temperature constant for one hour and collect the distillation products in asbestos wool. Separation of the distillation products from the water formed was accomplished by maintaining the asbestos wool at a temperature slightly higher than 100 $^{\circ}$C.

Table 37. Potential Tar Versus Volatile Matter in Untreated Coal. Results Obtained Under Laboratory Conditions (18).

Coal	Volatile Matter Percent Weight Dry Basis	Potential Tar gram/ton Dry Basis
1	4.8	223
2	5.5	251
3	6.5	84
4	6.9	140
5	7.0	558
6	7.0	223

Table 38. Potential Tar Versus Volatile Matter in Treated Coal. Results Obtained Under Laboratory Conditions (18).

Coal	Volatile Matter Percent Weight Dry Basis	Potential Tar gram/ton Dry Basis
1	2.9	139
2	3.4	5078
3	4.2	1172
4	5.1	112
5	5.6	1032
6	7.1	112
7	8.0	28
8	10.0	698
9	10.2	56
20	42.0	66

These findings do not agree with laboratory and road tests when South Wales anthracite was used as fuel (29). The results are given in Figure 121 in grams of potential tar evolved per ton of dry coal.

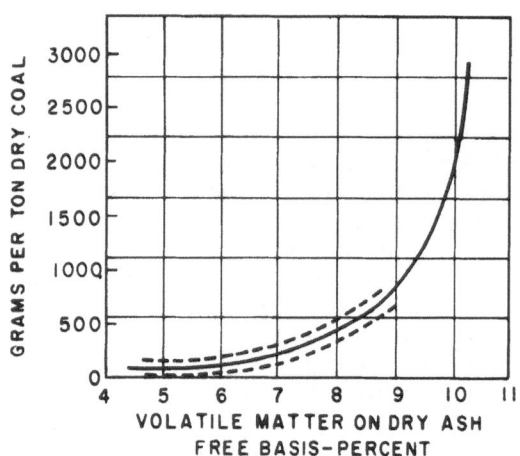

Figure 121. Potential Tar Versus Volatile Matter (29).

The points plotted for different samples of anthracite fall into a narrow band about a smooth curve becoming very steep when about 9% of volatile matter is exceeded. Road tests confirmed the general trend of the curve.

The American ASTM Designation D-271-70 defines volatile matter as those products (exclusive of moisture) which are given off as vapor when the coal is heated to 950 ^{o}C and this temperature is maintained for seven minutes. The loss of weight, minus the moisture content, is considered the volatile matter of the coal.

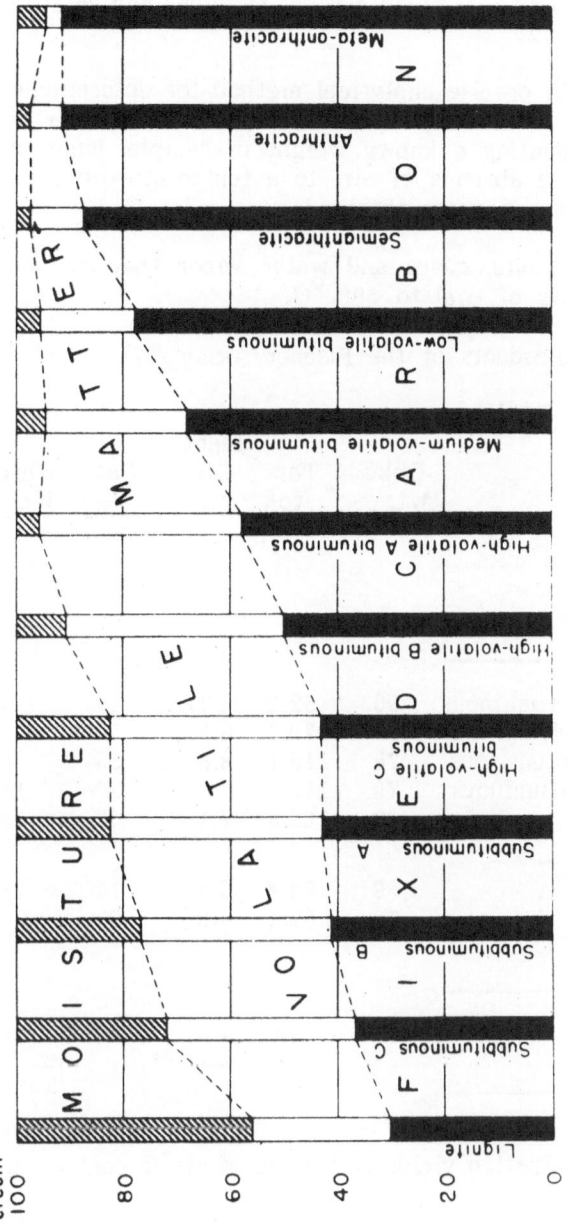

Figure 122. Classification of Coal (27).

165

In general, coal is classified according to its fixed carbon, moisture and volatile matter content as given in Figure 122.

The Fischer assay is an arbitrary but precise analytical method for determining the yield of products obtained from the distillation of organic substances in coal. The conditions consist of heating a known weight of sample, under a controlled rate of heating and in the absence of air, to a temperature of 500 °C then collecting and weighing the products obtained.

Table 39 lists the amount of tar, oils, gases and water vapor that can be expected when heating various ranks of coal to 500 °C.

Table 39. Distillation Products of the Fischer Assay (27).

A.S.T.M. CLASSIFICATION BY RANK

CLASS	GROUP	Coke Wt. %	Tar /ton	Light Oil /ton	Gas m^3/ton	Water Wt. %
I Anthracite	1. Meta-anthracite 2. Anthracite 3. Semianthracite					
II Bituminous	1. Low volatile bituminous 2. Medium vol. bituminous 3. High vol. A bituminous 4. High vol. B bituminous 5. High vol. C bituminous	90 83 76 70 67	32.0 70.4 115.1 112.9 100.6	3.7 6.3 8.6 8.2 7.1	49 54 55 56 50	3 4 6 11 16
III Subbituminous	1. Subbituminous A 2. Subbituminous B 3. Subbituminous C	59 58	76.4 57.4	6.3 4.8	74 63	23 28
IV Lignite	1. Lignite A 2. Lignite B	37	56.6	4.5	59	44

Tables 40, 41 and 42 list the carbonization yields of various kinds of coal when heated to various temperatures.

Table 43 shows the amount of tar and its specific gravity obtained from anthracite at various temperatures. It clearly indicates the considerable amount of tarry matter given up by a low volatile fuel such as anthracite. The table also shows that a considerable amount of tar (mostly the light oils) generated in the distillation zone of a gas producer is already converted at lower temperatures into noncondensable gases.

Figure 123 shows the effect of varying distillation temperature upon the tar yield from a bituminous coal. The enclosed part covers on an average, the distillation test results.

166

Table 40. Assay Yields from Carbonization of Dried Subbituminous Coal at Various Temperatures (27).

Temperature of Distillation, $^{\circ}C$

	300	400	500	600	700	800	900	1,000
Carbonization yields, moisture and ash free (weight %)								
. Char	98.2	85.7	70.4	63.0	60.9	60.3	59.5	59.0
. Water formed	0.9	4.6	9.1	10.3	10.9	10.5	10.5	10.5
. Tar, dry	0	4.9	9.1	9.5	9.3	9.2	9.2	8.9
. Light oil	0.7	1.0	1.4	1.6	1.6	1.3	1.4	1.7
. Gas	0.1	3.5	9.1	15.1	16.8	18.3	18.8	19.3
. Hydrogen sulfide	0	0.4	0.8	0.8	0.5	0.5	0.4	0.5
Total	99.9	100.0	99.9	100.3	100.0	100.1	99.8	99.9
Composition of assay gas, O_2-and N_2-free (volume %)								
. Carbon dioxide	50.0	57.6	20.2	18.7	15.7	12.0	10.9	10.0
. Illuminants	0	0.3	1.3	0.4	0.2	0.4	0.4	0.3
. Carbon monoxide	43.3	21.3	19.0	19.8	17.7	17.3	17.1	17.3
. Hydrogen	1.7	3.5	12.4	20.8	33.6	43.6	48.6	52.0
. Methane	3.3	16.4	41.4	39.8	32.4	26.5	22.8	19.9
. Ethane	1.7	0.9	5.7	0.5	0.4	0.3	0.2	0.5
Total	100.0	100.0	100.0	100.0	100.0	100.0	100.0	100.0

Table 41. Low Temperature Distillation of Raw Texas Lignite (27).

Temperature of Distillation, $^{\circ}C$	150	200	250	300	400	500
Assay yields, moisture and ash free (weight %)						
Char	99.5	99.2	98.0	93.8	74.6	62.5
Water	0	0	0	1.8	6.8	8.3
Tar	0	0	0	0.2	7.3	10.7
Light oil	0	0.2	0.7	1.4	1.6	1.9
Gas	0.5	0.7	1.4	3.0	9.4	16.0
Hydrogen sulfide	0	0	0	0.1	0.6	0.9
Total	100.0	100.1	100.1	100.3	100.3	100.3
Composition of assay gas, O_2-and N_2-free (volume %)						
Carbon dioxide	95.9	90.3	88.7	78.2	67.6	45.9
Illuminants	0	0.3	0.4	0.8	1.0	1.1
Carbon monoxide	0	6.5	8.8	12.7	13.3	10.3
Hydrogen	0	0	0	0.8	0.8	15.3
Methane	4.1	2.9	2.1	7.3	16.9	25.1
Ethane	0	0	0	0.2	0.4	2.3
Total	100.0	100.0	100.0	100.0	100.0	100.0

Table 42. Low, Medium and High Temperature Carbonization of Dried Texas Lignite (27).

Temperature of Distillation, °C

	300	400	500	600	700	800	900	1,000
Assay yields, moisture and ash free (weight %)								
Char	96.2	84.8	66.5	58.7	53.5	52.1	51.0	50.6
Water	1.0	4.1	8.0	10.0	10.3	10.3	10.5	10.4
Tar	0	3.9	10.5	10.8	10.9	10.8	10.9	11.0
Light oil	0.9	0.9	1.4	1.4	1.5	1.5	1.5	1.7
Gas	1.8	5.8	12.7	18.3	22.6	24.0	25.1	25.2
Hydrogen sulfide	0	0.4	1.0	1.0	1.2	1.2	1.2	1.2
Total	99.9	99.9	100.1	100.2	100.0	99.9	100.2	100.1
Composition of assay gas, O_2-and N_2-free (volume %)								
Carbon dioxide	79.0	72.3	44.5	33.2	25.0	22.0	19.7	18.3
Illuminants	0.4	1.0	1.4	1.1	0.7	0.5	0.4	0.4
Carbon monoxide	14.3	13.7	10.8	12.4	16.3	16.8	16.8	17.0
Hydrogen	0.8	1.0	14.9	22.9	32.2	38.2	42.6	45.8
Methane	5.3	11.9	26.7	28.4	24.9	21.6	19.8	17.9
Ethane	0.2	0.1	1.7	2.0	0.9	0.9	0.7	0.6
Total	100.0	100.0	100.0	100.0	100.0	100.0	100.0	100.0

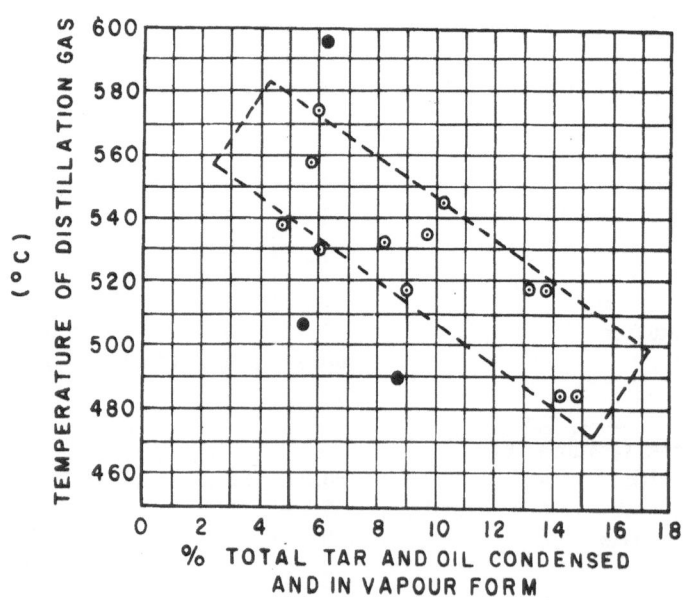

Figure 123. Total Tar and Oil in the Distillation Gas (41).

168

Table 43. Tar Yield Versus Temperature (41).

Temperature Degrees $^{\circ}$C	Tar, liter/ton	Specific Gravity of Tar
900	34.6	1.200
800	46.1	1.170
700	57.7	1.140
600	69.2	1.115
500	80.8	1.087
400	88.5	1.020

Working with samples from 222 different coals, Landers determined their proximate and ultimate analysis from the yields of low temperature (500 $^{\circ}$C) carbonization and the proximate and ultimate analysis of the char. Based on this work, equations were made which predict the amount of distillation products such as tar, light oils, char and gas as well as the heating value of the distillation gases and its volume per pound of carbonized coal. The results are given in Table 44.

Table 44. Prediction Equations for 500 $^{\circ}$C Coal Carbonization Yields (27).

Tar plus light oil yield, maf, wt % = -20.8954 + 0.00333 (Btu) -0.4624 (FC) + 2.6836 (H_2)

Char yield, maf, wt % = 32.1310 + 0.7815 (FC) + 0.2318 (O_2)

Gas yield, maf, wt % = 53.9549 - 0.00340 (Btu)

Heating value of gas, Btu/scf = -1395.94 + 0.1529 (Btu) -2.4101 H_2O (AR)

Gas volume, scf/lb, maf = 6.9377 - 0.000216 (Btu) -0.2849 (H_2) - 0.0884 (C/H_2)

Btu	Maf heating value, of fuel sample
FC	Fixed carbon, maf, weight percent
O_2	Oxygen content, maf, weight percent
H_2	Hydrogen content, maf, weight percent
C/H_2	Carbon to hydrogen ratio, maf, wet basis
H_2O (AR)	As-received moisture content, weight percent
maf	Moisture-ash free basis

The decomposition of peat takes place as follows (19):
At 120 $^{\circ}$C peat material begins to decompose with the formation of CO_2.
At 200 $^{\circ}$C decomposition process is fairly significant.
250-500 $^{\circ}$C maximum yield of tarry and gaseous substances
Above 850 $^{\circ}$C distillation products are mainly H_2 and CH_4.

The final products of distillation are within the following range:

coke	30-40%
tar	10-15%
gas	30-35%
water	20-25%

The distillation or carbonization of wood has been extensively examined because the process yields valuable products for the chemical industry such as phenols, light oils and charcoal (12,26,44,54,57,58).

Table 45 shows the chemical composition of wood divided into softwood and hardwood. The softwoods are: Pines, Firs and Redwood. The hardwoods are: Oak, Elm, Hickory, Walnut and Beech.

Table 45. Average Percent Chemical Composition of Soft Woods and Hardwoods (22).

	Softwoods	Hardwoods
Cellulose	42 + 2	45 + 2
Hemicellulose (xylan)	27 + 2	30 + 5
Lignin	28 + 3	20 + 4
Extractives	3 + 2	5 + 3

Lignin, cellulose and xylan act differently under thermal decomposition. Figure 124 shows the weight loss; i.e., the distillation products given up from lignin, xylan, cellulose and wood when heated to 500 ^{o}C.

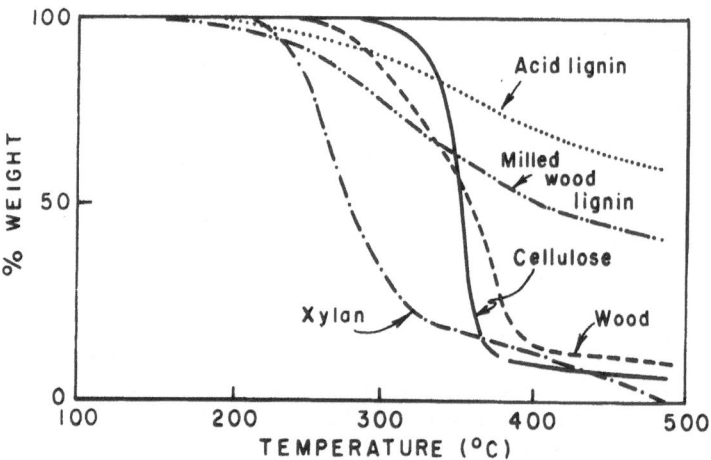

Figure 124. Thermogravimetry of Cottonwood and Its Components (44).

The carbonization products of wood can be grouped into four categories: tar, gases, pyroligenous acid and char. Upon further heating the tar is finally converted into noncondensable combustible gases and char. This last stage of the tar, usually called thermal cracking is not accomplished at all in an updraft gas producer because the tar vapors leave the gas producer without coming into contact with the hot partial combustion zone. The composition of the tar from wood as well as any fuel containing organic matter is extremely complex and not fully understood. There have been at least 200 major components identified, some of which are listed in Table 46.

170

Table 46. Constituents of Pyrolisis (Distillation) of Wood (12).

Carbon monoxide	Formic acid	Methacrylic acid
Carbon dioxide	Acetic acid	γ-Butyrolactone
Hydrogen	Propionic acid	n-Butyric acid
	Crotonic acid	iso-Butyric acid
Water	iso-Crotonic acid	Angelic acid
Tiglic acid	Δ^3-Hexenone-2	2-Methylfurane
Δ^2-Pentenoic acid	Methyl n-butyl ketone	3-Methylfurane
γ-Valerolactone	3,6-Octanedione	Dimethylfurane
n-Valeric acid	2-Acetylfurane	2,5-Dimethyltetrahydro-
Methylethylacetic acid	Cyclopentanone	furane
n-Caproic acid	2-Methyl -Δ^2 cyclopen-	Trimethylfurane
	tenone	5-Ethyl-2-methyl-4,5-di-
Isocaproic acid	Methylcyclopentenolone	hydrofurane
n-Heptoic acid	Cyclohexanone	Coumarone
Lignoceric acid	Methylcyclohexenone	Pyroxanthone
Furoic acid	Dimethylcyclohexenone	Benzene
Methyl alcohol		Toluene
Ethyl alcohol	Phenol	Isopropylbenzene
Allyl alcohol	o-, m, and p-Cresol	m-Xylene
Propyl alcohol	o-Ethylphenol	Cymene
Methylvinylcarbinol	2,4-Dimethylphenol	Naphthalene
Isobutyl alcohol	3,5-Dimethylphenol	1,2,4,5-Tetramethyl-
Isoamyl alcohol	Catechol	benzene
	Guaiacol	Chrysene
Formaldehyde	2-Methoxy-4-methylphe-	
Acetaldehyde	nol	Ammonia
Propionaldehyde	2-Methoxy-4-vinylphenol	Methylamine
Valeraldehyde	2-Methoxy-4-ethylphenol	Dimethylamine
Isovaleraldehlyde	2-Methoxy-4-propylphe-	Trimethylamine
Trimethylacetaldehyde	nol	Pyridine
Furfural	1,2-Dimethoxy-4-methyl-	3-Methylpyridine
5-Methylfurfural	benzene	Dimethylpyridine
Hydroxymethylfurfural	2,6-Dimethoxyphenol	
	2,6-Dimethoxy-4-methyl-	Methane
Methylal	phenol	Heptadecane
Dimethylacetal	2,6-Dimethoxy-4-propyl-	Octadecane
	phenol	
Acetone	Propylpyrogallol mono-	Eicosane
Methyl ethyl ketone	methyl ether	Heneicosane
Diacetyl	Coerolignol (or -on)	Docosane
Methyl propyl ketone	Euppittonic acid (or eupit-	Tricosane
Methyl isopropyl ketone	ton)	Furane
Diethyl ketone		
Ethyl propyl ketone		

It should be noted that the pyroligneous acid contains the water volatile constituencies given up by the fuel when heated. Since the pyroligneous acid and the tar form two liquid layers, there is a solubility equilibrium between them. Pyroligneous acid contains from 80% to 90% water. Table 47 lists products mostly found in the pyroligneous acid. It is worthwhile to mention that the pyroligneous acid also contains 7 to 12% soluble tar, made up mostly of constituents the same as those occurring in settled tar.

Table 47. Products Identified in the Pyroligneous Acid of Wood (26).

Formic acid	Pyromucic acid	Methyl ethyl ketone
Acetic acid	Methyl alcohol	Ethyl propyl ketone
Propionic acid	Allyl alcohol	Dimethyl acetal
Butyric acid	Acetaldehyde	Methylal
Valeric acid·	Furfural	Valero lactone
Caproic acid	Methylfurfural	Methyl acetate
Crotonic acid	Acetone	Pyrocatechin
Angelic acid	Pyroxanthen	Ammonia
Methylamine	Methyl formate	Isobutyl alcohol
Isoamyl alcohol	Methyl propyl ketone	Keto-pentatmethylene
α-Methyl β-keto-penta- methylene	Pyridine	Methyl pyridine

The gas given up when heating wood consists mainly of the noncondensable constituents but is also saturated with vapors of the more volatile liquids of the pyroligneous acid and carries a small amount of the less volatile constituents. The amount of each product varies depending on the species of wood and the conditions of distillation. However a rough estimate for the average yields from wood when heated to 350-400 °C are 38% charcoal, 9% total tar, 33% pyroligneous acid (without dissolved tar) and 20% gas (26). A more refined analysis is given in Table 48 which lists the products obtained from heating birch wood and birch bark up to 500 °C under atmospheric pressure within 4 hours.

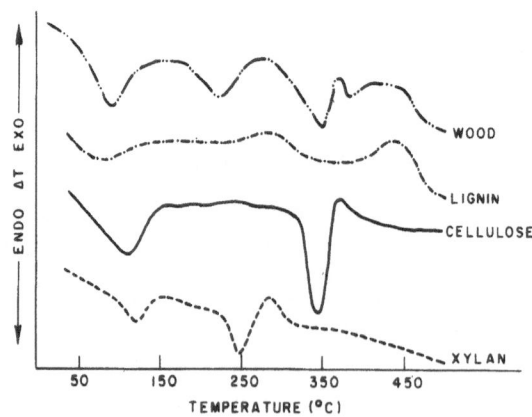

Figure 125. Differential Thermal Analysis of Cotton Wood and Its Components (44).

172

Table 48. Products of Carbonization of Birch Wood and Bark at Atmospheric Pressure up to 500 $^\circ$C for Four Hours (% oven dry weight) (26).

Product of Carbonization	Element, Substance, or Group of Substances	Outer Bark	Inner Bark	Wood
Charcoal	C	12.4	29.5	23.9
	H	0.4	1.0	0.8
	O	0.6	2.1	1.6
Ash		0.4	2.1	0.2
	Total	13.8	34.7	26.5
	Neutral oil	45.6	2.5	1.4
	Acids	0.8	0.4	0.4
	Phenolic compounds	7.5	2.1	1.6
Tar,	Insoluble in ether	1.1	0.4	0.2
waterfree	Precipitate on soda treatment	1.8	0.3	0.1
	Water soluble tar	0.7	0.8	0.9
	Error in analysis	0.2	0.9	0.0
	Total	57.7	7.4	4.6
	Water soluble tar	0.3	2.8	11.1
Tar water, excluding moisture of charring material, including water of tar	Acids calculated as acetic acid	1.2	5.8	9.7
	Methyl alcohol	0.2	1.0	1.3
	Water-soluble neutral compounds b.p., 95°C.	0.4	1.1	1.6
	Water, formed in the reactions	8.3	22.2	25.2
	Total	10.4	32.9	48.9
Gases	CO_2	5.5	16.9	11.8
	C_nH_{2n} (as C_2H_4)	1.1	0.3	0.5
	CO	2.0	4.5	5.1
	CH_4	2.7	2.1	1.8
	H_2 (remaining)	0.1	0.2	0.1
	Total	11.4	24.0	19.3
Total	Accounted for	93.3	99.0	99.3
	Unaccounted for	6.7	1.0	0.7
		100.0	100.0	100.0

The temperature at which the carbonization of wood is exothermic or endothermic is also of interest. Figure 125 shows the differential thermal analysis of wood and its components. In this test the wood sample and an inert substance which does not undergo any thermal reaction, are simultaneously and uniformly heated. From this test, the time function of the temperature difference between the wood sample and the control substance is determined.

The rate of combustible volatiles can be obtained through a thermal evolution analysis. This test utilizes a temperature programmed furnace combined with a flame ionization detector which responds in a predictable manner to the evolved gases. Figure 126 gives the results for carbonization of cottonwood. The graph shows that the maximum production of combustible volatiles is reached at 355 °C and ceases dramatically beyond this temperature.

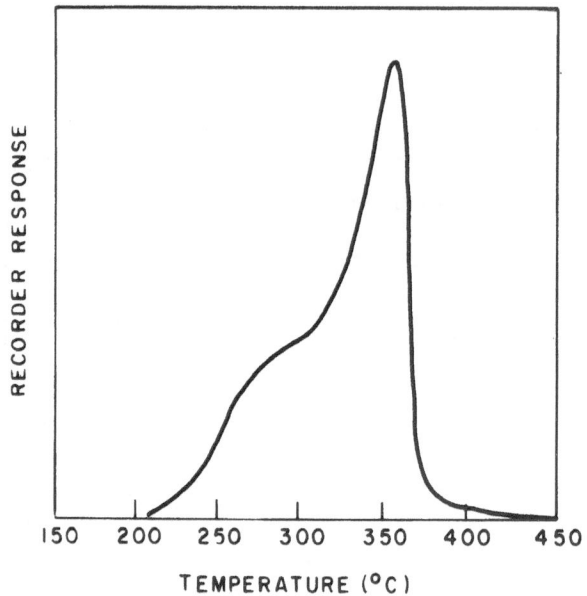

Figure 126.   Rate of Formation of Combustible Volatiles from Cottonwood Versus Temperature (44).

The outlined distillation of coal and wood helps to understand the complications involved in surpressing the tar production in a gas producer. Almost all known successful producers, portable or stationary, which have been used to drive an internal combustion engine used anthracite, charcoal or low volatile fuels which, under normal running conditions have little tendency to generate tar when gasified. The long history of gas producer practice, originally in large stationary units and later in portable units (for the sole purpose of driving a vehicle) has produced data and regulations about the tolerance level of tar in the raw gas. Terms such as tar free fuel, tar free gas or reports about gasifiers generating tar free gas are mostly misleading. Any gasifier fueled with a substance containing organic matter will generate tarry products. The best that can be achieved is a gas producer that, when brought up to its proper running temperature, generates an almost tar free raw gas. This gas can be used for driving an internal combustion engine with maintenance comparable to gasoline driven engines. The design of a gas producer requiring extensive cleaning of tar vapor from the raw gas is an unacceptable solution for two reasons. First, the extraction of tar vapor from gas is difficult and expensive. Second, the tar has by far the highest heating value of all combustible products obtained from gasification of organic matter including the fuel itself.

The heating value of the tar generally varies from 34 to 37 MJ/kg. The loss in energy through tar formation in the distillation zone is therefore considerable. Figure 127 shows the heat loss in percent versus tar content of the fuel in percent weight. The heating value of the tar was assumed to be 36 MJ/kg. The different lines on the graph represent various fuels with different heating values on a dry basis.

It is well known that even under favorable conditions the quality of the gas can change rapidly although the operation may be carefully controlled. This is especially true for the tar generation in a gas producer which depends on the temperature. Results of tests conducted on two large, stationary updraft gasifiers of 3.25 m and 2.6 m diameter are represented here. The tar content was found to vary at random between 5.72 and 27 g/m^3 of hot gas during a 90 hour period. Figure 128 shows the frequency of occurrence. The coal gasified had 30% volatile matter.

The carbonization of biomass under partial vacuum or with an inert carrier gas such as nitrogen is different from the process that takes place in the distillation zone of a gasifier. Although the mechanism is the same, the yield and composition

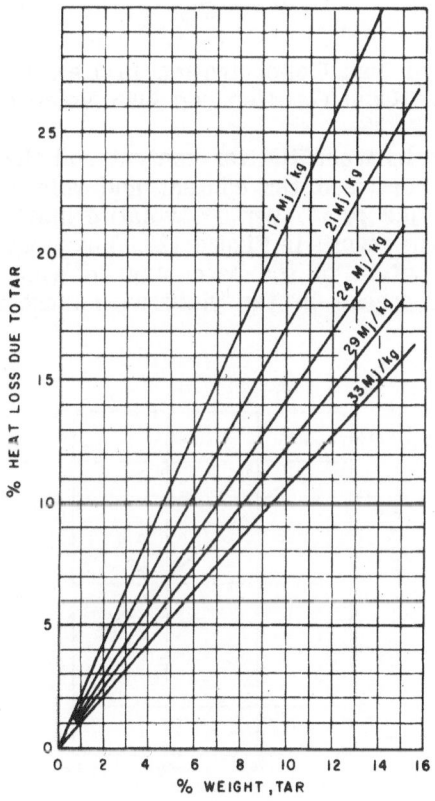

Figure 127. Heat Loss Through Tar Formation (41).

175

of the tar will be different. At least one author claims that under very unfavorable conditions the tar yield will be about 80% of what had been obtained under laboratory conditions (41). Data about the correlation between the tar content of the gas and fuel parameters such as ash content, size distribution and volatile matter are scarce and unreliable because they are difficult to obtain.

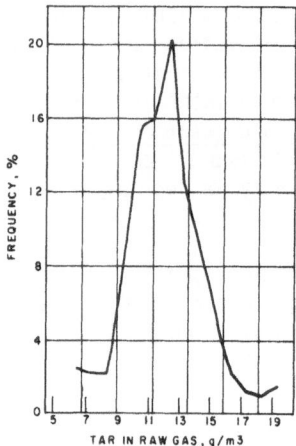

Figure 128. Frequency of Tar Concentration in Raw Gas Over a 90 h Time Period. The Coal Gasified had 30% Volatile Matter (16).

No correlation was found between the tar content and the ash content of coal. A correlation was found between tar content and volatile matter in the coal. An average variation of 0.9 g/m^3, per 1% volatile matter, has been reported. This agrees roughly with the results in Figure 123. The most important correlation was the effect on tar content of the percentage of fine material in the coal. Figure 129 shows the decrease in tar yield with increasing amounts of fine particles.

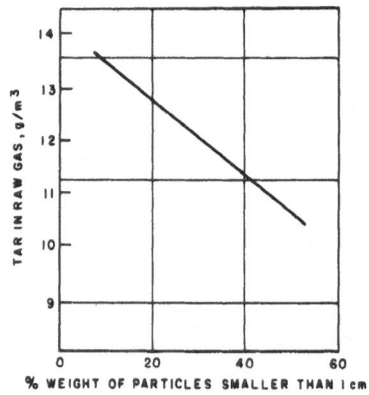

Figure 129. Tar Yield as a Function of Fine Particle Content in the Fuel (16).

176

It was found that the superficial velocity of the stream carrying away the tar had a strong influence on the tar production. A ten fold decrease of the superficial velocity decreases the tar production to about half the original figure.

It is also well known from low temperature carbonization that a slower rate of heating the fuel increases the tar yield.

Most conditions favorable for a high tar yield are satisfied in an updraft gas producer. This type of gasifier is an ideal tar generator because the gas stream from the oxidation zone passes through a long column of partly carbonized and green fuel at a continuously decreasing temperature. However, updraft gasifiers can handle high slagging and high ash fuel much better than downdraft and crossdraft gasifiers. They also have the most favorable efficiency and yield a gas with a high heating value. To overcome the serious tar generation in updraft gasifiers three methods have been proposed and tested.

In order to avoid the distillation products in the raw gas the French C.G.B. Producer (Figure 130) draws off the gas above the reduction zone through a funnel. However, this type of arrangement is suspectible to disturbances in the downward flow of the fuel.

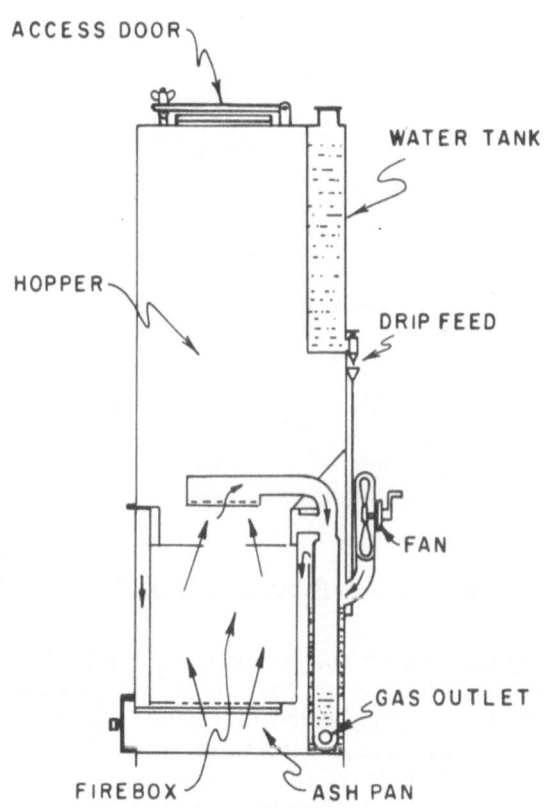

Figure 130.   French C. G. B. Updraft Gas Producer (21).

The American Suction Gas Producer (Figure 131) draws off the gas through a ring shaped gas collection chamber. The fuel column is therefore pinched at a height above the reduction zone. This design dating from 1902 is still applied in large modern gas producers (Figure 29).

A different approach is followed by the Duff Whitfield Gas Producer, Figure 132. The blast is introduced under the grate A. The gas exit is at D. E is a small steam blower which draws the gases given off from the surface of the fuel at F and delivers them at the lower part of the fuel bed through opening G. The same procedure is applied at points I, H and J. In each case, the distillation gases are taken from the distillation zone and forced up through the incandescent oxidation zone. This design utilizes the well-known fact of cracking the tar into noncondensable gases at high temperatures. The degree of decomposition into noncondensable gases is questionable. There does not appear to be any research on the question of how well the tarry vapors obtained from gasification of coal and biomass are cracked under the moderate conditions of 1 atm pressure and temperatures around 1,000-1,200 $^{\circ}$C.

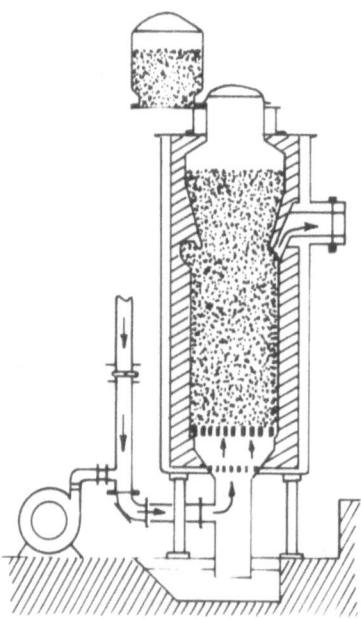

Figure 131. American Suction Gas Producers (34).

One rather interesting solution to the cracking of tar is applied in a large gas producer referred to as a 2-stage or dual mode gas producer. The particular gas producer, its schematic design given in Figure 133, has been operating for 30 years with no major repairs and is used to drive a 1,000 kW dual fueled-diesel engine. Its unique characteristic is that it can produce a completely tar-free gas. This unit was mainly used to produce charcoal and the producer gas was considered a byproduct.

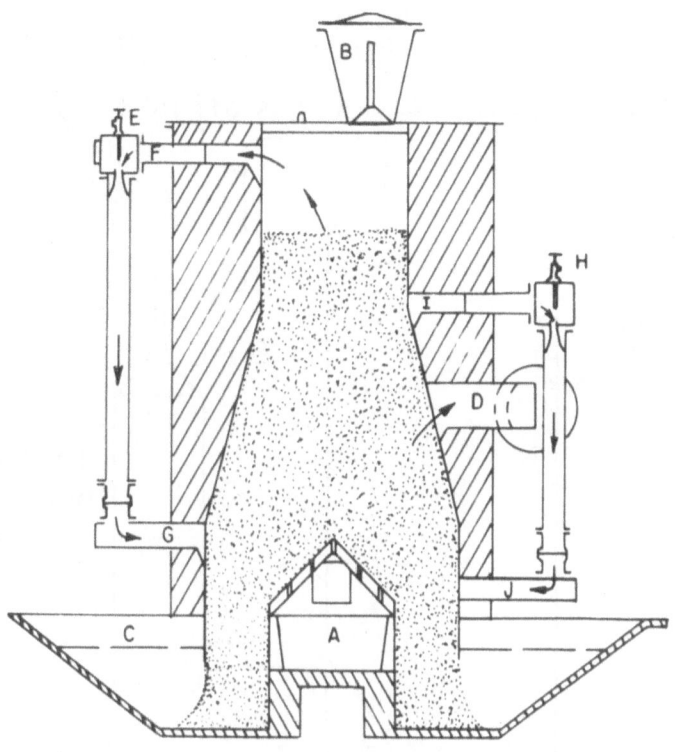

Figure 132. Duff Whitfield Gas Producer (34).

The flow from the top gas burner is split in such a way that approximately 1/3 of the gas moves up countercurrent to the fuel flow and 2/3 of the gas moves downward. The movement of the hot gas upward countercurrent to the fuel flow pyrolizes the wood and moves the volatile products upward. These volatile products are then mixed with incoming air and burned in excess air under controlled conditions such that all the heavier hydrocarbons are destroyed. The resulting noncondensable gas then enters the main body of the gas producer and the 2/3 that moves down undergoes reduction, thus producing the carbon monoxide. The offgas temperature is an indication of the progress of the endothermic reactions. The 1,000 kW, dual-fueled, diesel engine was manufactured by Societe Alsacienne de Constructions Mecaniques.

Because tar cannot be avoided in most cases, it is of interest to know how much tar will be carried out by the raw gas. This will determine the method of tar removal and the dimensioning of the scrubber or cyclone.

For instance, tar production in a Davy, single stage, fixed bed, updraft gas producer has been 0.065 kg/kg feed for bituminous coal and 0.081 kg/kg feed for venteak wood.

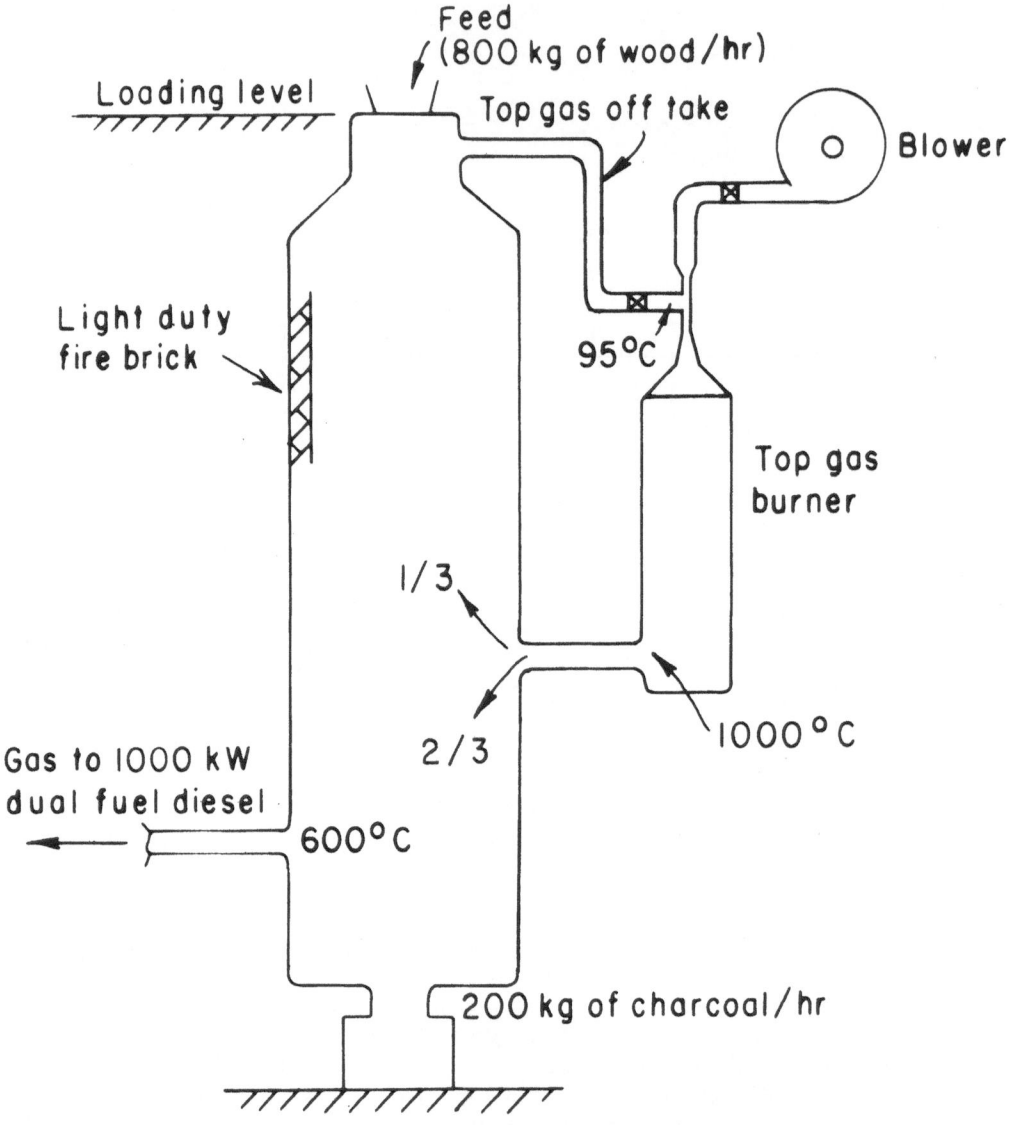

Figure 133. Dual Flow Gas Producer for Producing Charcoal and Powering a 1,000 kW Dual Fueled Diesel Engine (Delacott System, Distibois).

180

It has been shown that the raw gas needs to be cleaned and the degree of purification depends on the final use of the raw gas. The decision as to what kind of cleaning equipment should be installed between the gas producer and the burner or internal combustion engine should take into consideration the following:

1. Maintenance of the cleaning equipment, burner and I.C. engine.

2. Wear and corrosion on the I.C. engine, burner and purification system.

3. Pressure drop across the purification system and power input if purification is mechanically aided.

4. Air pollution.

5. Type of producer-engine system (portable, stationary, size).

6. Costs.

7. Availability of water.

8. Disposal of tar laden waste water.

Purification systems and the required condition of the raw gas for use in an internal combustion engine or burner differ considerably and will therefore be treated separately.

A. Purification of the raw gas for use in internal combustion engines:

Ideally only the combustible constituents CO, H_2, CH_4 and at ambient conditions noncondensable, higher hydrocarbons such as C_2H_6, C_4H_2 and C_2H_4 should reach the engine. In practice this goal is not attained. In order to extract the harmful and operational problem causing constituents such as water vapor, condensible higher hydrocarbons (tar, pitch, oils), mineral vapors in oxidized form and the corrosive agents NH_3, H_2S and HCN, a combination of different cleaning equipment in series or parallel are necessary. To what degree the gas should be purified is a difficult question and is almost always underestimated. The recent experience with producer gas driven, internal combustion engines is insignificant and has been mostly done on test units where the long term effects of impurities in the raw gas reaching the engine and economical considerations are of little concern. An exception are the extensive tests with producer gas driven trucks and tractors in Sweden since 1957. The vast European experience and elsewhere before and during World War II is mostly based on a narrow range of fuels such as wood, wood charcoal, anthracite and coke. The reported data based on low speed engines which are not any longer available to any extent, can not be readily transferred to high speed engines which are not as suitable for producer gas operation. Because the use of wood or charcoal for gas producers is rather questionable in Developing Countries and should be avoided in most cases, fuels that are more difficult to gasify and result in higher impurities need to be gasified. However, from the technical literature and personal contacts, there are reports of internal combustion engines running on producer gas for decades without any major repairs or considerable wear (60). The purification of the raw gas for stationary units is greatly simplified for a gas producer-engine water pump system where water is available. Weight and dimensions of the cleaning equipment are not so restrictive and simple inexpensive cleaning units can be built. For portable units mounted on automobiles, trucks or tractors the purification of the gas is not an easy task because of the specific requirements such as compactness and light weight. The available cleaning equipment can be classified into two categories:

    I. Units not mechanically or electrically aided.

    II. Mechanically or electrically aided units.

For small scale units the first type suffices, in particular with respect to Developing Countries where in most cases an external power source is not available. In some special cases such as gasification of high bituminous coal, rice husks and cereal straw in updraft gas producers, an electrostatic precipitator is an excellent solution to the tar problem and justifies the investment and additional power input. This judgment is based on very favorable results obtained during the 1930-1950 period in Italy and Germany with gasification of these problematic fuels in large updraft gas producers (60).

Non-mechanically aided units are: cyclones, fabric filters and scrubbers. All of them, although commercially available, can be designed and home made with the usual equipment necessary to build a gasifier. The common oil or fabric filter systems used in gasoline and diesel powered mobile equipment are by no means sufficient to clean the raw gas to the desired degree. Very few operational gas producer-engine systems around the world today display new concepts concerning the purification systems, except for units tested by the Swedish National Machinery Testing Institute which use fiberglass fabric filters for dry purification of the raw gas.

The remaining part of this chapter will deal entirely with the most common designs for cyclones, fabric filters and scrubbers. Some of the past gas cleaning systems will also be covered.

A cyclone is by definition a dust collector without moving parts in which the velocity of an inlet gas stream is tranformed into a confined vortex. The dust separation from the gas stream takes place through centrifugal forces. The suspended particles tend to be driven to the wall of the cyclone and are collected in an ash bin at the bottom. In almost all cases the cyclone is the first stage of cleaning the raw gas. It is usually located right after the gas exit. Cyclones are easy to build and inexpensive. They separate only coarse particles from the gas stream. One distinguishes between high efficiency and medium efficiency cyclones depending on the dimensions of the cyclone body. Ranges of efficiency for both types are given in Table 49.

Table 49. Efficiency Range of Medium and High Efficiency Cyclones (47).

Particle size (micron)	Medium (collection efficiency in %)	High
Less than 5	Less than 50	50-80
5-20	50-80	80-95
15-40	80-95	95-99
Greater than 40	95-99	95-99

There are some general recommended design criteria for the cyclones given in Figure 134. The height H of the main vortex should be at least 5.5 times the gas outlet diameter preferably up to 12 times. The cone serves the practical function of delivering the dust to a central point. The diameter of the apex of the cone should be greater than ¼ of the gas outlet diameter. The approach duct is usually round, therefore the round duct must be transformed to a rectangular inlet. The maximum included angle between the round and rectangular sections should not exceed 15 degrees. The optimum length of the gas outlet extension has been determined to be about one gas outlet diameter. This extension should terminate slightly below the bottom of the gas inlet (47,48).

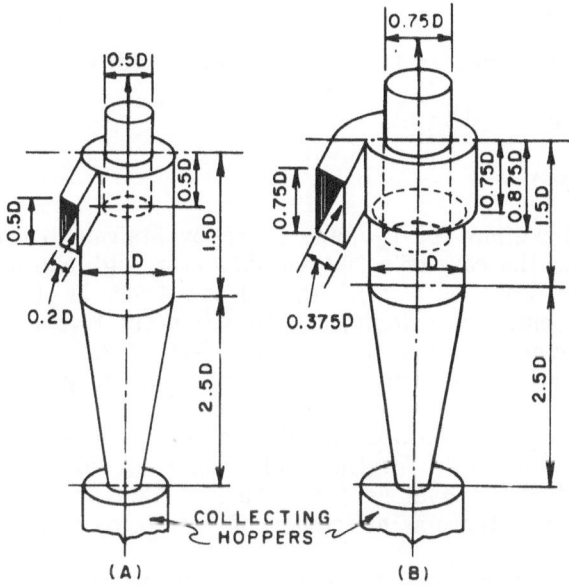

Figure 134. High (A) and Medium (B) Efficiency Cyclone (46,48).

The efficiency of a cyclone is highly dependent on the intake gas velocity. It is therefore advisable to locate the cyclone right after the gas exit, where the gas has its highest velocity because of the reduced area at and after the exit. This location of the cyclone also has the advantage of cooling down the gas through expansion before it reaches the subsequent purification units such as fabric filters and wet scrubbers which are more sensitive to high temperatures. There are several rules of thumb for both types of cyclones.

183

Cyclone efficiency increases with an increase in (14,15,47,48):

1. Density of the particle matter

2. Inlet velocity into the cyclone

3. Cyclone body length

4. Number of gas revolutions inside the cyclone (2-10 are normal for a high efficiency cyclone)

5. Particle diameter

6. Amount of dust, mg/m^3 (milligram/cubic meter of gas)

7. Smoothness of the cyclone wall

Cyclone efficiency decreases with an increase in:

1. Carrier gas viscosity

2. Cyclone diameter

3. Gas outlet diameter and gas inlet duct width

4. Inlet area

5. Gas density

There are standard designs for cyclones given by Stairmand (Figure 134) which are outlined here for the case of a 200 mm diameter, high and medium efficiency cyclone with a flow rate of 500 D^2 (m^3/h) and 1,500 D^2 (m^3/h) respectively. At these flows the entrance velocity is approximately 15.2 m/s for both types. The fractional efficiency curves (Figure 135) were obtained for gas streams at 20 ^oC and solid particles of density 2,000 kg/m^3 in the gas stream.

The cyclone efficiency and the pressure drop across a cyclone may be predicted without reference to a known fractional efficiency curve. The solution for the case of a high efficiency cyclone for a 10 kW (13.4 hp) gas· producer engine system with an overall thermal efficiency of 14% is as follows:

Producer gas energy that needs to be provided by the gas producer:

$$\frac{(10,000 \text{ J/s}) (3,600 \text{ s/h})}{0.14} = 257 \text{ MJ/h}$$

where a cool gas efficiency of 70% for the gas producer and an efficiency of 20% for the internal combustion engine or the generator are assumed.

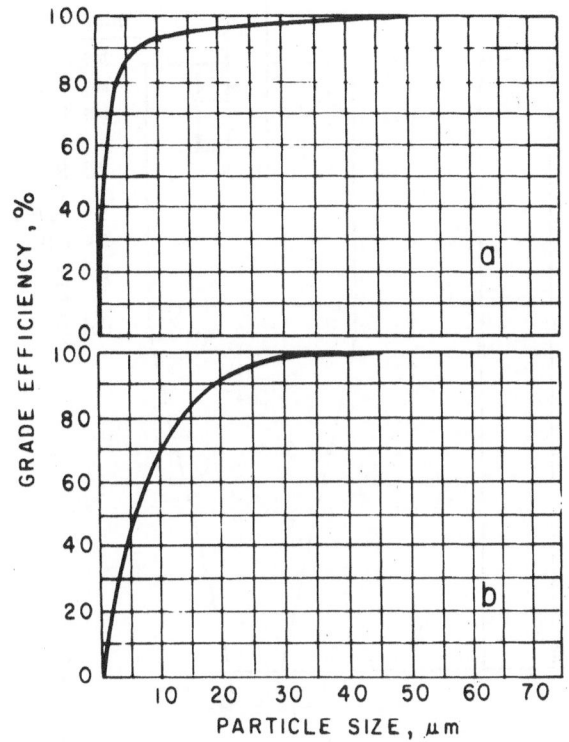

Figure 135. Stairmand's Fractional Efficiency Curve for Cyclone A and B in
Figure 134 (48).

Gas volume flow rate at cyclone inlet: Assumed heating value of the gas is
5.5 MJ/m^3 at inlet conditions to the engine of 25 $^{\circ}$C and 1 atm pressure.

Gas volume flow rate at engine manifold: 257 MJ/h/5.5 MJ/m^3 = 46.7 m^3/h.

Gas volume flow rate at cyclone inlet: (46.7) (573.16/298.16) = 89.8 m^3/h where
the inlet temperature of the raw gas into the cyclone is assumed to be 300 $^{\circ}$C.
The geometric proportions of the cyclone are the recommended ones as given
in Figure 136.

The parameter that is chosen first is the diameter of the pipe between the gas
producer and the cyclone. All other dimensions of a cyclone are based on this
chosen diameter.

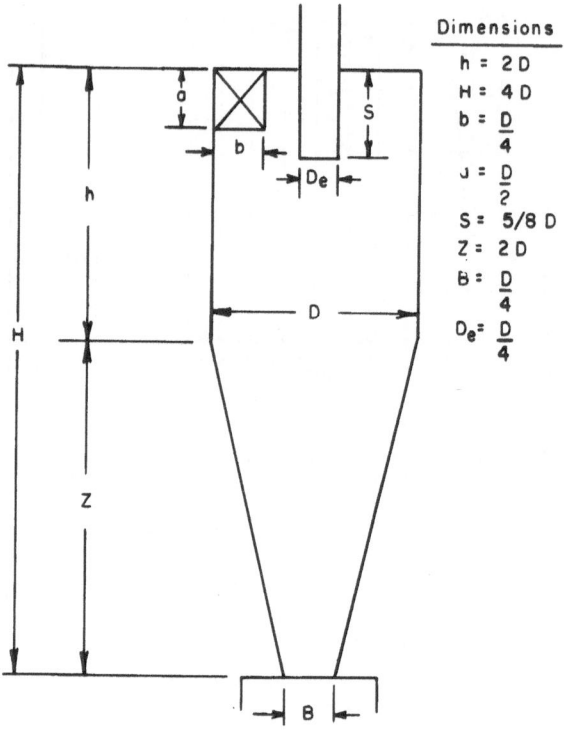

Dimensions

h = 2 D

H = 4 D

b = $\dfrac{D}{4}$

J = $\dfrac{D}{2}$

S = 5/8 D

Z = 2 D

B = $\dfrac{D}{4}$

D_e = $\dfrac{D}{4}$

Figure 136. Recommended Dimensions for a High Efficiency Cyclone (15).

Choice of pipe diameter between gas producer exit and inlet cyclone:

The conveying velocities in pipes are dependent upon the nature of the contaminant. Recommended minimum gas velocities are (15):

Contaminant	Velocity
Smoke, fumes, very light dust	10 m/s
Dry medium density dust (saw dust, grain)	15 m/s
Heavy dust (metal turnings)	25 m/s

A pipe of 3 cm inner diameter therefore allows a velocity of

$$\frac{89.8 \text{ m}^3/\text{h}}{(3{,}600 \text{ s/h})(3.14)(0.015\text{m})^2} = 35.3 \text{ m/s.}$$

The round pipe must be transformed to a rectangular inlet of width 3 cm and height 6 cm. Consequently the gas velocity at cyclone inlet will be:

186

$$\frac{89.8 \text{ m}^3/\text{h}}{3,600 \text{ s/h} \times 0.03 \text{ m} \times 0.06 \text{ m}} = 13.85 \text{ m/s.}$$

The final dimensions of the cyclone are given in Figure 137.

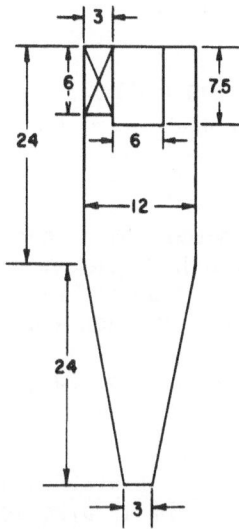

Figure 137.   Dimensions of a High Efficiency Cyclone for a 10 kW Gas Producer.

Prediction of the cyclone efficiency:

D_{pc} is called the particle cut size and defined as the diameter of those particles collected with 50% efficiency. D_{pc} may be predicted by the following equation:

$$D_{pc} = \sqrt{\frac{9 \mu b}{2 N_e v_i (\rho_p - \rho_g) \pi}} \tag{15,48}.$$

   b = cyclone inlet width, m

   μ = dynamic gas viscosity, kg/m-s

   N_e = effective number of turns in a cyclone.   Assume 5 for a high
        efficiency cyclone.

   v_i = inlet gas velocity, m/s

   ρ_p = actual particle density, kg/m^3

   ρ_g = gas density, kg/m^3 at inlet

The gas composition is assumed to be a typical volumetric analysis obtained from a downdraft gas producer fueled with wood chips:

$$CO = 28.7\%,$$
$$H_2 = 13.8\%,$$
$$CH_4 = 6.46\%,$$
$$C_2H_6 = 0.34\%,$$
$$N_2 + Argon = 44\%$$
$$CO_2 = 6.7\%$$

The gas viscosity at 300 $^{\circ}$C computed with respect to the various mole fractions and different viscosities as given in Table 49 equals 255.434 x 10^{-7} kg/m-s. The molecular weight, M, of the mixture is 24.79 which is the sum of the molar fractions of the molecular weight of each constituent as given in Table 50.

Assuming ideal gas behavior the density of the gas at atmospheric pressure and 300 $^{\circ}$C is:

$$\frac{24.79 \text{ kg/mol x } 1.01325 \text{ x } 10^5 \text{ N/m}^2}{573.16 \text{ K x } 8314.41 \text{ J/kg mol} \quad \text{K}} = 0.527 \text{ kg/m}^3$$

Particle density depends on the type of dust in the gas. It will be assumed as 2,100 kg/m^3 (20). Consequently the particle cut size is:

$$D_{pc} = \sqrt{\frac{(9)\,(255.434)\,(10^{-7})\,(0.03)}{(2)(5)(13.85)(2,100 - 0.527)(3.14)}} = 2.75 \text{ micron}$$

Knowing the particle cut size, D_{pc}, one can now predict the fractional efficiency curve for this particular cyclone with the help of Figure 138.

Example: The collection efficiency for 5 micron diameter particles is obtained as follows:

$$D_p/D_{pc} = \frac{5}{2.75} = 1.82.$$

The efficiency with which 5 micron diameter particles are collected is now obtained from Figure 138 as 75%. The complete fractional efficiency curve is given in Figure 139.

Table 49. Dynamic Viscosity of Producer Gas Constituents at Various Temperatures.

$$kg\ m^{-1}s^{-1} \times 10^7$$

Temperature °C	Air	N_2	O_2	CO_2	CO	H_2	CH_4	C_2H_4	C_2H_6	Water Vapor
0	172	166	192	137	166	84.1	102	94	86	—
20	181	175	203	146	175	88.2	109	101	92	95.6
40	191	184	213	156	185	92.2	115	108	98	103
60	200	193	223	165	194	96.1	121	115	104	110
80	209	201	233	173	202	100.0	127	122	109	117
100	217	209	243	182	211	104	133	128	115	125
150	238	229	266	203	231	113	147	142	129	143
200	257	247	288	222	251	122	160	156	142	161
250	275	265	309	241	269	130	173	169	155	179
300	293	282	329	259	287	139	184	181	166	197
350	309	298	348	276	304	147	196	192	176	215
400	325	313	367	293	320	154	207	202	184	233
450	340	328	385	309	336	162	217	212	195	251
500	355	342	402	324	352	169	227	222	204	269
550	369	355	419	339	367	177	237	—	—	287
600	383	368	436	354	382	184	246	—	230	306
650	396	381	452	368	396	191	256	—	—	327
700	409	393	468	382	410	198	265	—	249	348
800	433	417	500	408	437	211	283	—	269	387
900	457	440	530	434	464	223	300	—	283	424
1000	479	461	559	459	490	235	316	—	299	456

Table 50. Molecular Weight of Producer Gas Constituents.

A	39.944
CO_2	44.011
CO	28.011
C_2H_6	30.070
CH_4	16.043
N_2	28.016
H_2O	18.016
O_2	32.000
H_2S	34.082
SO_2	64.066
H_2	2.016
Inert	28.164 (Molecular weight of constituents of air treated as inert)

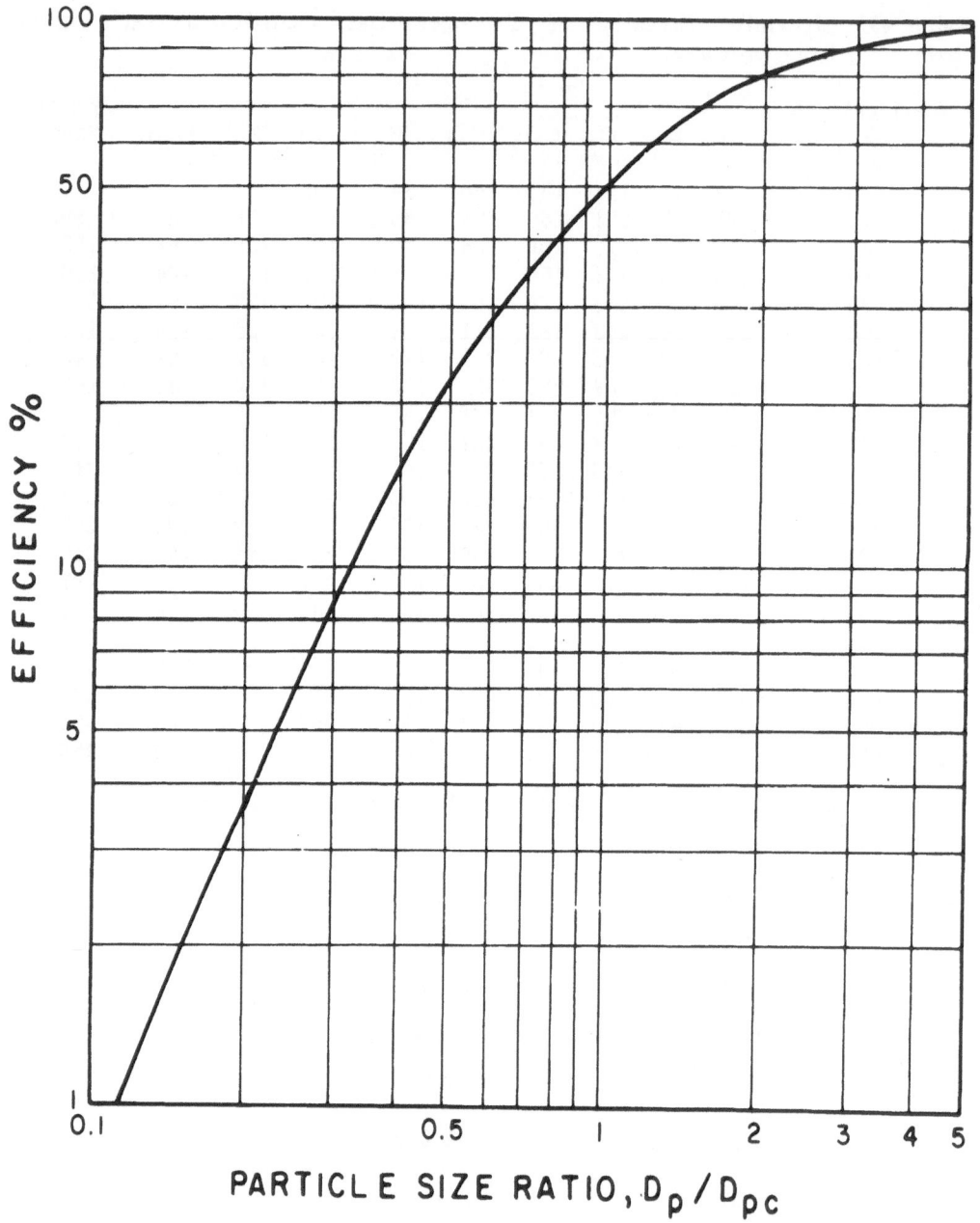

Figure 138. Particle Size Ratio D_p/D_{pc} Versus Collection Efficiency (33).

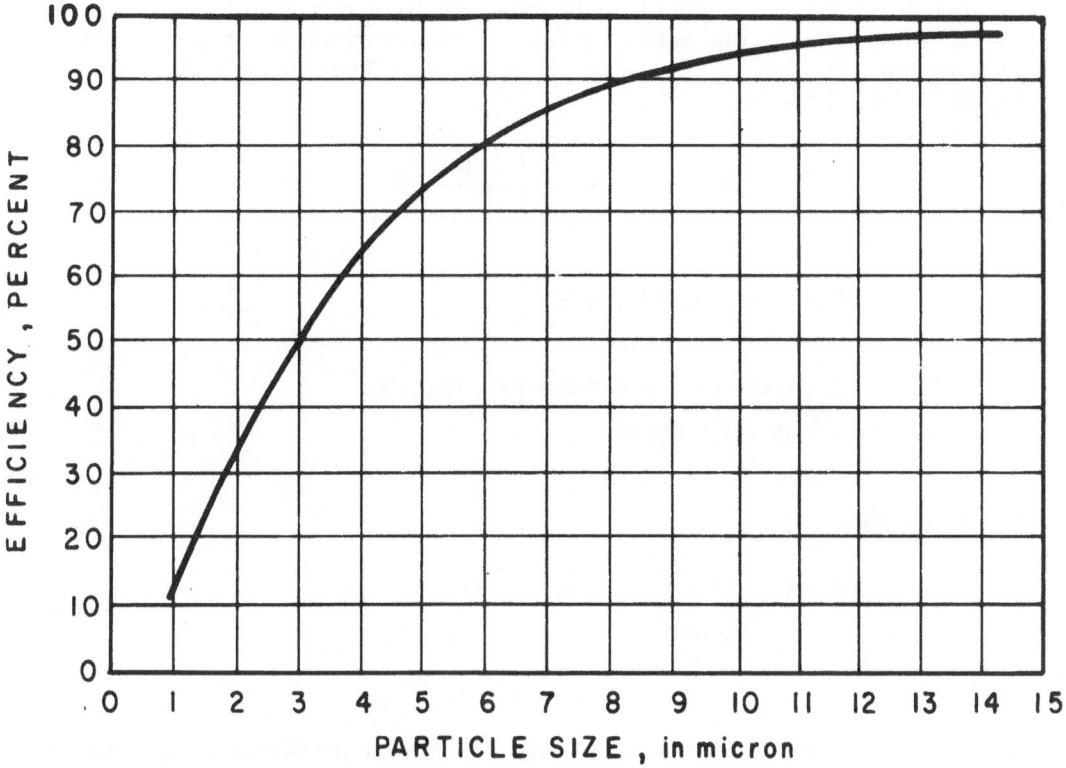

Figure 139. Fractional Efficiency Curve for Cyclone in Figure 137.

The outlined method was devised by Lapple. Its accuracy has been compared with additional experimental data and manufacturer's efficiency curves for cyclones. All results compared favorably with the original curve of Lapple. The maximum deviation noted was 5%.

Because the natural suction of an internal combustion engine has to overcome the pressure drop in the entire system, it is important to keep the pressure drop across the cyclone as small as possible. The pressure drop may be approximated by (33):

$$\Delta P = \frac{(6.5)\,(\rho_g)\,(U_i^2)\,A_d}{D_e^2}$$

U_i = Gas inlet velocity, m/s
A_d = Inlet duct area, m^2
D_e = Diameter of the cyclone exit duct, m
ρ_g = Gas density, kg/m^3

In this example:

$$\Delta P = \frac{(6.5)\,(0.527)\,(13.85^2)\,(0.06)\,(0.03)}{(0.06)^2} = 328.54 \ \ Nm^{-2} = 33 \ mm \ H_2O$$

The equation shows that the pressure drop depends on the square of the velocity. Consequently, there will be a four fold increase in the pressure drop if the inlet velocity is doubled. The largest pressure drop that is allowable at maximum engine load should be used to dimension the cyclone. Maximum efficiency should not be aimed at when dimensioning a cyclone. The task of a cyclone is to separate the coarse particles in the gas stream and prevent sparks from entering any cloth filter that may be part of the cleaning system. It should be noted that the pressure drop in the cyclone is only a small fraction of the total pressure drop through the entire gas producer-engine system. The main contributors to the total pressure drop are the tuyeres, the fuel bed and the scrubber and fabric filters.

The wall friction in the cyclone is a negligible portion of the overall pressure drop. The pressure drop is entirely due to the vortex and design of the gas inlet and outlet. From the equation for ΔP it is obvious that the collection efficiency of a cyclone strongly depends on the inlet velocity of the gas. The gas velocity at full engine load can be about ten times as high as at idling engine speed. It is therefore inevitable that the cyclone will work with a relatively unfavorable inlet velocity at a low engine load. This is demonstrated in Figure 140. The gas producer and cyclone were mounted on a diesel tractor (40 hp at 1,500 rpm).

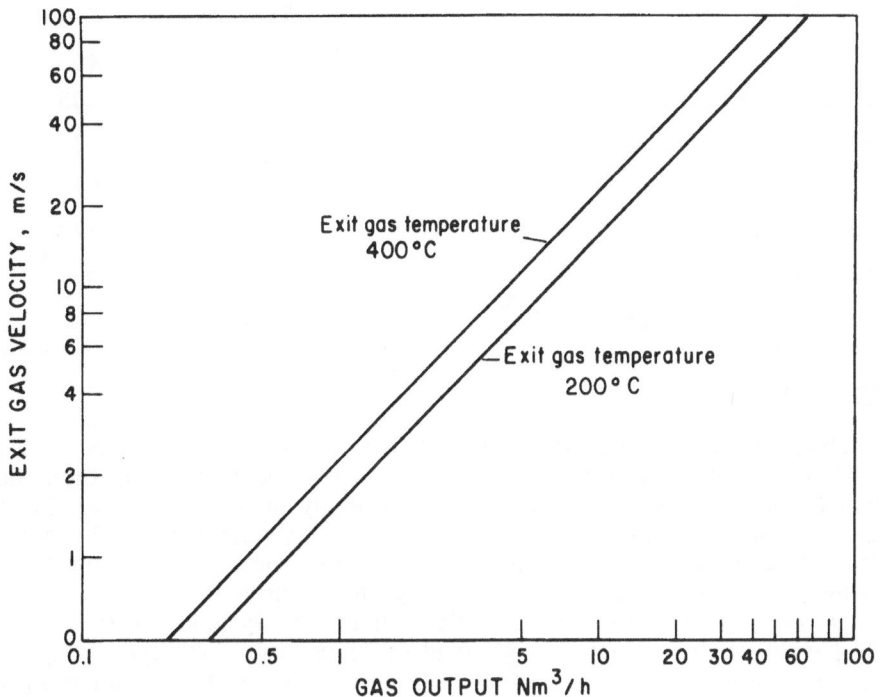

Figure 140. Gas Output Versus Exit Velocity at Two Different Gas Exit Temperatures (35).

Actual dynamometer and road tests with a diesel tractor and truck respectively gave the following performance of a cyclone:

Diesel tractor: Dynamometer (35)

Total test time: 4 h

rpm: 1,625

Wood consumption: 21 kg/h

Amount of gas: 40.3 Nm^3/h

Dust content of raw gas: 2.3 g/Nm^3

Degree of separation in cyclone: 67%

193

Truck: Road Test (35).   Test #

	1	2	3
Distance covered: km	342	479	305
Average speed: km/h	57	47	57
Gas consumption: Nm^3/h	57	63	68
Cyclone inlet velocity: m/s	10.6	11.5	12.5
Separation: %	53	63	54
Dust content of raw gas: g/Nm^3	2.1	2.9	2.4

Wet Scrubbers:

Wet scrubbers have been widely used for many decades in stationary as well as portable units. They can be built in a bewildering array and their design, dimensioning and construction is outlined thoroughly in Reference 14. In almost all cases the wetting agent is water and/or oil. Scrubbers have the ability to remove gaseous pollutants and solid particles while cooling the gas at the same time which makes them ideal for stationary units where the degree of removal is only limited by the pressure drop throughout the system. During the 1930 to 1950 period, engineers put a great deal of effort into the design of compact, efficient wet scrubber systems for portable units. Some of the past designs are presented at the end of this chapter because of lack of new developments for the very few existing gas producer engine systems. The following representation will closely follow Reference 14 and various descriptions of past systems. Scrubbers for small scale gas producer engine systems will be categorized as follows: Plate scrubbers, packed bed scrubbers, baffle scrubbers, impingement and entrainment scrubbers. In practice combinations of the above types have been used in an amazing variety.

Plate scrubbers: It consists of a vertical tower with one or more plates mounted transversely inside the hollow tower. The scrubbing liquid is fed in at the top and flows downward from plate to plate. Plate scrubbers are divided into three categories according to the method of feeding the gas through the plates and the downward flow of the water. Figures 141 to 143 show the various systems. The bubble cap tray is of the cross flow type. The gas bubbles through the holes and out of slots or caps while the water flows across the plates, down a downcomer and across the next plate in the reverse direction. The purpose of the cap is to disperse the gas effectively and keep the liquid from falling through the plates. The reader is referred to Reference 14 for detailed calculations on the performance of bubble cap scrubbers.

The wet impingement scrubber is similar to the sieve plate scrubber. The gas passes through many jets and impinges on plates located above the perforations. Collection of particles is due to impaction of the jets on the collecting plate and on the droplets sprayed and accelerated by the emerging gas stream. The collection may be only increased a few percent by adding additional plates or increasing the pressure drop. The impingement baffle is usually placed at the vena contracta of the gas jets formed by the perforation, several millimeters

194

above the plate. The standard operating pressure drop is 4 mbar per plate and can be increased to obtain better efficiencies. Water consumption varies between 0.13 to 0.27 liter per m^3 gas per plate. The design of the tower is the same as in bubble cap scrubbers.

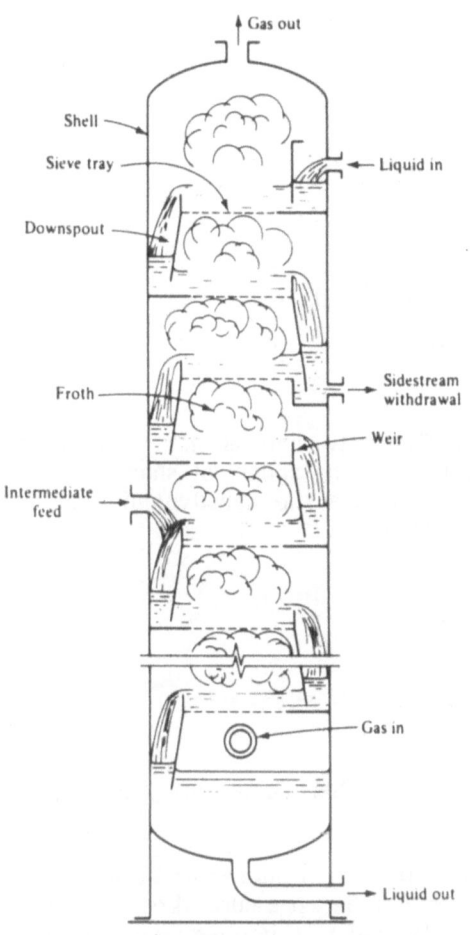

Figure 141.   Plate Column (53).

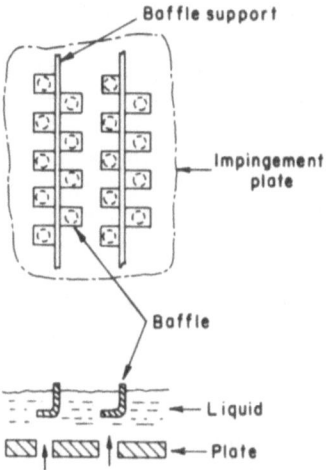

Figure 142.   Impingement Plate (14).

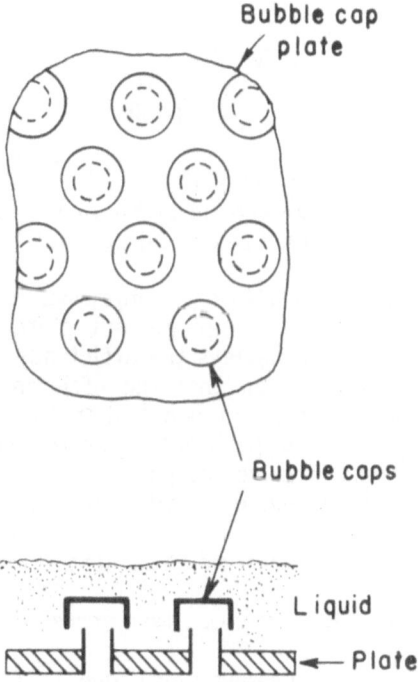

Figure 143.   Bubble Cap Plate (14).

In practice the gas flow rate, Q_G, through the column is known and the perforation diameter, d_h, as well as the number, n, of holes in a plate can be arbitrarily chosen by the designer. The gas velocity, u_o, through the perforations as well as the particle diameter, d_{p50}, collected with 50% efficiency may be computed provided the viscosity, μ_G, of the gas and the actual particle density, ρ_p, is known or fairly well approximated. In practice u_o is higher than 15m/sec and is computed from:

$$u_o = \frac{4Q_G}{n\pi d_h^2}$$

The diameter, d_{p50}, is given as:

$$d_{p50} = -0.0825 + \left(0.0068 + \frac{1.382 \times 10^8 n \mu_G d_h}{\rho_p Q_G} \right)^{\frac{1}{2}}$$

However, the actual particle density will fluctuate and is not well known. Moreover, ash particles do not even come close to the spherical shape assumed in these equations. In this case it is better to use the aerodynamic particle size d_{pa50} collected with 50% efficiency given as:

$$d_{pa50} = \left(\frac{1.37 \, \mu_G \, n d_h^3}{Q_G} \right)^{\frac{1}{2}}$$

The particle size is given in microns whereas all other values are expressed in centimeters, grams and seconds.

Figure 144 shows the marginal increase in efficiency if more plates are added to a standard impingement plate scrubber at 4 mbar pressure drop per plate. Figure 145 displays the efficiency curve of a typical wet impingement scrubber. It can be seen that the efficiency in collecting particles in the lower range, 1-3 microns, is much higher than for cyclones. The pressure drop across any part of the purification system must be of concern because the only force present to drive the gas is the natural suction of the engine. The pressure drop per plate can be divided into three main components:

1. Pressure drop across the dry plate.

2. Pressure drop across the wet plate, mainly due to the water depth.

3. Pressure drop due to friction within the system.

196

The dry plate pressure drop which is mainly due to the jet exit can be approximated by:

$$\Delta P = \frac{0.81 \, \rho_G \, Q_G^2}{n^2 g \, d_h^4}$$

The distance between the baffle and the plate should be larger than the hole diameter, d_h, when applying this equation.

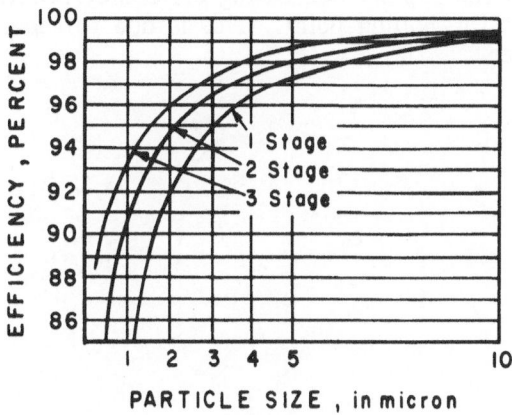

Figure 144. Standard Impingement Efficiency at a Pressure Drop of 4 mbar per Stage (SLY Catalog, 1969) (14).

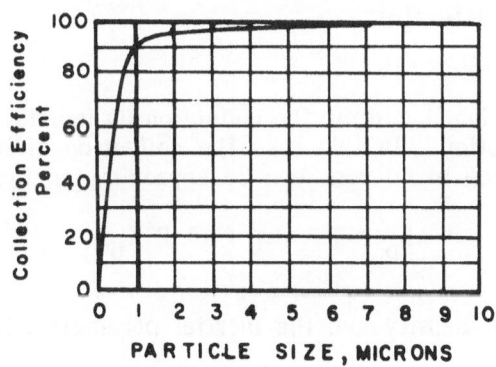

Figure 145. Grade Efficiency Curve for Wet Impingement Scrubber (Stairmand) (14).

A different design is shown in Figure 146. The gas passes through a sieve-like plate with 1-3 mm perforations with a high velocity, 17-35 m/sec. This high velocity is necessary in order to prevent the water from weeping through the perforations. A pressure drop of 2.5-10 mbar per plate is normal. Sieve plate scrubbers can be built with or without downcomers as in bubble plate scrubbers. Important factors influencing the performance are the perforation diameter, d_h, the foam density, F, and the gas velocity, u_h, through the perforation. The foam density, F, is defined as the ratio of the clear liquid height, l, to the total foam height, h. Clear liquid height, l, is the height of the liquid flowing on the plate when no gas is passed through the plate. Total foam height, h, is the height of the bubbling liquid over the plate when gas is introduced through it. Collection efficiency is, in general, good for particles larger than 1 micron. An important fact is the rapidly decreasing efficiency practically going to zero with temperatures at the boiling point. This is due to repulsion of particles by the evaporating water.

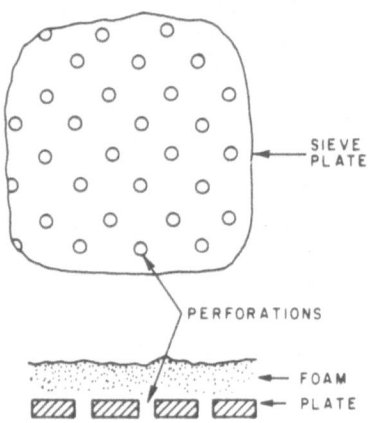

Figure 146. Sieve Plate (14).

For particles larger than 1 micron, the collection by inertial impaction dominates over diffusion collection. In this case the collection efficiency for a particle of diameter, d_p, given in microns can be approximated by:

$$E_p = 1 - \exp{(-40\ F^2 K_p)}$$

where F is the foam density and the inertial parameter, K_p, equals:

$$K_p = \frac{\rho_p\ d_p^2 u_h}{9\ \mu_G d_h}$$

The foam density, F, is in the range of 0.35 to 0.65.

Figure 148 shows the collection efficiency as a function of the hole velocity, u_h, for various gas and liquid flow rates combined in a generalized collection efficiency curve versus the generalized parameter $F^2 K_p$. Another approach is shown in Figure 147. In this case the aerodynamic particle size diameter, d_{pa50}, that is collected with 50% efficiency is plotted versus the hole velocity, u_o. Curves are given for various representative foam densities, F, of 0.4 and 0.65, respectively. The symbols, u_h and u_o, are synonymous symbols for hole velocity.

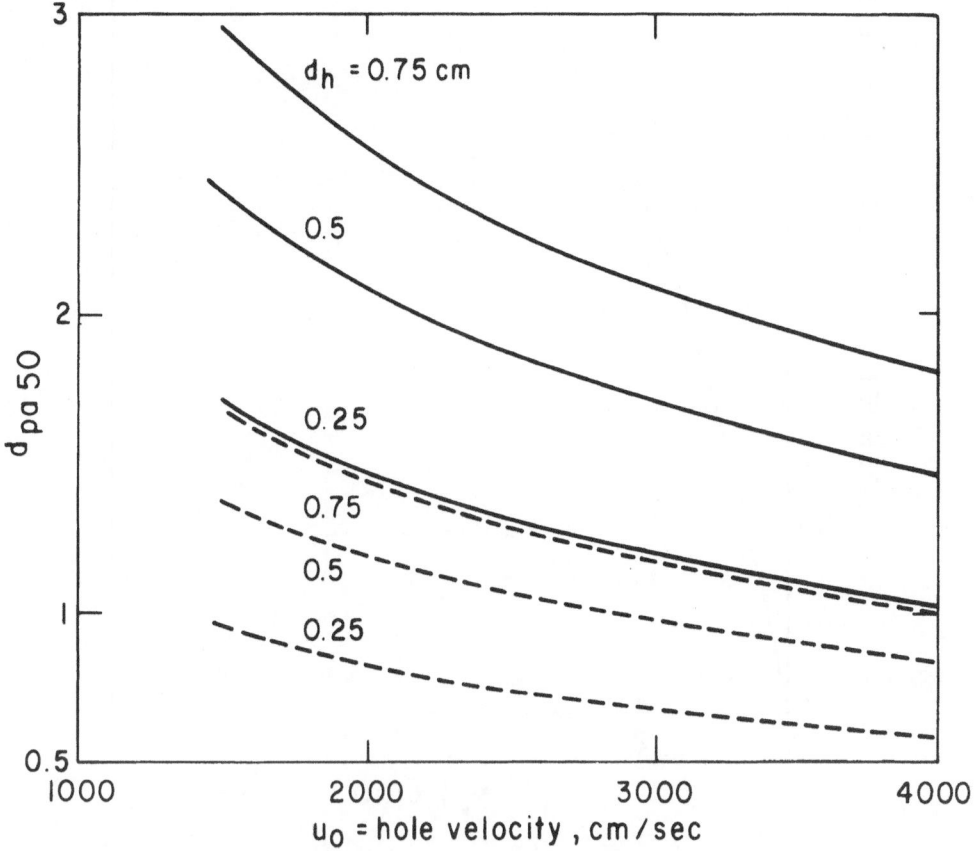

Figure 147. Predicted Sieve Plate Performance d_{pa50} Versus u_o with F and d_h as Parameters. For Impaction Only (14).

—— F = 0.4;   - - - - F = 0.65; $u_G = 1.8 \times 10^{-4}$

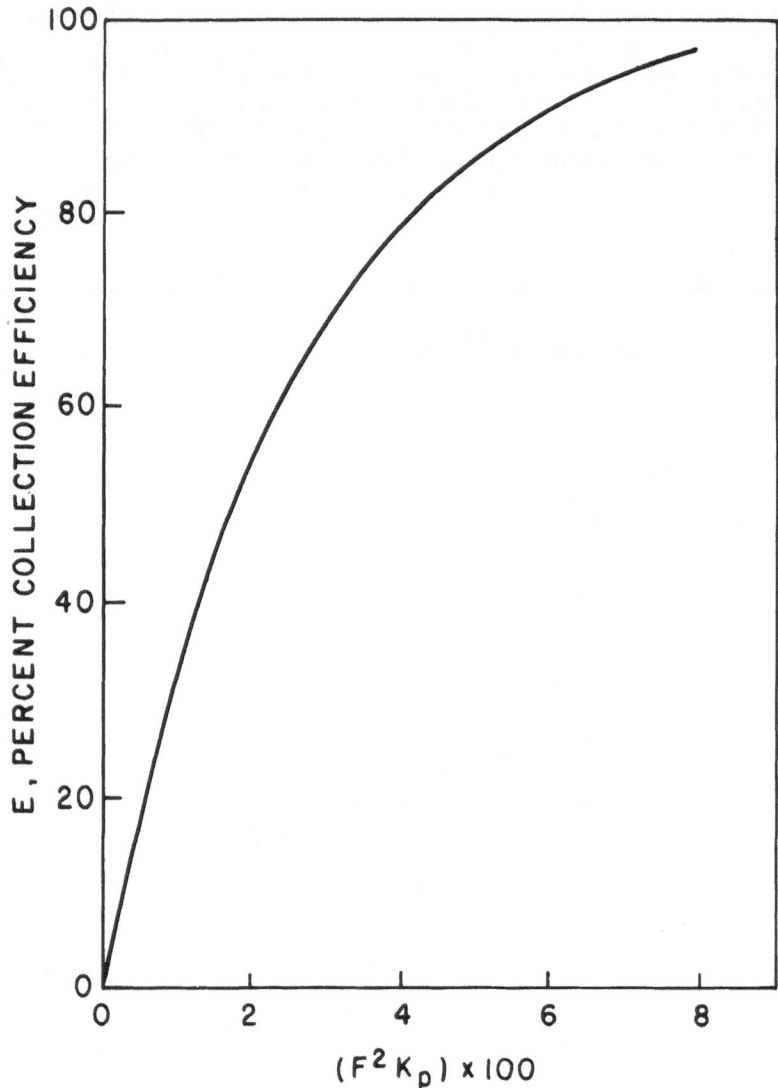

Figure 148. Collection Efficiency Versus Generalized Parameter in a Sieve Plate Column. After Taheri and Calvert (14).

200

The smallest allowable plate diameter, d_c, for wet plate scrubbers such as bubble cap, impingement and sieve types can be approximated by:

$$d_c = \Psi \, (Q_G \sqrt{\rho_G})^{\frac{1}{2}}$$

It is also true that d_c depends on several operating diameters but to a good approximation is constant for a given tray geometry and tray spacing. The values for Ψ are 0.0162 and 0.014 for bubble cap and sieve plates, respectively (dimensions $m^{\frac{1}{4}} \, hr^{\frac{1}{2}} \, kg^{\frac{1}{4}}$). They are based on a liquid specific gravity of 1.05 and 61 cm tray spacing. Figure 149 gives the correction factor if a different spacing than 61 cm is used. The height of a plate scrubber above the top tray consists of a space for liquid disengagement and is typically 0.6 and 0.75 meter. The column height below the bottom tray should be 0.6 to 1.0 meter in addition to the tray spacing to allow space for the gas stream inlet and some depth of liquid to provide a seal on the liquid outlet.

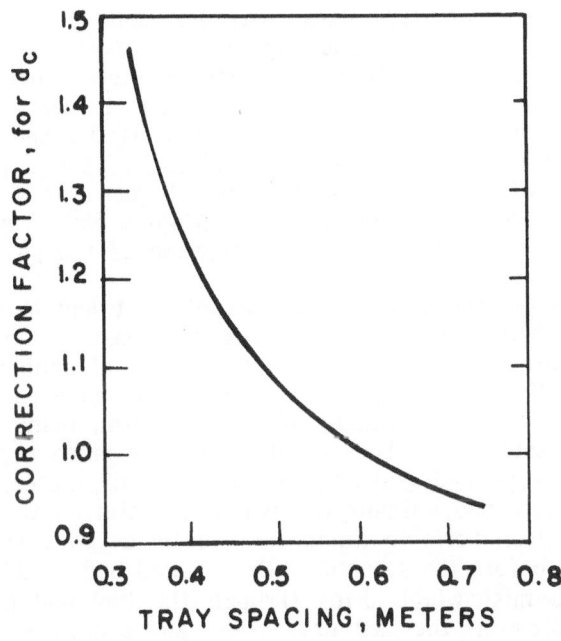

Figure 149. Correction Factors for Tray Spacings Different from 61 cm (14).

Another wet scrubber of interest for gas producers, the packed bed column, is commonly used for absorbtion of gas but less commonly used for particle separation. The packed bed scrubber typically consists of the packing, liquid

201

distributor plates, a support grate for the packing and ports for gas and water as shown in Figure 150.

Commercially available packing materials are shown in Figure 151. Their purpose is to break down the liquid flow into a film with high surface area. The design and shape of the material determines the pressure drop through the column, overall surface area available for water-gas contact and, most important, the homogenuity of the packing. They are usually made out of metal, ceramics, plastics or carbon.

Packings are by no means restricted to the commercially available materials. They can be "home made" materials. The systems of the past and some recent research units have used a wide range of material and shapes such as steel wool, metal turnings, sisal tow, wood wool, wood chips or blocks, saw dust, charcoal, coke, gravel, crushed rocks, sand, cork, and procelain marbles. Unfortunately, very little is reported about their performance and maintenance characteristics as well as reliability. Whether home made or commercial, a proper packing is most important. Improper packing causes channelling and lowers the collection efficiency considerably. Friable packings such as soft coke are not as suitable because of their physical properties. Soft coke soon chokes up the scrubber. Packings may break up through careless placement and then cause severe blocking problems at the absorber base. The packing material must be perfectly clean and free from dust. For home made packings it is best to grade them according to size and start with a layer of coarse material (3-5 cm diameter) separated by a liquid distribution plate from the fine material layers (1-3 cm diameter). If the bed is not packed evenly the washing water will tend to take the easiest path and leave unwetted areas through which the gas will pass. The water distributor at the top of the column must give an even distribution over the full cross section. Figure 152 displays an adequate and inadequate liquid distribution system. A good water distribution is quite often not achieved because it requires restriction of the gas flow.

Packed bed scrubbers are able to remove pollutant and harmful gases to any desired degree. The limiting factors are the cost and the pressure drop. Typically, a pressure drop of 10 cm H_2O per meter of column can be expected for the commercially available packings. For home made packings the pressure drop depends mainly on the grading of the material, being extremely high in case of sand and saw dust. The hot dust and tar laden producer gas has the tendency to clog a packed bed scrubber if too fine material is used for the bottom layers. A continuous liquid flow is not essential for this type of scrubber. It can be operated on periodic liquid flow or on a dry basis if the condenser (cooler) is located before the scrubber. Dry packed bed scrubbers are commonly used to eliminate mist which drips through the bed and is collected at the bottom. Some systems in the past used a wet and a dry bed scrubber in series, to first remove undesirable gases and vapors and then separate the entrained mist in a dry bed scrubber. It is sometimes necessary to replace the coarse material such as wood blocks or crushed rock in the dry scrubber with finer material such as saw dust and fine gravel if impurities in the gas are particularly persistent. The sizing of packed bed scrubbers is outlined in detail (14,53) for the commercially available packings. For home made packings no data can be given and the best system must be found by error and trial.

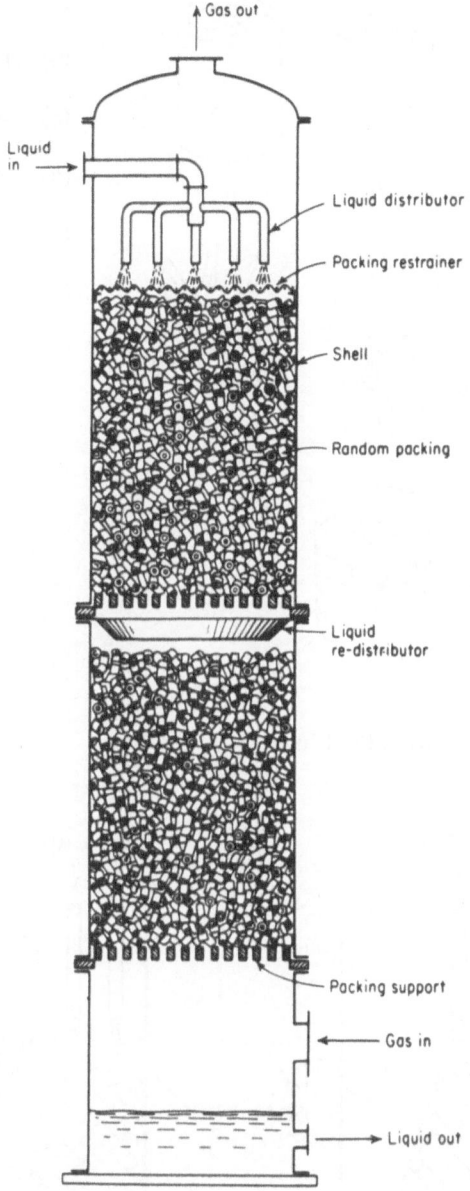

Figure 150. Packed Bed Scrubber (53).

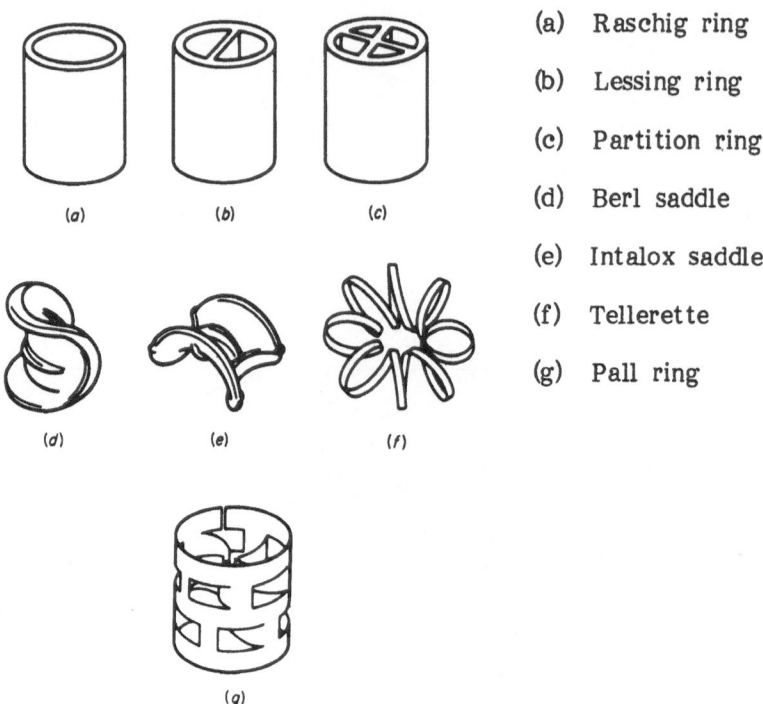

(a) Raschig ring

(b) Lessing ring

(c) Partition ring

(d) Berl saddle

(e) Intalox saddle

(f) Tellerette

(g) Pall ring

Figure 151. Random Tower Packings (53).

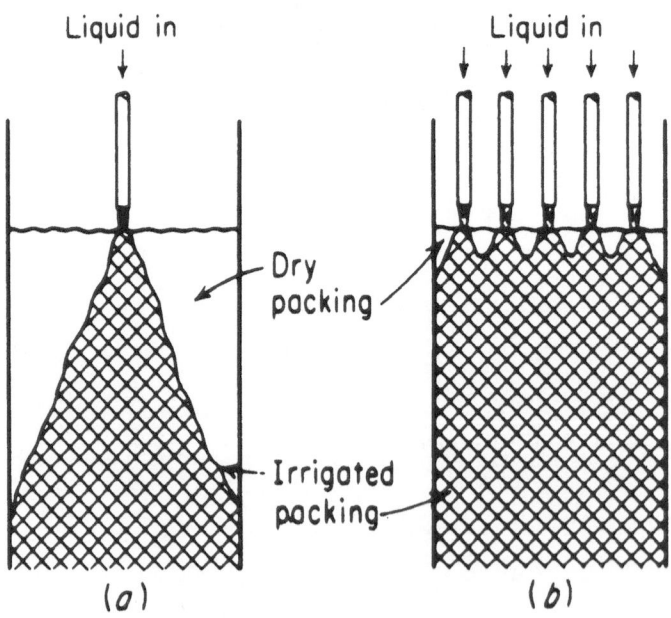

Figure 152. Inadequate (a) and Adequate (b) Liquid Distribution Systems (53).

The necessary amount of water to clean and cool the gas entering the packed bed is about 3-5 liter per B.H.P.-hour, based on practical experience with coke bed scrubbers. A 20-hp engine would therefore require a barrel of water every three hours to clean and cool the gas. The water should be reasonably clean for packed bed scrubbers, whereas sieve-plate scrubbers have a good ability to use dirty water as a washing liquid.

A type of wet scrubber operating with a stationary water bed and no packing material is shown in Figure 153. This so-called impingement, entrainment or self-induced spray scrubber is most frequently used for particle collection of several microns diameter. The pressure drop may range from 10-50 cm of water column. The principle of operation is based on particle collection by multitudes of drops generated through the gas flow below or past the water surface. This class of scrubber has some important advantages such as no clogging or blocking of the unit can occur under heavy dust load. The system is always well irrigated and does not require a continuous water flow if evaporation is not excessive. The spray is self-induced by the gas stream without employing mechanical devices or spray orifices. Circulating water is used without purification, excessive build up of solids being avoided by purging the settled solids and adding clean water. The sensitivity to + 25% changes in the gas flow is minor. Water consumption is low, between 0.03 and 0.67 liter per m^3 gas depending on the temperature of the gas and the allowable concentration of the slurry. Concentration is usually kept at 5-10%.

Self-induced spray scrubbers have been widely used in the past for stationary and portable units. They were commonly employed as the first stage of a wet, packed-bed scrubber to separate the coarse particles out of the gas stream before it enters the packing. Figures 154 and 155 show more recent designs that are commercially available.

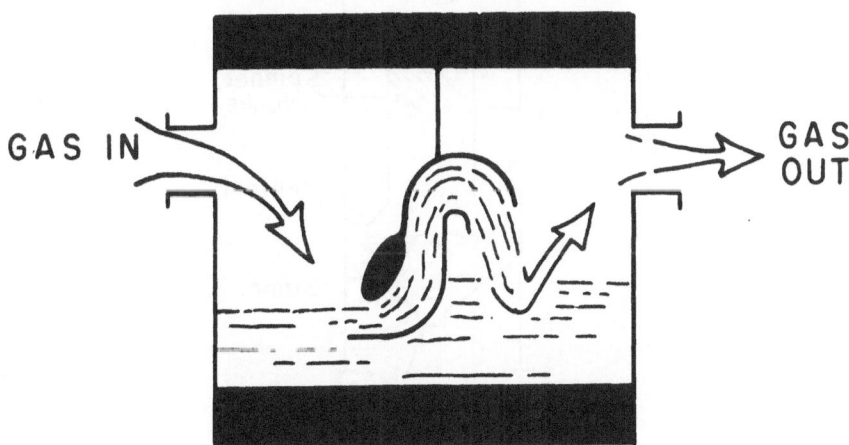

Figure 153. Self Induced Spray Scrubber (14).

Baffle scrubbers operating on a dry or wet basis are frequently used. They are designed to cause changes in gas flow direction and velocity by means of solid surfaces. Louvers and wall plates are examples of surfaces which cause changes

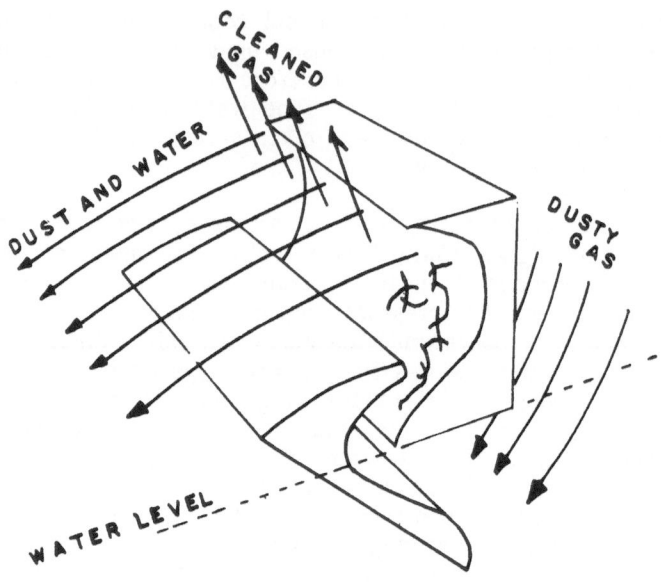

Figure 154. "Rotoclone" Type N Precipitator (14).

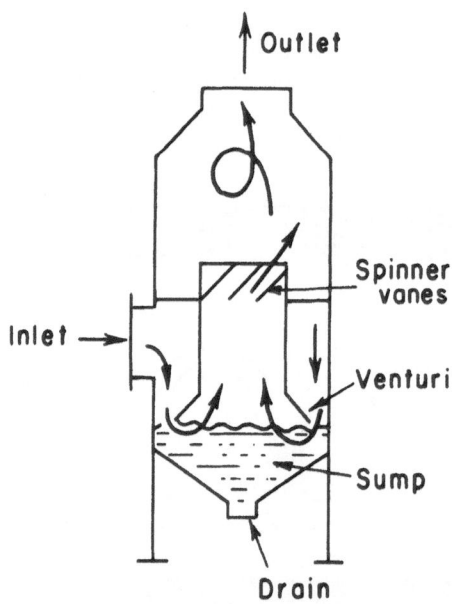

Figure 155. Turbulaire Type D Gas Scrubber (Joy Manufacturing Co.) (14).

in the main flow direction. If the material separated is liquid, it runs down the collecting surfaces into a collection sump. If solid, the separated particles may be washed intermittently from the collecting surfaces. Figure 156 shows the general principle involved for a baffle scrubber. Figures 157 to 159 are typical examples of commercially available units. Most baffle scrubbers are used for collecting water drops and mist produced in wet scrubbers. Their efficiency is good down to particle drop size of 5 micron. However, if a large number of baffles are required for good efficiency, the pressure drop can be considerable. This is shown in Figures 159 and 160 for a baffled wall collector with various spacing and height of the baffles.

It should be pointed out that self-induced spray and baffle scrubbers, although simple devices, are not very effective for small particles. The pressure drop in both devices can be so high that the manifold suction of the engine is not capable of drawing the gas through the gas cleaning system.

A simple and very effective method of removing solid matter from a gas stream is to filter it through cloth or some other porous material. Several types of natural and synthetic fabrics such as cotton, dacron and fiberglass are commercially available and their characteristics concerning cost, permeability, durability, resistance to certain acids and temperature are given in Table 51.

Fabric filters have a very high collection efficiency, in excess of 99%, which can not be matched by any other purification system previously discussed. Their pressure drop is within the range of 5-20 cm of water column. The overall pressure drop is caused by a combination of fabric resistance which is primarily a function of air flow, structure of the fabric and resistance due to the dust accumulated on the fabric surface. Usually less than 10% of the total resistance is attributed to the clean fabric resistance.

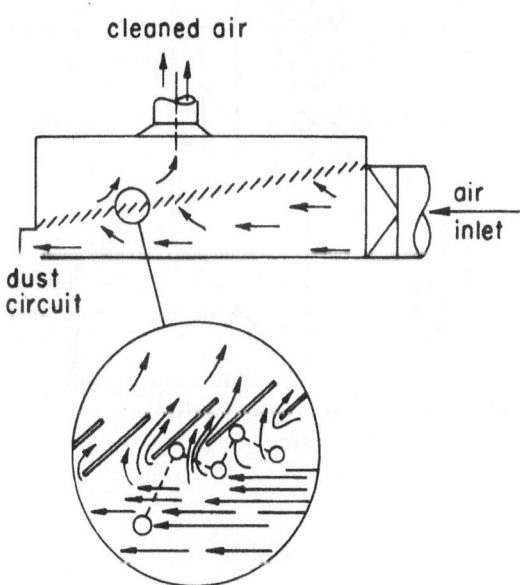

Figure 156. Louver Impingement Separator (A.I.H.A. 1968) (14).

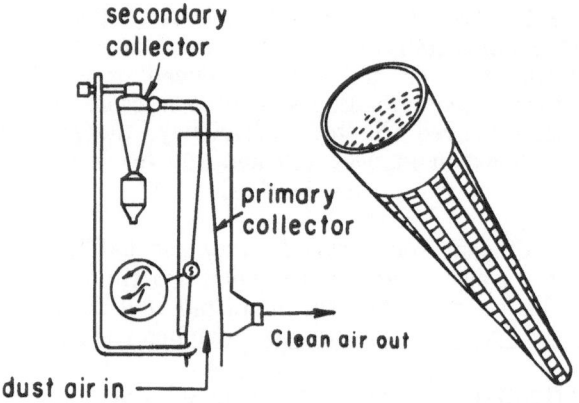

Figure 157.   Conical Louver Impingement Separator (A.I.H.A. 1968)   (14).

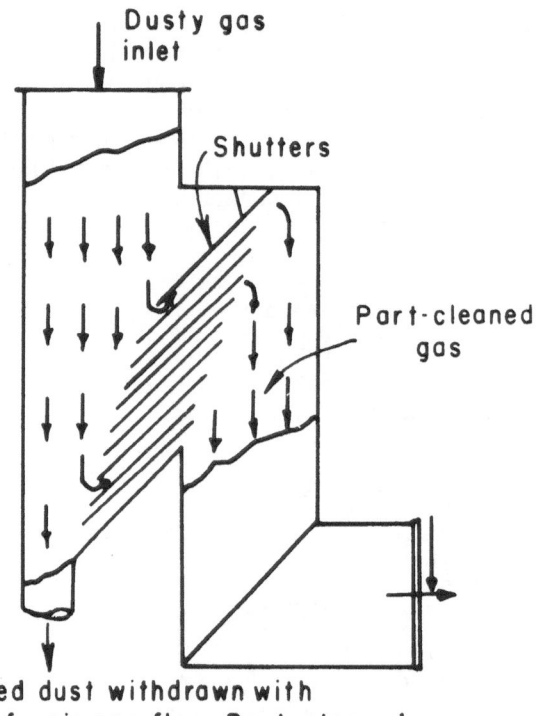

Collected dust withdrawn with
20 % of main gas flow. Part-cleaned
gas to main cyclone.

Figure 158.   Shutter Type Collector (Stairmand, 1951) (14).

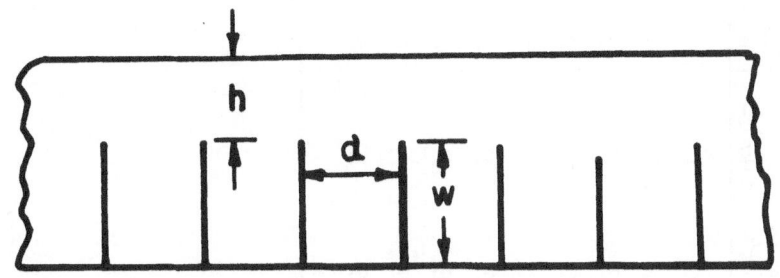

Figure 159. Collector With One Baffled Wall (Calvert and Hodous, 1962) (14).

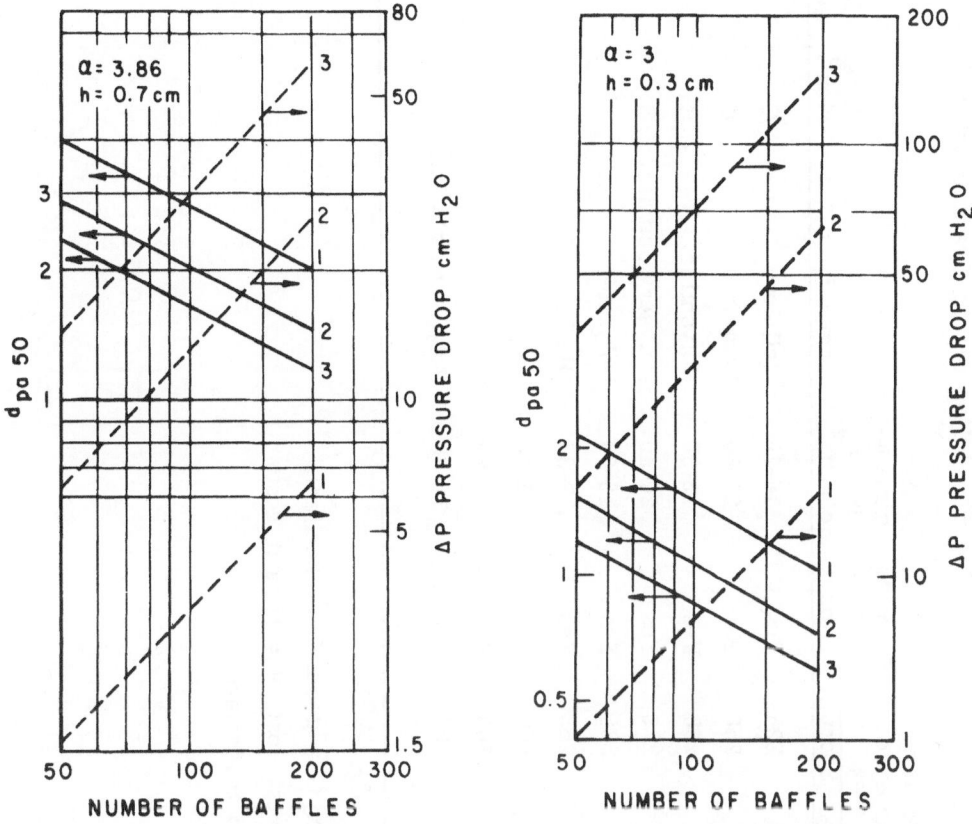

Figure 160. Predicted Secondary Baffles Device Performance, $d_{pa\,50}$ Versus Number of Baffles, with u, α and h as Parameters (14).

Table 51. Filter Fabric Characteristics (55).

Fiber	Operating exposure (°F) Long	Operating exposure (°F) Short	Supports combustion	Air permeability [a] (cfm/ft²)	Composition	Abrasion [b]	Mineral acids [b]	Organic acids [b]	Alkali [b]	Cost rank [c]
Cotton	180	225	Yes	10-20	Cellulose	G	P	G	G	1
Wool	200	250	No	20-60	Protein	G	F	F	P	7
Nylon [d]	200	250	Yes	15-30	Polyamide	E	P	F	G	2
Orlon	240	275	Yes	20-45	Polyacrylonitrile	G	G	G	F	3
Dacron [d]	275	325	Yes	10-60	Polyester	E	G	G	G	4
Polypropylene	200	250	Yes	7-30	Olefin	E	E	E	E	6
Nomex [d]	425	500	No	25-54	Polyamide	E	F	E	G	8
Fiberglass	550	600	Yes	10-70	Glass	P-F	E	E	P	5
Teflon [d]	450	500	No	15-65	Polyfluoroethylene	F	E	E	E	9

[a] cfm/ft² at 0.5 in. w.g.

[b] P = poor, F = fair, G = good, E = excellent.

[c] Cost rank, 1 = lowest cost, 9 = highest cost.

[d] Du Pont registered trademark.

The openings between the threads of the cloth are many times larger than the size of the particles collected. Consequently, a new filter has a low efficiency and low pressure drop initially. Their disadvantages are their short life and temperature sensitivity. At temperatures above 300 ^{o}C fabric filters can not be used. From previous discussions it should be clear that at full load, gas temperatures in excess of 300 ^{o}C can be expected at the cyclone exit.

Moreover, the hygroscopic material, condensation of moisture and tarry adhesive components found in producer gas may cause crust caking or plugging of the fabric. Fabrics can burn if readily combustible dust, as in producer gas, is being collected. This may explain why their use has not been as widespread in the past although their efficiency is excellent even in the submicron range where wet scrubber systems are totally ineffective.

Almost all fabric filters are either envelope or cylindrical shaped as shown in Figure 161.

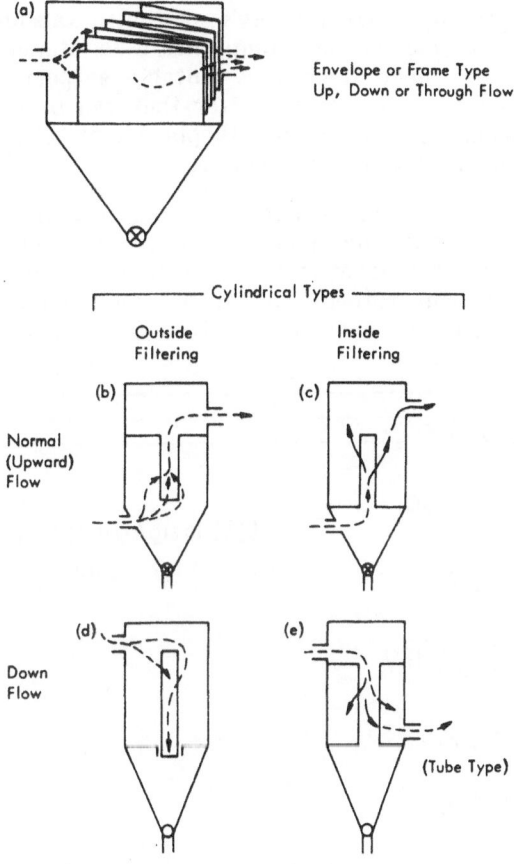

Figure 161. Configuration of Fabric Filters (33).

The design and operation of fabric filters depend primarily on the air-to-cloth

211

ratio used. This means the volume of gas passed through one unit of surface area of the cloth in one hour. This ratio may be as low as 30 m³/m²-h or as high as 600 m³/m²-h. If the gas contains smaller particles which are difficult to collect one should choose a low air-to-cloth ratio which also keeps the pressure drop down but requires large cloth areas. More detailed information about fabric filters and their industrial applications are contained in References (10) and (33). Recent tests with fabric filters as part of the purification system of small portable gas producers on tractors and trucks have been carried out by the National Machinery Testing Institute, Uppsala, Sweden and their findings are presented below.

The highest temperature allowable for textile filters such as cotton or wool is about 120 °C. This is below the dew point of many of the tarry vapors and oils in the gas stream. Their use is therefore not recommended.

The development of synthetic fibers during the 1930-1950 period was not at a level that would have permitted use of them as part of the filter system. On the other hand, the very few cloth filters tested performed so badly that they were not used commercially. With today's synthetic fabrics, the situation is totally different. The reasonably high temperature resistance and low moisture absorption of these fibers make them very suitable as part of the purification system. The performance of a fiberglass filter that can be used at temeratures as high as 300 °C depends primarily on its placement within the purification system and the performance of the condenser.

In general, the fabric filter should be placed immediately after the cyclone. Its performance and life depend on the type of gasifier, the ambient air conditions, the fuel moisture content, the specific gasification rate and how the vehicle is driven. In any case, temperatures should be kept below the point where considerable aging of the material occurs as given in Table 52 from the Swedish test series.

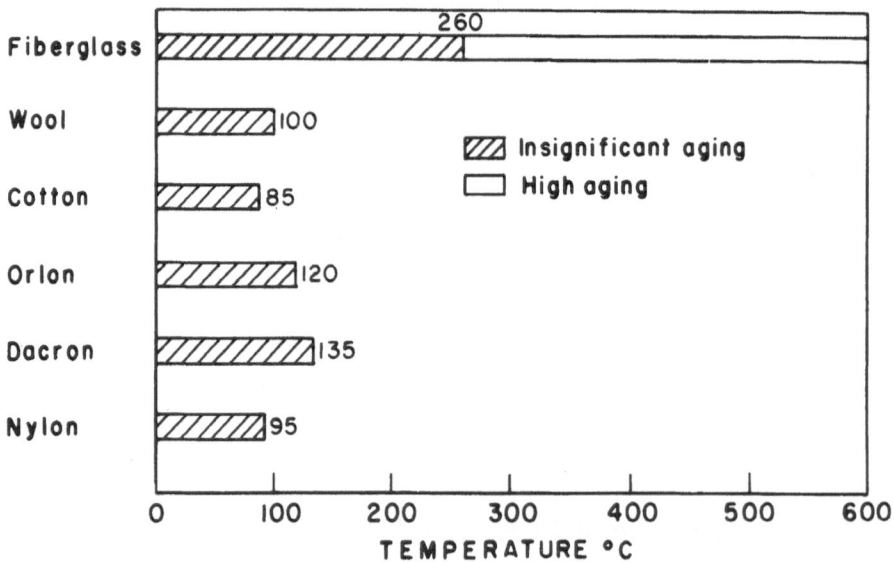

Table 52. Temperature Resistance of Various Fabric Filters (35).

212

Blowing the gasifier too hard results in sparks or glowing particles passing the cyclone and burning through the fabric filter. On the other hand, low loads or cold weather favor fast cooling of the gas before it reaches the fabric filter and therefore causes condensation of water, tar and oil vapors at the fabric surface. This usually chokes off the gas supply to the engine. Figure 162 shows the principal parts of the purification system for tractors. Figure 164 is a plan view of the cleaning system for trucks. In both cases a patented Bahco fiberglass filter of 2.33 m^2 total area was used. Its dimensions are given in Figure 163. The rectangular container is 538 mm x 355 mm x 625 mm.

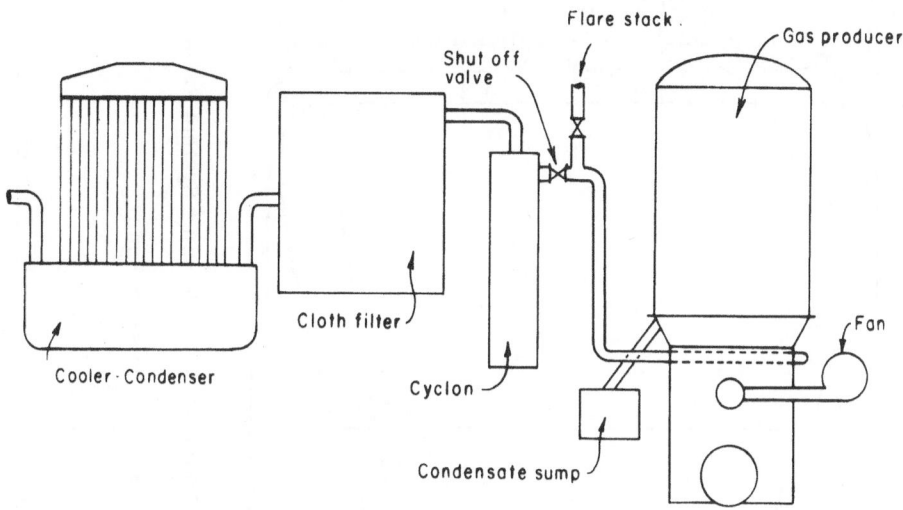

Figure 162.   Purification System For Tractors (35).

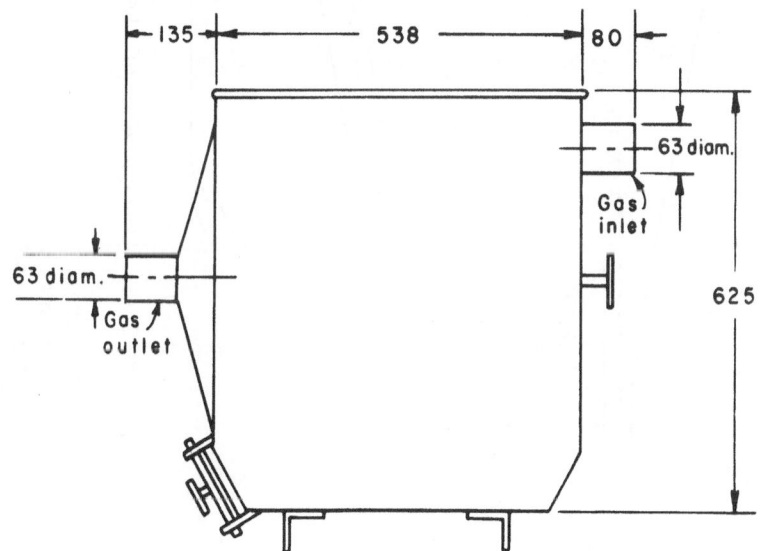

Figure 163.   Bahco Cleaner Side View (35).

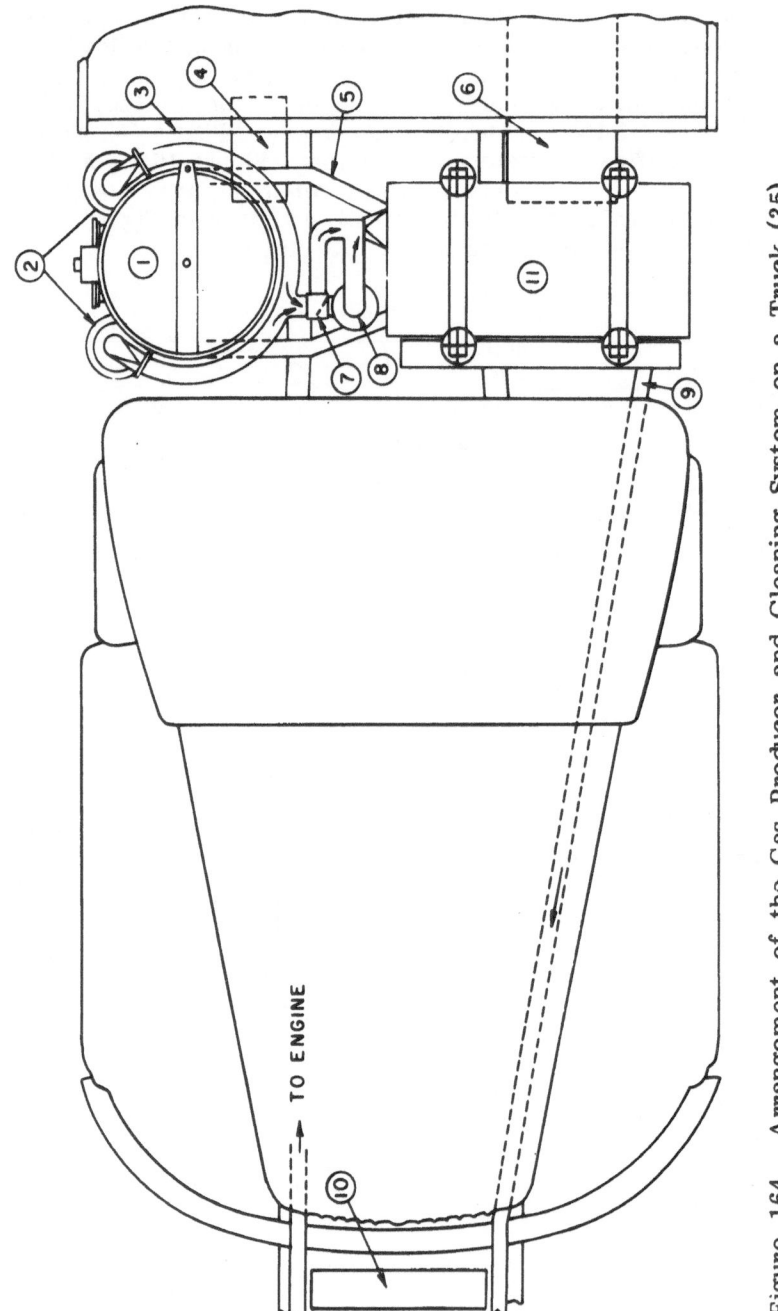

Figure 164. Arrangement of the Gas Producer and Cleaning System on a Truck (35).

1. Generator
2. Cyclones
3. Front wall of truck
   cargo box
4. Condensate Sump
5. Frame
6. Fuel Tank
7. Valve
8. Precooler
9. Gas Lines

10. Gas Cooler

11. Fabric Filter

TO ENGINE

214

The air-to-cloth ratio, the pressure drop across the fabric filter and the temperature after the cloth filter as a function of the generator load over the entire range of possible gasification conditions are the most important design parameters. Data obtained from actual trials with tractors are shown in Figure 165.

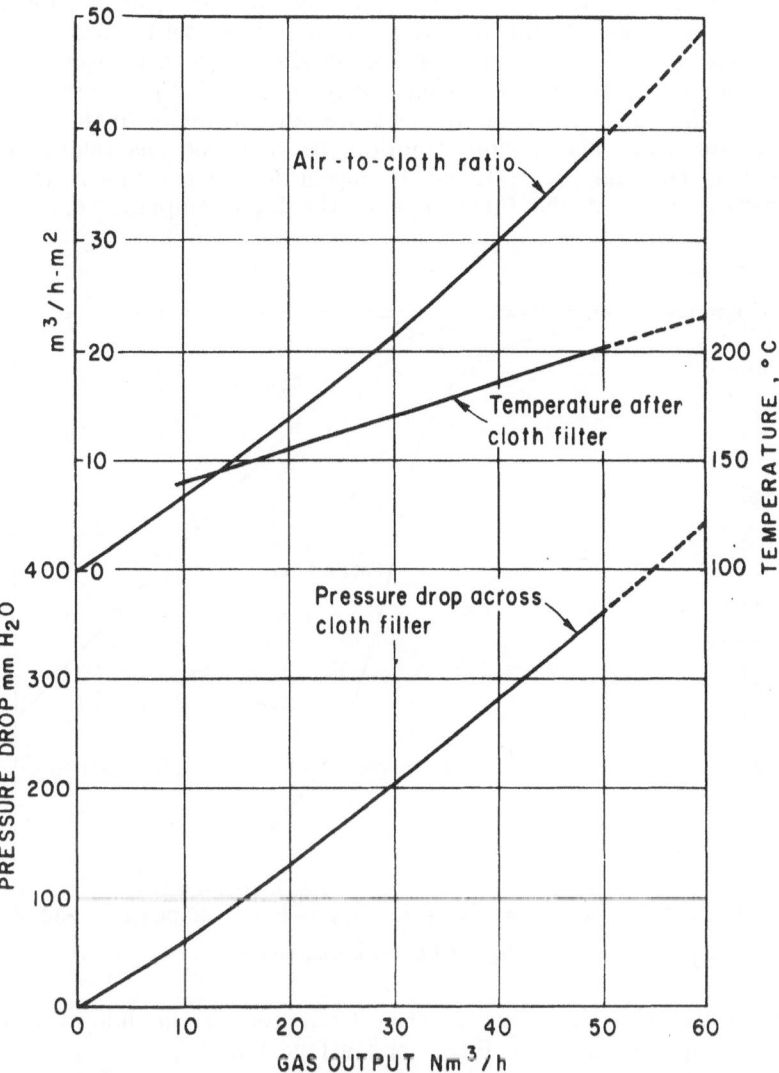

Figure 165. Pressure Drop, Temperature at Filter Exit and Air-to-Cloth Ratio as a Function of the Load (35).

A comparison of the collection efficiency of the Bahco Fabric Filter with a cyclone is shown in Figure 166. This data was obtained with a wood and charcoal gas producer as shown in Figure 74. The results clearly show the superiority of a fabric filter over a cyclone in the below 5 micron range. In addition, the ideal combination of a cyclone with a fabric filter in series is demonstrated. It can not be denied that, although the combination of a cyclone-fabric filter-condenser gas cleaning system is most effective in collecting a wide range of particles, there have been problems with such an arrangement. Most troublesome is the fact that producer gas leaving the gasifier is either already saturated with tar vapors or close to saturation. Any drastic temperature drop across the cyclone and the fabric filter leads therefore automatically to tar condensation at the fabric surface. The obvious solution, to keep the gas temperature high between cyclone and the fabric filter, is impracticable because it would result in a fast deterioration of the fabric due to the high temperatures.

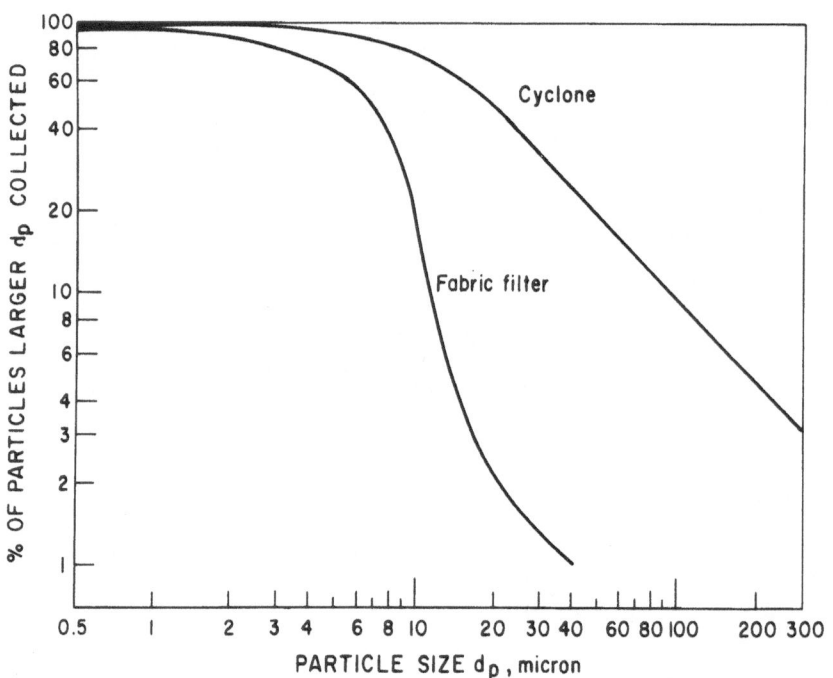

Figure 166. Accumulation Curve of Dust Collected in the 0.5 to 300 Micron Range for a Cloth Filter and a Cyclone (35).

High percentages of particles in the range 0.2-3 micron have been reported occasionally. In particular, silica dust is especially troublesome being extremely fine and highly abrasive. The reader may recall that ashes of biomass fuels may have a high percentage of SiO_2, up to 93% of the ash in the case of rice hulls. With the previously described arrangement, the usual dust content of 1-5 g/Nm^3 under normal running conditions has been reduced to 0.3 mg/Nm^3. This remaining dust concentration is well below the limit of 10 mg/Nm^3 that has been considered as safe for an internal combustion engine.

This last part of this chapter describes some of the past systems used in portable and stationary units. Although their efficiencies are not as good as what can be done today with more advanced filter materials and purifiers, they are useful for stationary units. They also demonstrate well how different particle and vapor collection methods such as collection by diffusion, impaction, and gravity have been combined in a purfication system. Their advantage lies in the fact that they can be home made, are simple to build and easy to repair and maintain.

Figure 167 shows the combination of a packed bed scrubber and self-induced spray scrubber. The packing was coke and about 3 liters of water were used per BHP-h to clean and cool the gas.

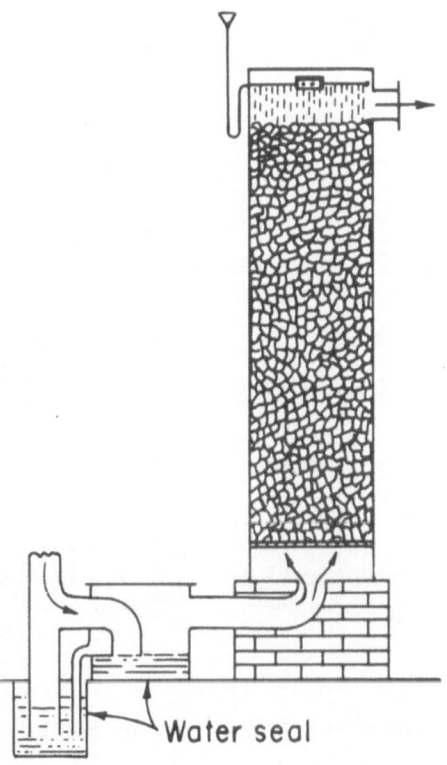

Figure 167. Combination of Packed Bed Scrubber and Self-induced Spray Scrubber for Stationary Units (17).

The purification system used in the Brush-Koela Plant is shown in Figure 168. Both elements in the first scrubber are packed with steel wool, instead of wood wool because the high temperatures damaged the wood wool. In the first wet

217

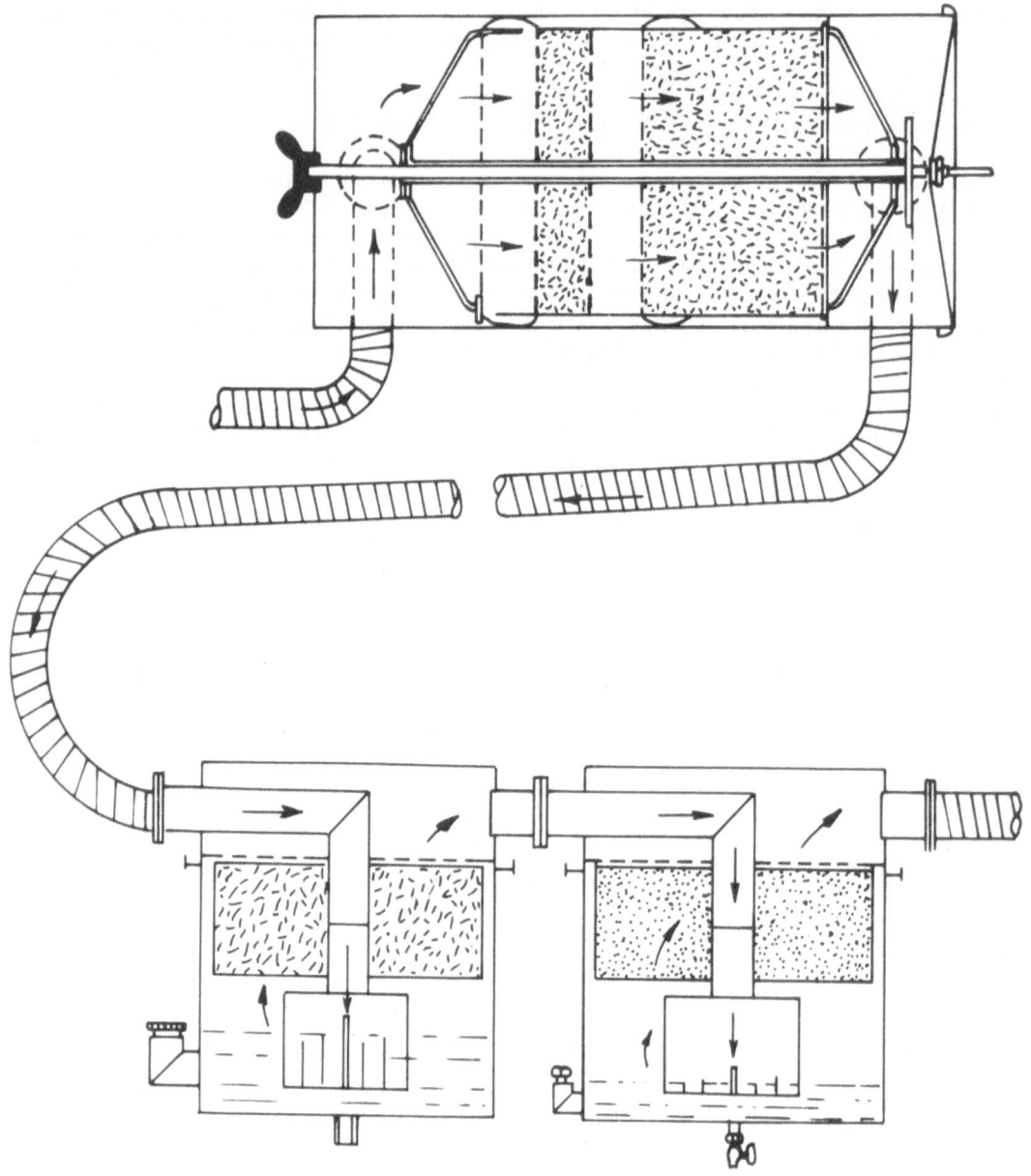

Figure 168.   Purification System of Brush-Koela Plant (2).

scrubber the gas is drawn through a solution of sodium carbonate in water and a filter bed of rusty steel wool. In the last filter the gas is passed through an oil bath and then through a pack of porcelain rings. The oil entrained with the gas was actually used as upper cylinder lubricant for the engine. The permissible mileage between cleaning and servicing the system was given to be between 320 km and 1600 km. An interesting design feature is the vertical slots in the wet scrubbers. The water and oil seals are automatically broken when starting or idling. During normal operation all the slots in the skirt of the filter distributor are submerged, but with reduced suction the fluid level outside the distributor falls. This exposes the ends of the longer slots and enables sufficient gas to be drawn through by by-passing the fluid.

Figure 169 shows a dry gas filter used successfully in buses with an excellent separation for fine particles. The gas was obtained from high grade charcoal of 2% ash content and 90% fixed carbon. The size of the cylinder is 30 cm diameter and 1.5 m long. The packing had to be removed after only 100 km or 200 km when two parallel cylinders were used. Both coir and cotton waste have been found to provide an excellent separation for fine particles. The gas entered the chamber at 115 ^{o}C and was cooled down in the purification process to 50 ^{o}C.

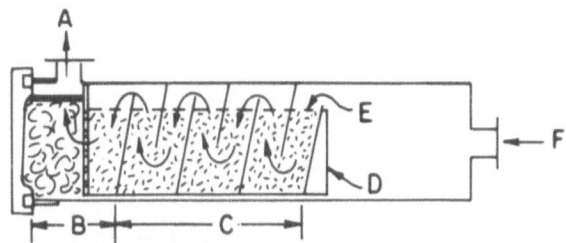

Figure 169.   A Combination of Baffle and Dry Packed Bed Scrubber (20).

A: Gas Outlet (50°C); B: Cotton Waste; C: Coir or Sisal Tow; D: Removable Tray; E: Cooling Chamber; F: Inlet (115°C).

A combination of wet oil scrubber and dry filter was frequently used. Oil is superior to water and experiments carried out with charcoal ash showed that it floated for a long time in water, but in oil it sank immediately. The gas, after passing through the oil well, continues through the vanes of a swirl plate which separates oil and gas. The speed and direction are then changed by a series of perforated plates packed with coir and cotton waste. In practice, cleaning was required after 800 km (20).

Under normal conditions, a combination of a dry and wet scrubber system was sufficient for the old units. However, the manufacturer realized the human element in driving a gas producer fueled automobile, in particular in Third World Countries where a large number of these units were sold and therefore, most purification systems had a so-called safety filter installed ahead of the carburetor. Its only purpose was to stop the gas flow and shut down the engine in case a careless driver did not clean the filter system frequently. They were usually made out of fine fabric or metal cloth as shown in Figures 170 and 171. A more sophisticated one is shown in Figure 172. It was used with considerable success on a 38 seat bus working in a tropical, sandy seaport. Its dimensions are 75 cm length by 30 cm diameter. Gas enters at the base and is diffused by a perforated cone over the oil through which it passes at high velocity. It is then slowed down to allow the oil to separate before the gas reaches the coir and cotton waste trays.

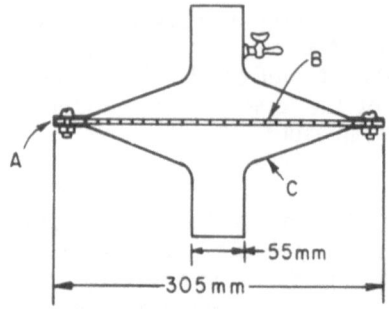

A. Rubber packing
B. 250 mesh gauze
C. Metal frame

—55mm—

—305mm—

Figure 170. Membrane Safety Filter (13).

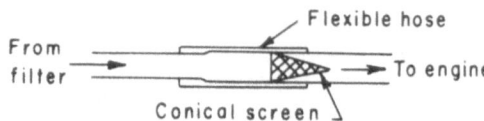

From filter

Flexible hose

To engine

Conical screen

Figure 171. Cone Safety Filter (51).

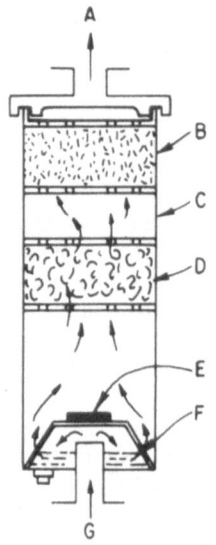

A. Gas Exit at $27^{\circ}C$

B. Cotton Waste

C. Metal Cylinder

D. Coir

E. Lead Weight to hold cone down

F. Oil bath level

G. Gas inlet at $49^{\circ}C$

Figure 172. Oil and Fabric Safety Filter (20).

The dry paper filter for carburetors on today's cars and trucks is probably suitable as a safety filter for producer gas.

A four-stage filter is shown in Figure 173. The gas first expands in the expansion chamber which also serves as a dust bin. After having passed through the main sisal pack, it travels up the outer annular space. It then enters the annulus between the two inner cylinders and passes through the oil. Here the suction of the engine causes the oil seal to be made and broken continuously, giving rise to a self-induced spray effect. Upon emerging from the oil, the gas passes through the final sisal tow pack, which retains any excess oil without impairing the oily vapors present in the gas. Practical separation of up to 98% has been achieved with this unit.

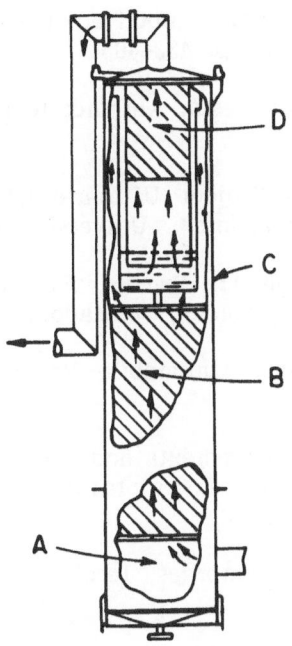

Figure 173.   Four Stage Filter (3).

A: Dust Box; B and D: Sisal Tow; C: Outer Cylinder.

There have been many more purification systems on the market for small-scale stationary and portable gas producers and their design and performance has been well documented in the literature.

In general, the past systems performed reasonably well under normal running conditions and high-grade fuel. It is particularly difficult to find an optimal gas cleaning system for automotive gas producers. In such units, compactness and lightness are important. In stationary units, the use of large packed bed columns with an adequate liquid flow can be employed. The efficiency may be considerably improved by using additional blowers to overcome the pressure drop in large columns instead of relying only on the natural suction of the engine manifold.

CHAPTER VI

1. Allcut, E. A., Producer Gas for Motor Transport, Automotive and Aviation Industries, v 89, n 4, 1943, pp 38-40,42,44,60.

2. Anonymous, An Improved Producer, Automobile Engineer, v 30, May 1940, pp 147-148.

3. Anonymous, Gas Producers: Modified Government Plant, Automobile Engineer, November, 1942, pp. 433-464.

4. Anonymous, Gas Producer Tests, Automobile Engineer, v 31, n 417, 1941, pp 418-420.

5. Anonymous, Gasification Project Ultimate Chemical Analysis Log, Agricultural Engineering Department, University of California, Davis, 1979.

6. Anonymous, Generator Gas The Swedish Experience from 1939-1945, Solar Energy Research Institute, Golden, Colorado, SERI/SP 33-140, January 1979.

7. Anonymous, Producer Gas, Automobile Engineer, v 26, n 352, 1936, pp 475-478.

8. Bailie, R. C., Current Developments and Problems in Biomass Gasification, Sixth Annual Meeting, Biomass Energy Institute, Winnipeg, Manitoba, Canada, October, 1977.

9   Berg, Torsten, Svavel i generatorgas, Jernkont. Annal., v 114, n 5, 1930, pp 213-272.

10. Billings, C. E. and J. Wilder, **Handbook of Fabric Filter Technology,** Vol. 1, GCA Corporation, 1970.

11. Bowden, A. T., Discussion on Bench and Field Tests on Vehicle Gas Producer Plant, Institution of Mechanical Engineers, Australia, v 148, 1942, pp 65-70.

12. Bugge, G., **Industrie der Holzdistillationsprodukte,** Theodor Steinkopff Company, Leipzig, East Germany, 1927.

13. Campbell, J. L., Gas Producers: An Outline of the Compulsory Government Tests in Australia, Automobile Engineer, v 32, n 422, 1942, pp 156-158.

14. Calvert, Seymour, et al., Wet Scrubber System Study, Volume I: Scrubber Handbook, U.S. Department of Commerce, NTIS PB-213 016, August, 1972.

15. Danielson, J. A., **Air Pollution Engineering Manual,** U.S. Department of Health, Education and Welfare, Cincinnati, Ohio, 1967.

16. DeGraaf, J. E., A Note on Variations in Producer-Gas Quality, Iron and Steel Inst. Journal, v 157, October 1947, pp 183-190.

17. Dowson, J. E. and A. T. Larter, **Producer Gas,** Longmans Green and Company, London, England, 1907.

18. Edgecombe, L. J., The Determination of Potential Tar in Anthracite and Fuels Containing Small Amounts of Tar, Fuel Science and Practice, v 19, n 9, 1940, pp 201-203.

19. Ekman, E. and D. Asplund, A Review of Research of Peat Gasification in Finland, Technical Research Centre of Finland, Fuel and Lubricant Research Laboratory, Espoo, Finland.

20. Enever, W. F., Gas Conditioning, Some Notes on Cleaning and Cooling Equipment, Automobile Engineer, v 33, n 436, 1943, pp 199-200.

21. Goldman, B. and N. C. Jones, The Modern Portable Gas Producer, Institute of Fuel, v 12, n 63, 1939, pp 103-140.

22. Goldstein, I., Wood Technology, Chemical Aspects, Am. Chem. Soc., 172nd Meeting, San Francisco, August 31, 1976.

23. Goss, J. R., et al., Transient and Steady-State Temperature Fluctuation in a Downdraft Gas Producer, Meeting of American Society of Agricultural Engineers, Pacific Region, Hilo, Hawaii, March, 1980.

24. Gumz, W., **Vergasung fester Brennstoffe,** Springer-Verlag, Berlin, West Germany, 1952.

25. Harahap, F., et al., Survey and Preliminary Study on Rice Hull Utilization as an Energy Source in Asian Member Countries, Development Technology Center, Institute of Technology, Bandung, Indonesia, March 1978, pp 193-236.

26. Hawley, L. F., **Wood Distillization,** The Chemical Catalog Company, New York, 1923.

27. Hendrickson, T. A., **Synthetic Fuels Data Handbook,** Cameron Engineers Inc., Denver, Colorado, 1975.

28. Holman, J. P., **Thermodynamics,** Third Edition, McGraw-Hill Company, New York, 1980.

29. Hurley, T. F. and A. Fitton, Producer Gas for Road Transport, Proceedings of the Institution of Mechanical Engineers, v 161, 1949, pp 81-97.

30. Jenkins, B., Downdraft Gasification Characteristics of Major California Residue-Derived Fuels, Ph.D. Thesis, Department of Agricultural Engineering, University of California, Davis, 1980.

31. Lees, B., Ammonia, Hydrogen Cyanide and Cyanogen in Producer Gas, Fuel, v 28, n 5, 1945, pp 103-108.

32. Lees, B., Particle Size Distribution of the Dust in Producer Gas, Fuel, London, v 28, n 9, 1949, pp 208-213.

33. Marchello, J. M. and J. J. Kelly, **Gas Cleaning for Air Quality Control,** Marcel Dekker Inc., New York, 1975.

34. Marks, L. S. and S. S. Wyer, **Gas and Oil Engines and Gas Producers,** Chicago American School of Correspondence, 1908.

35. Nordström, O., Redogörelse för Riksnamndens för ekonomisk Försvarsberedskap forsknings- och försöksverksamhet på gengasområdet vid Statens maskinprovninger 1957-1963, (from) Overstyrelsen for ekonomisk forsvarsberedskap, Sweden, January, 1962.

36. Orr, C., **Filtration, Principles and Practices, Part I,** Marcel Dekker Inc., N.Y., 1977.

37. Partridge, J. R., Manitoba Crops as an Energy Source, Sixth Annual Conference Biomass Energy Institute, Winnipeg, Manitoba, Canada, October 13, 1977.

38. Panasyuk, V. G., Zh. Priklad Khimitt, v 39, n 590, 1957, pp 813.

39. Payne, F. A., et al., Gasification-Combustion of Corncobs and Analysis of Exhaust, American Society of Agricultural Engineers Summer Meeting, San Antonio, Texas, Paper #80-3025, 1980.

40. Peart, R. M., et al., Gasification of Corn Cobs in a Producer Gas Generator, Third National Conference and Exhibition on Technology for Energy Conservation, Tuscon, Arizona, January, 1979.

41. Rambush, N. E., **Modern Gas Producers,** Van Nostrand Company, New York, 1923.

42. Redding, G. J., The Effect of Fuel Moisture Content on the Quality of Gas Produced from the Gasification of Crop and Forest Residues, Master's Thesis, Department of Agricultural Engineering, University of California, Davis, 1979.

43. Schläpfer, P. and J. Tobler, **Theoretische und Praktische Untersuchungen über den Betrieb von Motorfahrzeugen mit Holzgas,** Schweizerische Gesellschaft für das Studium der Motorbrenstoffe, Bern, Switzerland, 1937.

44. Shafizadeh, F., et al., **Thermal Uses and Properties of Carbohydrate and Lignins,** Academic Press, New York, 1976.

45. Spiers, H. M.,, Technical Data on Fuels, Sixth Edition, British National Committee of the World Power Conference, London, 1961.

46. Stairmand, C. J. and R. N. Kelsey, Chemistry and Industry, 1955, pp 1324.

47. Stern, A. C., **Air Pollution,** Academic Press, New York, 1974.

48. Strauss, W., **Industrial Gas Cleaning,** Pergammon Press, New York, 1975.

49. Takeda, S., Development of Gas Engine (I), The Bulletin of the Faculty of Agriculture, Mie University, Tsu, Japan, n 58, 1979, pp 137-141.

50. Takeda S., Research on Gas Engine (II), Annual of Institute of Tractor Research and Testing, Mie University, Tsu, Japan, n 3, 1979, pp 19-36.

51. Taylor, G., Gas Generators Capture European Interest, Automotive Industries, v 82, n 1, 1940, pp 22-25.

52. Telford, W. M., Some Notes on the Design of Mobile Producer Gas Units, Gas and Oil Power, v 36, September, 1941, pp 179-181.

53. Treybal, Robert, **Mass-Transfer Operations,** McGraw-Hill Co., New York, 1980.

54. Tsoumis, G., **Wood as Raw Material,** Pergammon Press, New York, 1968.

55. U.S. Department of Health, Education and Welfare, Control Techniques for Particulate Air Pollutants, Washington, D.C., 1969, pp 4-172.

56. Vigil, S. and G. Tchobanoglous, Thermal Gasification of Densified Sewage Sludge and Solid Waste, Water Pollution Control Federation Conference, Las Vegas, Nevada, October, 1980.

57. Wenzl, H. F. J., **The Chemical Technology of Wood,** Academic Press, New York, 1970.

58. Wise, L. E., **Wood Chemistry,** Reinhold Publishing Corporation, New York, 1944.

59. Wyer, S. S., **A Treatise on Producer Gas and Gas Producers,** Hill Publishing Company, 1906.

60. Personal Communication with Eldon Beagle, Consultant to FAO, Sacramento, 1981.

CHAPTER VII: INTERNAL COMBUSTION ENGINES

One of the most attractive applications of producer gas is its use in internal combustion engines for power or electricity generation. Although producer gas can be combusted in gas turbines, this chapter is concerned only with reciprocating internal combustion engines commonly referred to as Diesel and Otto engines. One of the not so well known facts about Diesel and Otto engines is their ability to run on fuels other than what they were designed for with very little modifications. However, there are many questions that should be looked into before attempting to run an internal combustion engine with an alternative fuel such as producer gas. A producer gas-air mixture as delivered to the combustion chamber is certainly inferior in some respects to the gasoline-air or diesel fuel-air mixture for which the engine has been designed. The chemical and physical properties of producer gas as compared to these mixtures are so different that a thorough evaluation of the following topics is necessary in order to understand the operational differences:

1. Actual efficiency of the engine

2. Power output on producer gas operation and engine modification

3. Engine wear and long-term effect on engine

4. Engine exhaust

Most reported data specifically about engine performance and long-term effects have been published before 1945 except for the report of the Swedish National Machinery Testing Institute from 1957-1963. Our worldwide search for operational automotive units as well as experimental ones revealed less than 100.

Although the search did not contact all operators and new projects are being frequently initiated, it appears there are very few operational units worldwide. In addition, the research reports and operational experience accumulated in the last decade is totally insignificant when compared to the thousands of papers written about the subject prior to 1950 and the more than a million units that had been built and operated up to 1950. There is no data available about small-scale gas producer-engine systems operated with fuels other than wood, charcoal, coal and coke on a commercial basis. However, bench test experiments with automotive gas producers fueled with corncobs can be traced back as early as 1948.

As pointed out in Chapter IV, the cold gas efficiency of a gas producer may be 70% under favorable conditions. A further loss must be taken into account when converting the cold gas energy into mechanical power by means of an internal combustion engine. The performance of I.C. engines is usually given in terms of their volumetric, indicated thermal and mechanical efficiencies. The volumetric efficiency, η_v, is of chief interest as a measure of the performance of the cylinder-piston-valve assembly as a gas pumping device. It is defined as the mass of fresh mixture which passes into the cylinder in one suction stroke divided by the mass of this mixture which would fill the piston displacement at inlet density. The indicated thermal efficiency, η_t, is the ratio of actual work done by the pistons to the heat supplied by the mixture. Finally, the mechanical efficiency, η_m, is the ratio of the power developed by the piston to the actual power obtained at the shaft. Some authors prefer to call the product of indicated thermal efficiency and mechanical efficiency the brake thermal efficiency.

The product $\eta_v \cdot \eta_t \cdot \eta_m$ is the actual efficiency of the internal combustion engine and this efficiency combined with the cold gas efficiency η_c of the gas producer yields the overall efficiency, $\eta_s = \eta_c \cdot \eta_v \cdot \eta_t \cdot \eta_m$, of the entire gas producer-purification-engine system.

The actual conversion efficiency of internal combustion engines varies widely with design, size and running conditions. A rather conservative figure is 25% for diesel engines and 15% for spark ignition engines, when operated on their respective fuels. In general, one can assume a better indicated thermal efficiency under producer gas operation, since the combustion of the gas is more complete and the flame temperature is considerably lower. The mechanical efficiency under producer gas operation will be lower due to the induction of the gas-air charge. Mechanical losses are in general caused by friction of the bearings, pistons and other mechanical parts. In addition, the engine is providing all the suction that is necesary to overcome the total pressure drop in the gas producer, purification system and piping. This latter fraction can be considerable and is one of the major causes of reduced power output besides the power drop due to the lower energy density of the producer gas-air mixture. A pipe diameter of 35 mm - 50 mm has been suggested for 2.5 liter engines.

Since the resistance within the piping system increases with the third power of the mean velocity of the gas, it is obvious that a considerable loss in power output can be expected when using too small pipes, long connections and a complicated arrangement with many elbows. A detailed comparison of efficiencies of internal combustion engines operated with producer gas versus gasoline or diesel operation can not be presented due to insufficient data and the wide variation in the composition of the producer gas-air mixture. For instance, the water vapor in the producer gas will play a significant role in assessing the combustion fuel-air ratio, indicated thermal efficiency, volumetric efficiency and detonation limits in spark ignition engines. It is known that water vapor slows down combustion, decreases the flame temperature and increases time losses unless ignition is properly advanced as humidity increases.

The indicated thermal efficiency of a spark ignition engine operated with producer gas is shown in Figure 174. The engine was operated under various compression ratios from 4.91 to 15.7.

Comparable tests with a naturally aspirated air-cooled 14 hp diesel engine have been reported in Reference 27 and some of the findings are shown in Figures 175 and 176. The graphs indicate that volumetric and indicated thermal efficiency are higher under producer gas operation and favored by high engine rpm.

Although engine efficiency considerations are worthwhile from an economical point of view, much more attention should be paid to the unavoidable drop in engine power output associated with producer gas operation. The fundamental difference between gasoline or diesel oil and producer gas operation lies in the unsteady gas composition and the lower energy density of the air-gas mixture.

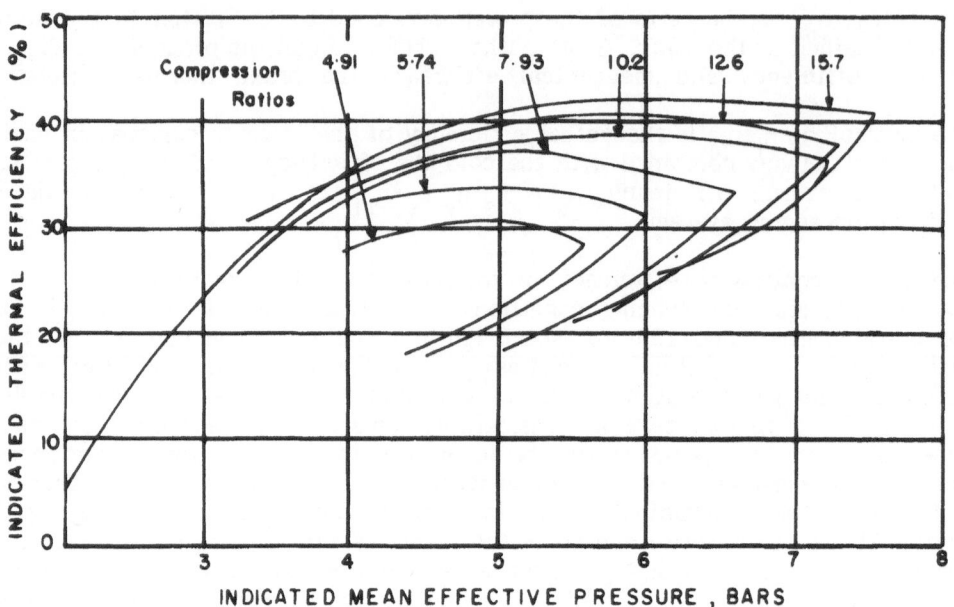

Figure 174. Indicated Thermal Efficiency of a Spark Ignition Engine at Various
Compression Ratios (41).

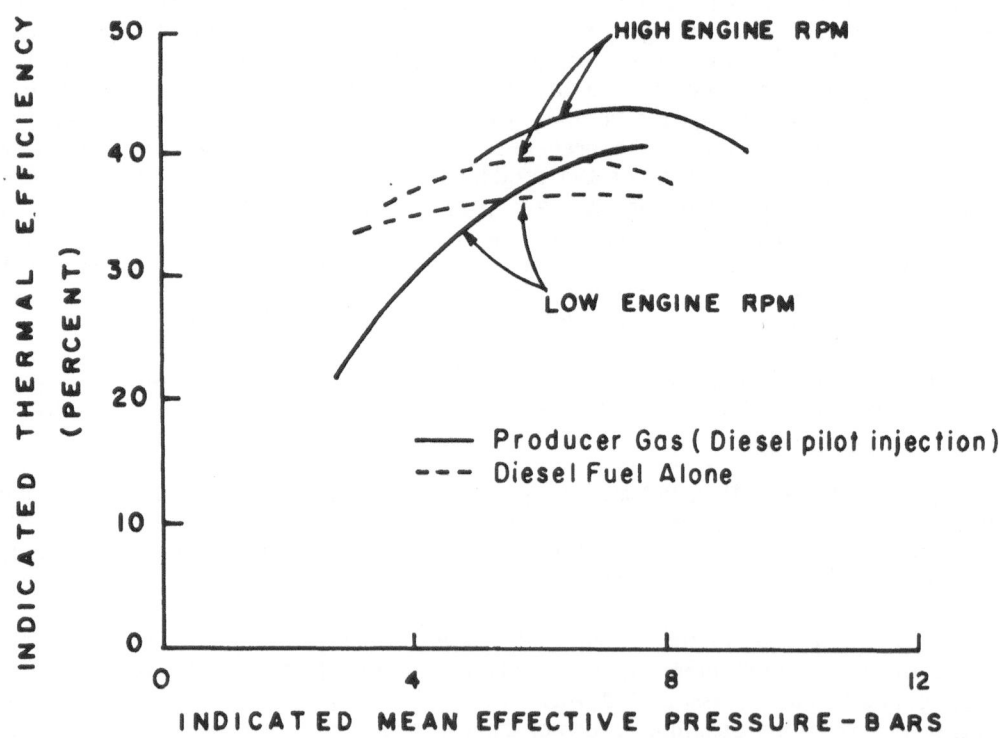

Figure 175.   Indicated Thermal Efficiency of Dual-Fueled-Diesel Engine (27).

228

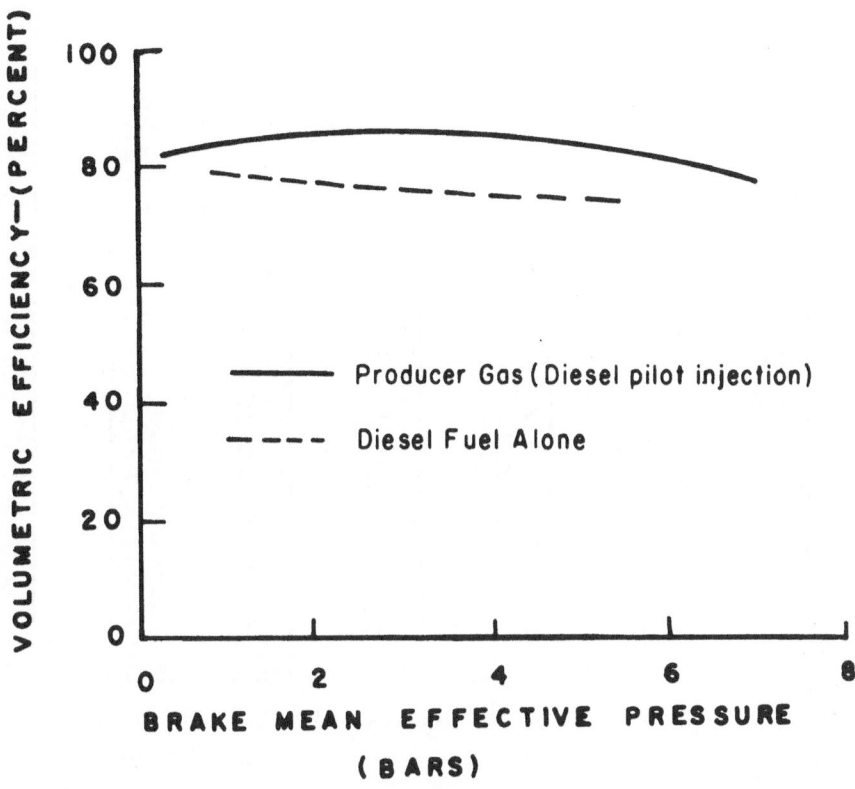

Figure 176. Volumetric Efficiency of Dual-Fueled-Diesel Engine (27).

For instance, a typical gas composition of 4.5% CO_2, 27% CO, 14% H_2, 3% CH_4 and 51.5% N_2 has a lower heating value of 5.7 MJ/m^3 at normal ambient conditions of 15 °C and 1 atm.

The stoichiometric gas-air mixture has an energy density of 2.5 MJ/m^3 compared to 3.5 MJ/m^3 for gasoline-air and 3.3 MJ/m^3 for diesel-air mixtures. Assuming no change in efficiency, a gasoline engine operated on producer gas will suffer a power drop of 29%. Taking into account the usual lower mechanical efficiency and the wide range of producer gas quality, a power drop from 40% up to 70% can be expected. Figure 177 shows possible gaseous fuels used in internal combustion engines and the energy density of their stoichiometric mixtures relative to the gasoline-air mixture.

The large number of automotive gas producers before and during World War II did not stimulate the development of a special gas producer engine. The reasons should be sought in the war conditions and the uncertain future of automotive gas producers. However, extensive research has been done on how to recover most conveniently most or all of the power loss. An internal combustion engine should meet certain design criteria for a possible conversion to producer gas operation.

229

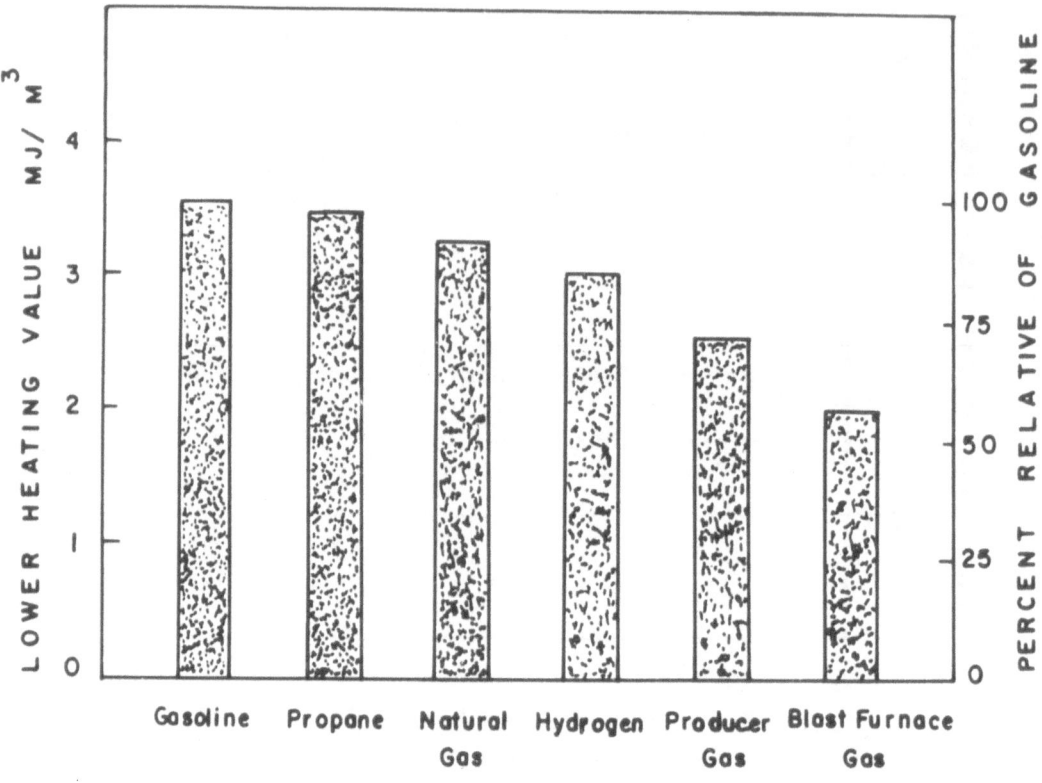

Figure 177. Typical Calorific Values for Some Fuel-Air Stoichiometric Mixtures at 1 Atmosphere and 15 °C (3).

There is a significant difference between a diesel and spark ignition engine with respect to its suitability for producer gas. Diesel engines operate on the compression-ignition principle, drawing in a full unthrottled charge of air during the intake stroke. A compression ratio between 12 and 20 is used to achieve a high air temperature at the end of the compression stroke. Just before top dead center, the diesel-air mixture is sprayed into the combustion chamber and the fuel burns almost immediately without any spark ignition. This will not be the case with a producer gas-air mixture. In fact, a diesel engine can not be operated on producer gas alone because the gas-air mixture will not ignite at the prevailing compression temperature and pressure. Spark ignition engines do not have this disadvantage and can be operated on producer gas alone without any pilot injection of gasoline. This is certainly very convenient when considering electricity generation in more remote areas or areas inaccessible for long periods over the year. In general, low speed engines with a large inertial mass, large piston displacement and large combustion space have a great advantage over today's high speed light and compact engines. Large intake valves with appropriate opening timing and good aerodynamically designed and built induction pipes will make the difference between a poorly functioning unit with low power output and a smooth running one with a power output closer to that with the normal fuel.

230

Since the conversion of diesel and spark ignition engines to producer gas operation differs so much, it is best to treat both cases separately.

Conversion of a gasoline engine to producer gas:

Today's compression ratio for spark ignition engines lies within the range 5 for industrial and tractor engines and 10 for premium gasoline passenger cars. The expected power drop for an unalternated engine will be about 40%. There are four alternatives to recover part or all of the power loss:

1. No modifications of the engine. In this case recovering of the power loss means driving the engine at a higher speed on a continuous basis.

2. Supercharging or turbocharging the engine.

3. Supercharging or turbocharging the engine and supercharging the gas producer.

4. Increasing the engine compression ratio.

5. Dual fueling the engine.

The use of an unalternated gasoline engine for producer gas operation is appealing from an economical point of view and technically sound. This approach is, in particular, beneficial in cases where an existing unit is operated on half load most of the time and the full power requirements are not crucial. Examples are engine-water pump systems and electricity generation. This approach will certainly not work for tractors and trucks which depend heavily on a full power output for a considerable part of their running time. There is a considerable diversion of opinion as to what extent the recovery of the power is actually useful. One should clearly distinguish between efficiency and power output. The actual efficiency of the gasoline engine will be only slightly affected or may be even better for producer-gas operation. It therefore makes sense to anticipate the expected power drop in a new installation and choose a larger engine to meet the power output requirements and extend the life of the unit.

In case an already installed gasoline engine is converted to producer gas operation and it is necessary to recover at least some of the power lost, supercharging (or turbocharging) the engine is one technically feasible method. Supercharging the engine was done during the 1940's on a commercial basis and is therefore not new in connection with automotive gas producers (7,22,25,28,31). There are some problems related to this method. The supercharger would be required to deliver the gas-air mixture into the existing unaltered engine at a differential pressure of 100,000 N/m^2 (1 atm) to achieve the equivalent of a compression ratio increase from 5 to 10. This is difficult to achieve with a centrifugal type compressor and a positive displacement pump seems to be more appropriate. The power consumption of turbochargers is considerable and when such a device is installed the power should be provided by a turbine driven with the exhaust gas and not taken off from the shaft. Other problems related to turbocharging producer gas are the excessive wear on the equipment due to moisture and dust in the mixture. This method will probably not have appreciable application in Third World Countries.

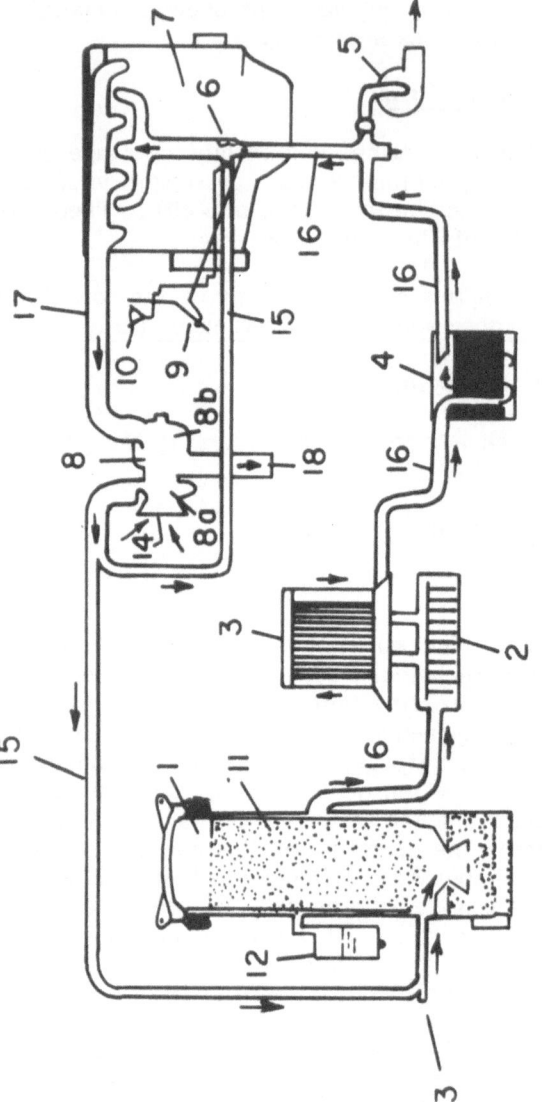

Figure 178. Diagram of a Wood-Gas Producer Plant for Motor Vehicles, with
Turbocharging the Gas Producer (22).

1. Imbert producer
2. Baffle-plate settling filter
3. Cooler
4. Fine filter
5. Starting fan
6. Mixing nozzle for producer gas and air
7. Motor
8. Turbocharger
8a. Blower
8b. Gas turbine

9. Gas pedal
10. Air lever
11. Wood load
12. Seepage water trap
13. Ignition aperture
14. Air inlet to blower
15. Combustion air pipe
16. Wood-gas pipe
17. Exhaust gas pipe from
    engine
18. Exhaust to atmosphere

Supercharging the plant as a whole avoids the excessive wear or clogging up of the compressor, because only air is compressed before it enters the air intake to the gas producer as shown in Figure 178. However, the entire plant will be under pressure instead of under suction which could cause some problems related to the safety and health of the driver.

The usual suction drive of automotive gas producers prevents poisonous gases from leaking through bad joints and fittings during normal operation. This is not the case in supercharged plants.

In summary, supercharging the engine or the gas producer is one method to recover most or even all of the power loss without alternating the internal combustion engine. This is shown in Figure 179 for two gasoline engines with a compression ratio of 4.5 and 7.5, respectively. How reliable supercharging of producer gas is remains to be seen and much more research and road trials must be done before any conclusive answer can be given.

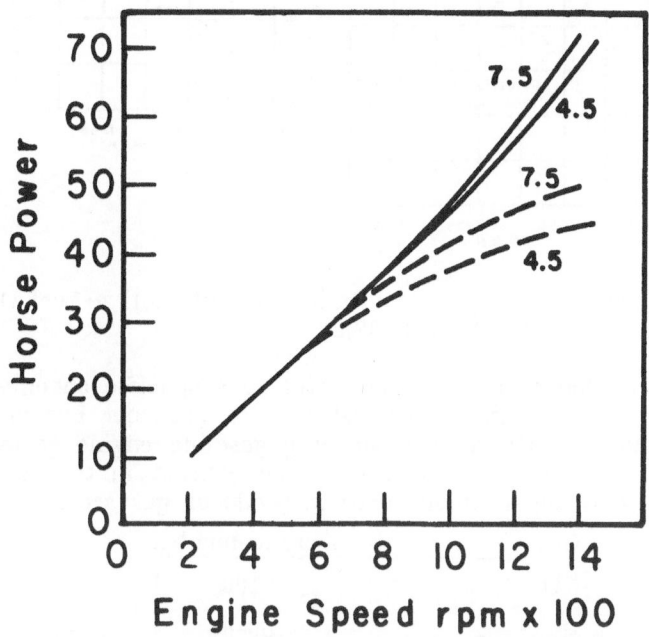

Figure 179. Power Increase of a Gasoline Engine with Unchanged Compression Ratio 1 to 4.5 and of a Gasoline Engine with Increased Compression Ratio 1 to 7.5 as a Function of the Speed, Both Driven by Wood Gas (22).

- - -   Without supercharging
———   With supercharging

Installing high pressure pistons in a gasoline engine is another technically feasible method to recover some of the power loss associated with producer gas drive. The advantages and disadvantages are best understood by looking at the thermodynamic behavior of gasoline and producer gas under high pressure and temperature. In normal combustion after the flame is ignited at the spark plug, the flame front travels across the chamber compressing the unburned gas ahead of it. The gas ahead of the flame spontaneously ignites under normal combustion conditions, resulting in a high-pitch knocking sound. It is well known that the occurence of knocking is closely related to the octane number of the fuel and the engine compression ratio. Figure 180 shows this trend in compression ratio and automotive fuel octane number over the years in the United States. When using producer gas as a fuel the compression ratio of a gasoline engine, usually limited by the danger of knocking, can be increased considerably, due to the higher octane number of producer gas. It can not be emphasized enough

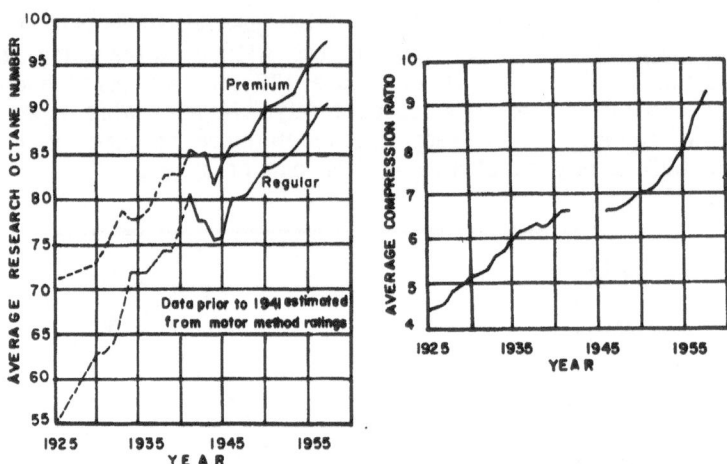

Figure 180. Trends in Compression Ratio and Automotive-Fuel Octane Number in the United States (37).

that the term producer gas does not refer to a specific chemical composition as is the case for gasoline. The behavior of producer gas under conditions prevailing in the combustion chamber of a gasoline engine varies considerably due to the wide range of possible hydrogen content in the gas. The octane numbers for various gases which occur in producer gas are given as:

Gas	Octane number
CO	105
H_2	60-66
CH_4	105

The table indicates the unsuitability of H_2 as a fuel in high compression engines. On the other hand, hydrogen in the producer gas is necessary to achieve a high heating value and even more important to increase the flame speed of the gas-air mixture and therefore to decrease the time the mixture needs for complete combustion. This is a very important fact and one of the reasons why producer gas is more efficiently used in low speed engines. Flame speeds of various gases as a function of their concentration in a gas-air mixture are given in Figure 182. From the graph it is clear that the flame speed of hydrogen is about ten times that of CH_4 or CO. The graph also indicates the flame speed of a representative producer gas-air mixture. Comparing this graph with

234

Figure 181 which shows the flame speed of gasoline as a function of the mass ratios, it is obvious that gasoline engines and in particular high-speed engines can not perform as well on producer gas-air mixtures.

Although the cited flame speeds of various gases and the producer gas-air mixture are based on laboratory tests and the actual flame speed in an engine is probably a magnitude higher, there still remains a large difference between a gasoline-air mixture and a producer gas-air mixture. Consequently, the ignition must be advanced to allow the flame to cross the combustion zone before the piston reaches the top center. This means a loss in area of the indicator diagram and therefore a loss in power and efficiency. The location of the spark plugs and the shape of the combustion chamber as well as the fuel-air ratio also have a pronounced effect on flame speed and ignition timing. Probably the most influencial factor on the ignition advancement is the volumetric hydrogen fraction in the producer gas. Depending on the fuel and the mode of running, this fraction amounts to from 2% to 20% of the producer gas. Hydrogen production in the gas producer depends primarily on the moisture content of the feed material and the partial combustion temperature and can, therefore, change drastically even when using the same type of fuel. Furthermore, the hydrogen content limits the increase of the compression ratio. The optimum ignition advance as a function of hydrogen content in the producer gas is shown in Figure 183 and is compared to gasoline operation.

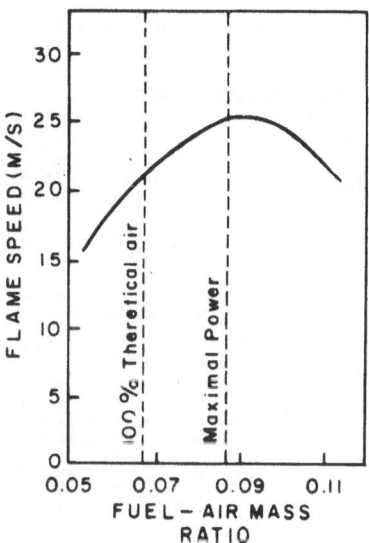

Figure 181. Representative Flame Speed of Gasoline as a Function of Its Mass Fraction in Air (36).

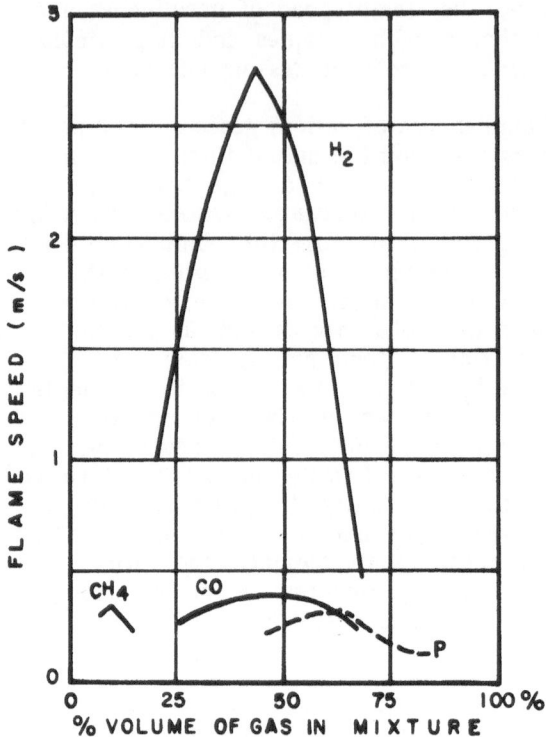

Figure 182. Flame Speed of Various Gases as a Function of Their Volume Fraction (7,26,32).

P = 4.4% CO_2; 29.1% CO; 10.2% H_2; 56.3% N_2

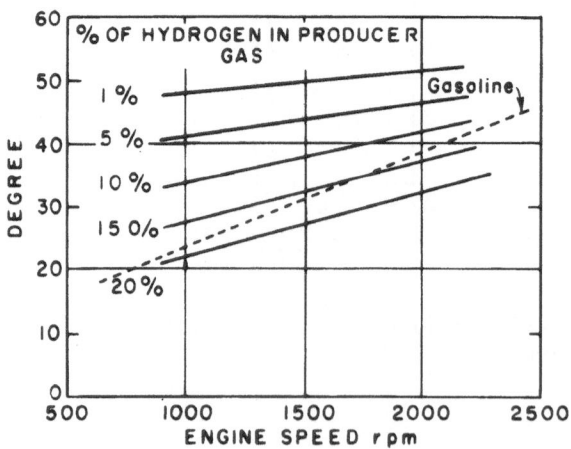

Figure 183. Hydrogen Content of Producer Gas and Ignition Advancement.

236

Although the ignition timing depends not only on the hydrogen content, the curves show quite well how difficult it is to operate a gasoline driven engine with a fixed ignition advancement over such a wide range of possible gas compositions. The ignition advancement for gasoline driven engines is usually 5-15 degrees and some compromise needs to be made regarding optimal performance of the engine. The graph also indicates that with a high hydrogen fraction, 15%-20%, the ignition advancement required using gasoline and producer gas are roughly the same.

A wide range of gas compositions with a seemingly uniform fuel is shown in Table 53 which lists average values and the range obtained with a charcoal fired crossdraft gas producer. In order to achieve maximum performance at various loads the ignition advancement had to cover a range of 35 to 57 degrees (Figure 184).

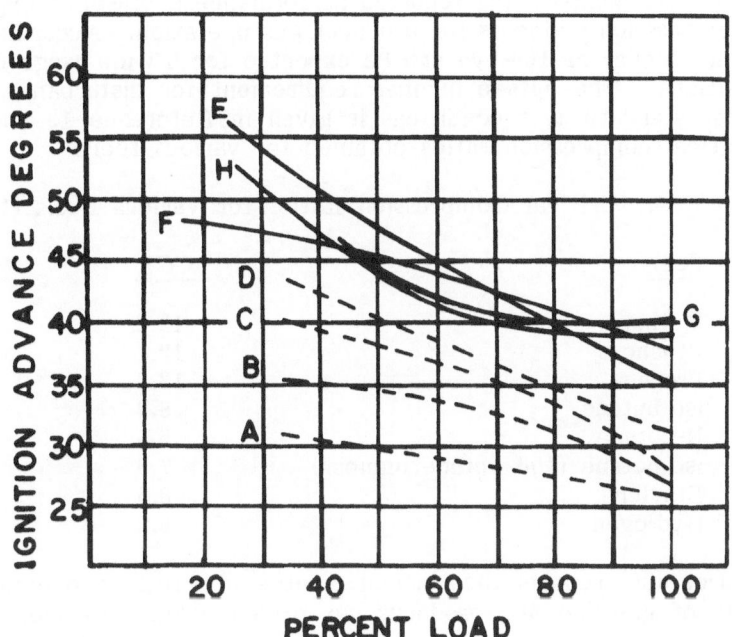

Figure 184.   Ignition Advance as a Function of Load for Gasoline and Gas Drive (24).

A,E	900 rpm
B,F	1200 rpm
C,G	1500 rpm
D,H	1800 rpm

———— Charcoal   - - - - Gasoline

Table 53. Gas Composition in a Charcoal Fired Crossdraft Gas Producer (24).

Gas	Average	Range
CO_2	1.8	0.8 to 4.1
O_2	1.4	0.1 to 2.3
H_2	5.2	0.3 to 13.0
CH_4	1.8	0.0 to 7.0
CO	28.2	21.3 to 30.4
N_2	62.0	52.7 to 67.9

It is illustrative to compare the reported performance of gasoline engines fitted with high compression cylinders for producer gas operation. According to Table 54, an octane number of 100-105 can be expected for a wide range of producer gas compositions. The octane number requirement for disturbance free combustion under standard test conditions is given in Reference 13 and Table 54 lists the critical compression ratios obtained for various fuels.

Table 54. Critical Compression Ratios for Various Fuels (13).

Fuel	CCR
Methane	12.6
Ethane	12.4
Propane	12.2
Iso-butane	8.0
N-butane	5.5
Iso-octane (100 octane number)	7.3
Ethylene	8.5
Hydrogen	8.2

Another source (26) reports the critical compression ratio and octane number requirements of gasoline and producer gas driven engines as shown in Figure 185.

All engines tested before 1950 had compression ratios of 5 or slightly lower. Summarizing all the data about increasing the compression ratio of gasoline engines contained in the reviewed literature, the following can be concluded:

Bench tests were conducted with compression ratios increased up to 16.2, but commercially built automotive gas producers usually operated at compression ratios between 6.5 to 7.5 (12,33). The increase in power output due to increased compression ratios is much more pronounced at higher heating values of the gas. It needs to be pointed out that a higher heating value of the gas is in almost all cases associated with a higher hydrogen content and not with increased methane or carbon monoxide generation. This applies in particular to the most common fuels used, such as charcoal, wood, and high grade coals. On the other hand, high hydrogen content is determimental for high compression ratios. It is therefore technically not feasible and uneconomical to increase the compression ratio above 10. In addition, because of the asymptotic behavior of the thermal efficiency versus compression ratio curve the gain in power from higher compression ratios decreases rapidly. Each increase in the compression ratio will result in higher friction which offsets some of the gains. A low hydrogen content can not be guaranteed in downdraft gas producers and is also not desirable for several other reasons such as lower heating value and flame speed. Hydrogen contents above 10% may require a highly retarded ignition timing at high compression ratios and most of the power gained will be lost. Besides the pure thermodynamic considerations there are some problems concerning the life and ease of operation related to very high compression ratios. High compression engines are much more difficult to start, making hand starting almost impossible. The strain and wear on pistons and the ignition system is considerably greater and have resulted in malfunction and short life of the equipment.

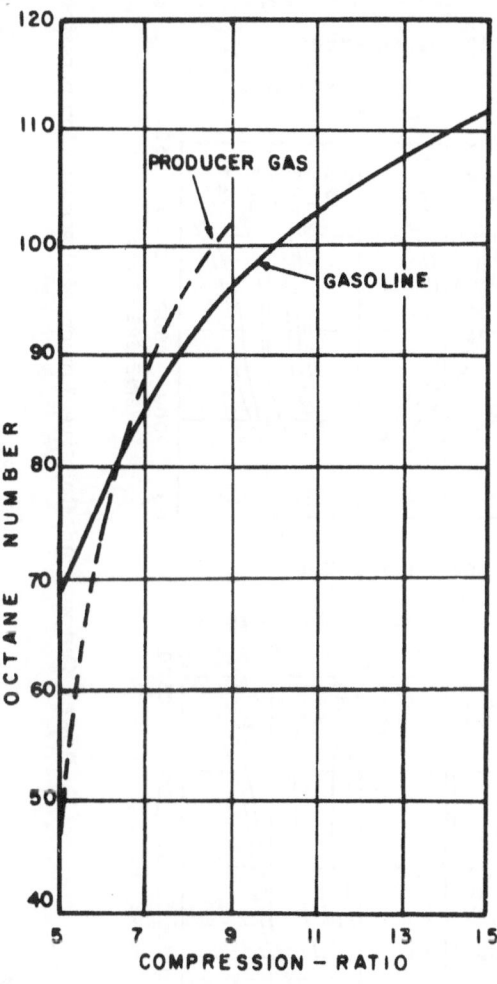

Figure 185.   Octane Number Requirement for Disturbance-Free Combustion at Various Compression Ratios (26).

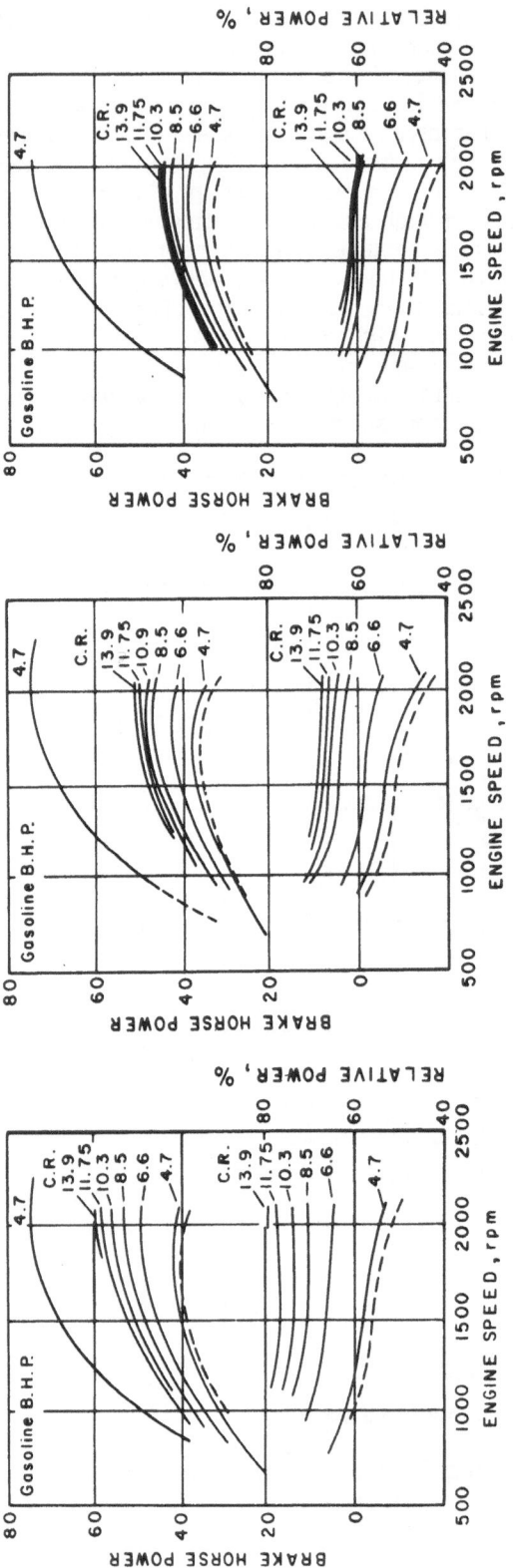

Power Output and Relative Power at Various Compression-Ratios.

Figure 186
Gas Heating Value 6.9 MJ/Nm³
Gasoline-Air Mixture 4.1 MJ/Nm³
Producer Gas-Air Mixture 2.8 MJ/Nm³

Figure 187
Gas Heating Value 6 MJ/Nm³
Gasoline-Air Mixture 4.1 MJ/Nm³
Producer Gas-Air Mixture 2.5 MJ/Nm³

Figure 188
Gas Heating Value 4.5 MJ/Nm³
Gasoline-Air Mixture 4.1 MJ/Nm³
Producer Gas-Air Mixture 2.3 MJ/Nm³

Once converted, the gasoline engine can no longer be operated with gasoline for extended periods. This is a very important point because gasoline was usually used to start the engine and supply the necessary suction for the gas producer air blast. As already pointed out, the initially generated gas from a gas producer cannot be used to drive the engine and must be flared off behind the cyclone in order to avoid serious damage to the purification system and the engine. In the case of a high compression engine, the gas producer must be equipped with a blower to provide the necessary power for the air blast during the first few minutes of operation.

We are not aware of any long-term tests conducted with modern high speed gasoline engines with compression ratios of 10, and our limited knowledge about engines manufactured in Third World Countries does not allow any final conclusions concerning the alteration of the compression ratio for these gasoline engines. Figures 186 to 188 show the power output of a gasoline engine with an initial compression of 4.7 when fueled with producer gas of various heating values at compression ratios from 4.7 to 13.9. The dotted curves represent the performance of an engine at 4.7 compression ratio when fitted with pistons giving the same friction losses comparable with those measured with the high compression pistons. As shown in the graphs only 80% of the gasoline power can be achieved under most favorable conditions. It is obviously difficult to say how much power loss is attributed to bad design of induction pipes, mixing valves and piping and how much is due to the fact that a gasoline engine is designed to run on gasoline and not on producer gas.

How much the power loss varies with engine type, good or bad induction piping, mixing valve design, generator type and compression ratio is best shown in Figure 189. Attention should be given to curves J and D which show the difference in power output obtained from the same engine but with different induction pipes and mixing valves. The observed increase in power of 15% is considerable and clearly indicates that the loss in induction pipes and engine manifold designed for gasoline-air mixtures is not negligible.

There are differences in controlling the producer gas-air mixture compared to the gasoline-air mixture in actual driving. The gasoline air-mixture is automatically adjusted by the carburetor and controlled with the accelerator. The only manual device is the choke for cold start. In producer gas driven automobiles or stationary units one has the choice between automatic, semi-automatic or hand controlled operation. Which system should be used is a matter of convenience and level of training. Finding the correct gas-air ratio is more difficult in producer gas driven engines for two reasons: 1. The gas composition will change over a run, sometimes drastically; and 2. The power output curve as a function of the percent theoretical air has a very pronounced sharp peak unknown in gasoline operation (Figure 190). This means the correct mixture is more difficult to adjust and a seemingly marginal change in the opening of the air intake valve can cause a significant power drop.

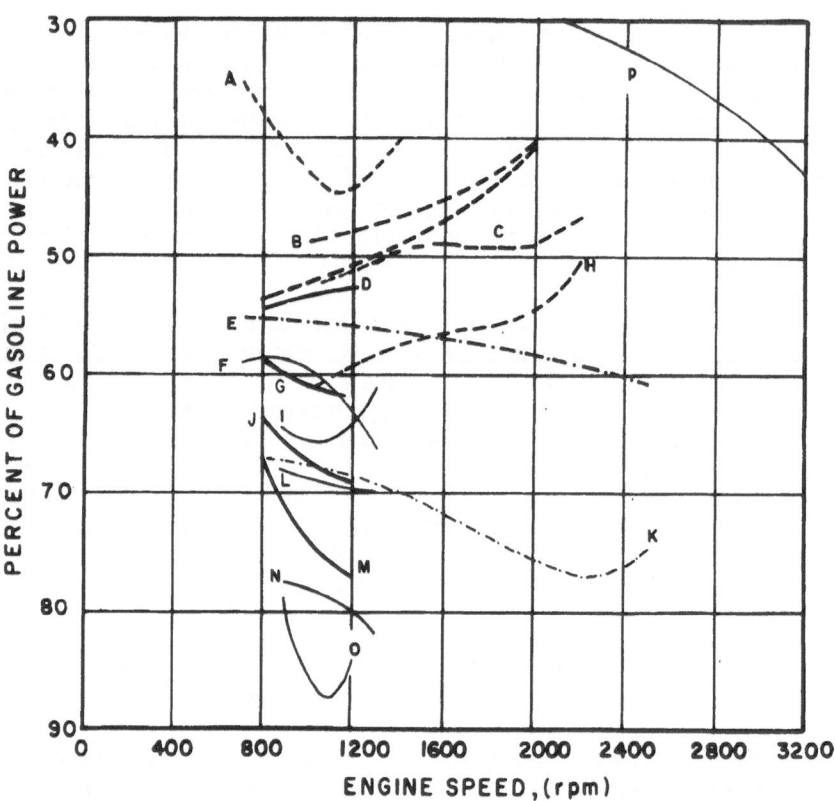

Figure 189. Power Drop in Converted Gasoline Engines (32).

Curve	Engine	Compression Ratio	Gas Producer
A	Arbenzmotor	3.9	Pava
B	Chevrolet 50 PS	4.22	Widegren
C	Ford 40 PS	5.22	Widegren
D	Bussing 90 PS (ill designed mixing valve and engine manifold)	5.6	Imbert
E	Chevrolet 30 PS	4.64	Imbert
F	Hanomag R 28	4.98	Imbert
G	Bussing 90 PS (manifold ill designed, well designed mixing valve)	5.6	Imbert
H	Ford 40 PS	7.0	Widegren
I	Kamper 52 PS	5.17	Imbert
J	Bussing 90 PS (well designed manifold and mixing valve)	5.6	Imbert
K	Chevrolet 30 PS	6	Imbert
L	Kamper 52 PS	6.89	Imbert
M	Bussing 90 PS	8.2	Imbert
N	Kamper 52 PS	8.89	Imbert
O	Kamper 52 PS (Supercharged)	6.89	Kromag
P	D.K.W. (Two stroke engine)	5.88	Oberbexbacher

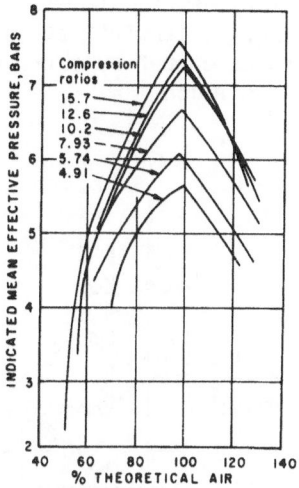

Figure 190. Indicated Mean Effective Pressure Versus Percent Theoretical Air (20).

One of the early mixing valve designs is shown in Figure 191. It consists of two separate butterfly valves for the control of air and producer gas. The gas valve is operated with the accelerator pedal while the air valve is hand operated. Although the design is simple, it is rather effective and allows a good control of the power output, provided the operator gets used to the new driving style.

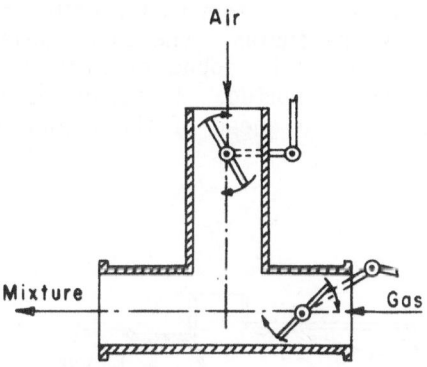

Figure 191. Hand-Operated Mixing Valve.

Several attempts have been made to make the gas-air mixture control more convenient. One design, a modified form of the first one, is shown in Figure 192. Throttle "a" is mechanically linked to the air throttle "b" and both are controlled by the accelerator. Throttle "c" is separately operated by hand to adjust the mixture ratio for maximum power conditions with full throttle. This design did not work very well in a dual fueled diesel engine because the partial closure of the gas throttle resulted in an enrichment of the mixture associated with an increase in exhaust smoke.

243

The small orifice hole, "d", provides the necessary air at idling where pilot diesel oil was injected. At idling, throttle "b" was completely closed and throttle "a" only slightly opened to provide enough suction for some gas flow to prevent the gasifier from cooling off too much.

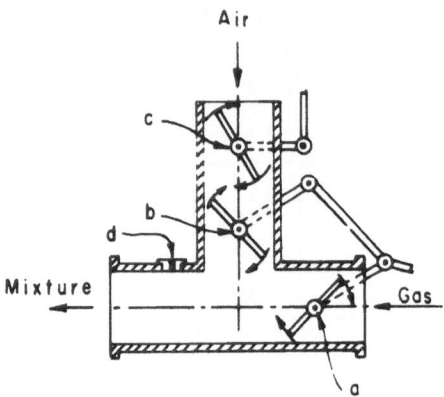

Figure 192. Semi-Automatic Mixing Valve (41).

A fully automated mixing valve is shown in Figure 193. The gas and air flow is fully controlled from the accelerator pedal by a butterfly valve. The air enters the mixing valve through the flap valve at high velocity which guarantees a turbulent mixing. The opening of the flap valve is governed by an increase in the depression inside the piping system that usually occurs with an increased demand in producer gas by the engine. The characteristic dimensions of such a mixing valve must, of course, be found by a trial and error method, but satisfactory performance was reported. It should be noted that the gas-air mixing arrangements are located ahead of the engine carburetor and do not replace this device.

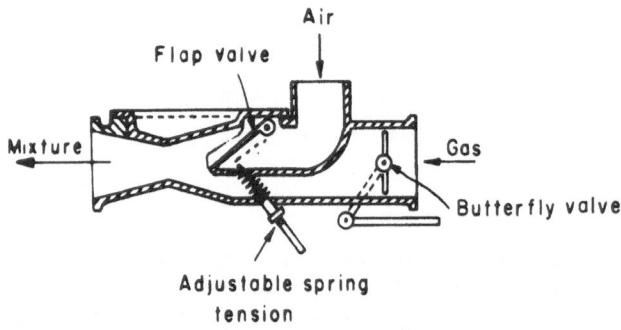

Figure 193. Fully Automated Mixing Valve (41).

Another more recent design for a small engine and a theoretical treatment on its performance is given in Reference 35.

The most widely used method to increase the power output and improve the convenience of driving the automotive gas producer was to dual fuel the gasoline engine whenever necessary. Dual fueling an engine means the simultaneous injection of the gas-air mixture with small amounts of gasoline mixed to the gas-air stream in the carburetor. The degree of dual fueling depends on the engine load and how gasoline independent the producer-engine set needs to be. Three general methods were in wide use:

1. Dual fueling on a continuous basis meaning a small amount of gasoline was continuously injected into the gas-air mixture in the carburetor.

2. Starting the engine on gasoline and, after the gas producer was working properly, switching over to producer gas-air mixture operation.

3. Dual fueling the engine only when additional power was needed on hills or under heavy load and the gas producer could not provide the additional power.

Figures 194 and 195 show the arrangements for option 2. Details of the idling air valve and load air valve are shown in Figures 196 and 197. The system is semi-automatic and similar to the one in Figure 192. The change over from gasoline to producer gas was accomplished by a screw valve.

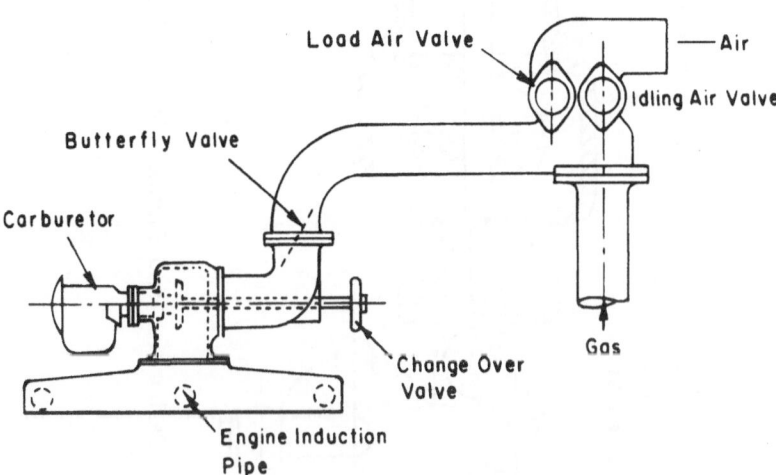

Figure 194. Carburetor and Gas-Air Mixing Arrangements for Gasoline Engines (41).

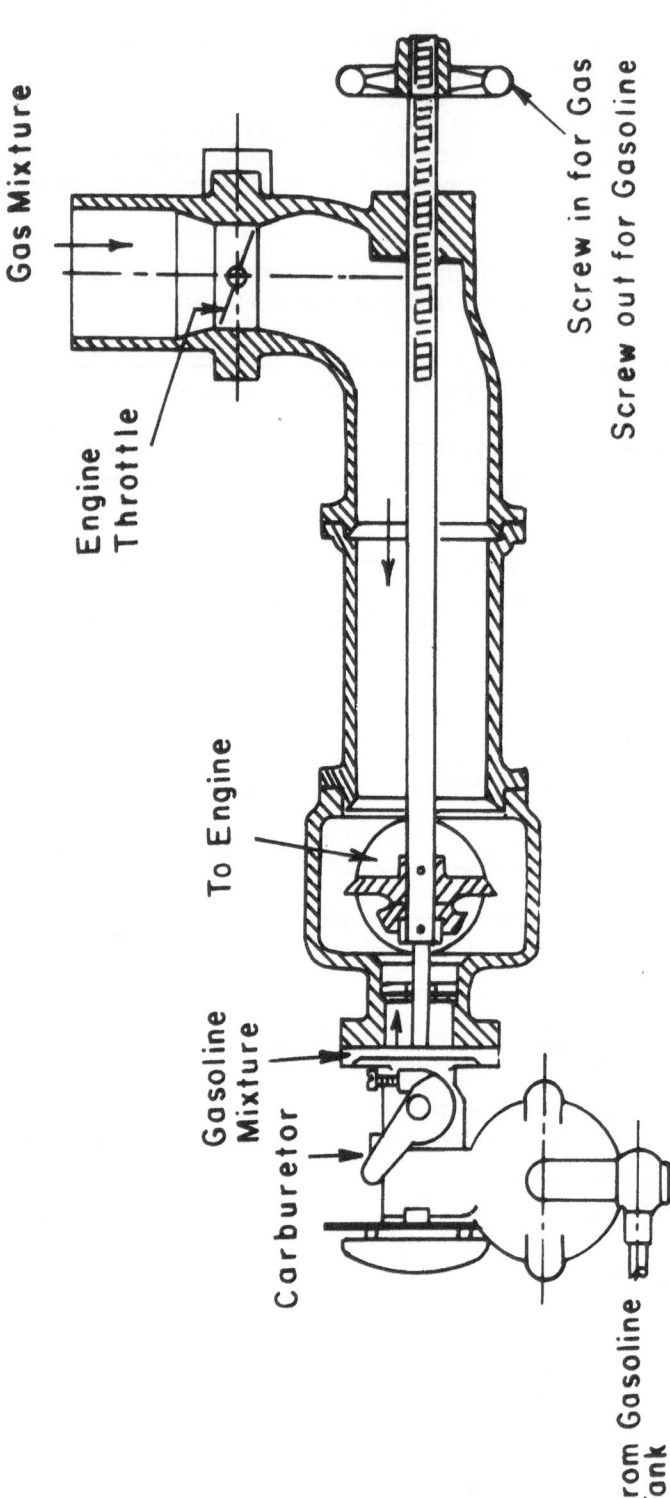

Gas Mixture

Engine
Throttle

Screw in for Gas

Screw out for Gasoline

To Engine

Gasoline
Mixture

Carburetor

From Gasoline
Tank

Figure 195. Detail of Carburetor and Engine Throttle (41).

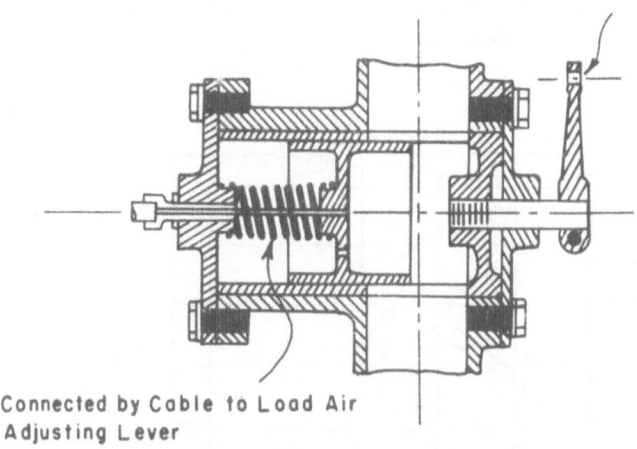

Connected to Engine Throttle

Connected by Cable to Load Air
Adjusting Lever

Figure 196. Load Air Valve Double Barrel Type. Right Barrel is Connected to
the Engine Butterfly Throttle (8).

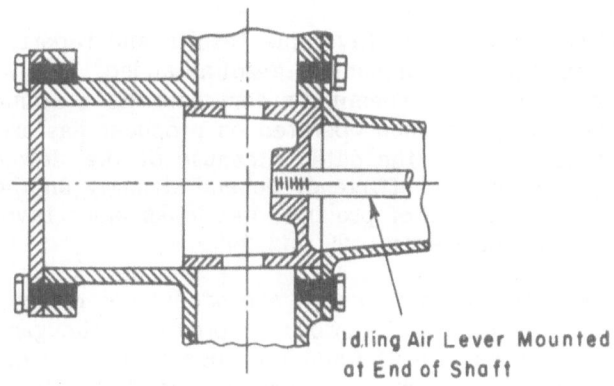

Idling Air Lever Mounted
at End of Shaft

Figure 197. Idling Air Valve (8).

It is not possible to recover all the power loss by dual fueling the engine.
However, there are many reasons why dual fueling is a good compromise between
gasoline savings, convenience and ease of operation. For instance, buses of the
Highland Transport Company equipped with the H.M.L. producer on a trailer
were dual fueled with 15% of the gasoline used before conversion. The benefits
were an increased power output of 28% and maintenance of the standard bus
schedule. The rather elaborate injection system provided the bus with gasoline
only when it was needed on hills or acceleration, but not during downhill driving
or idling (38). How much power can be gained through addition of gasoline is
not just a matter of theoretical calculations because the gasoline-producer gas-air

mixture will behave differently in various combustion chambers. As indicated in Figure 198 about 87% of the original power can be restored when adding 27% of the gasoline used before conversion.

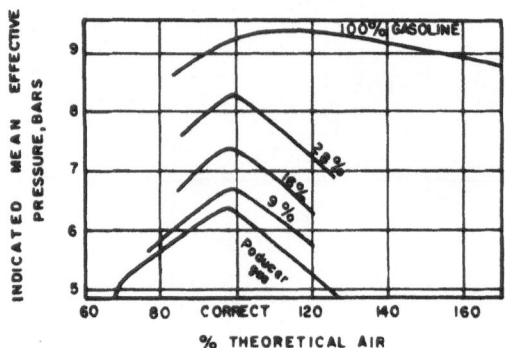

Figure 198. Power Output for Various Amounts of Gasoline Added (20).

Others (1,17) did not obtain such favorable results and report 60% to 70% of the original power for the same amount of gasoline added. The wide discrepancy in reported power losses and seemingly contradicting statements about the efficiency of gasoline engines when operated on producer gas are mostly due to different methods of reporting the data. Because of the slowness of producer gas combustion, power loss and efficiency depend strongly on the engine speed. At lower engine speeds the use of producer gas looks more favorable compared to gasoline as indicated in Figures 199 and 200.

It is difficult to judge the performance of a producer gas driven gasoline engine without relating the results to the producer providing the gas. There was a certain need to test and compare the major European makes, since gas producer-engine sets did not always stand up to the promises made by their manufacturers. The task of comparison was undertaken by the Subcommittee on Producer Gas of the Associate Committee on Substitute Fuels for Mobile Internal Combustion Engines of the National Research Council of Canada. The report (1) describes stationary and road tests to measure power, economy, ease of handling, durability and other characteristics under various conditions.

The published results are unique in the sense that all gas producers were tested with the same engine type and truck. The reported data are instructive and precise. A classification of the tested producers is given and although the makes were not revealed and gas producers labeled with letters, it is not difficult to conclude that a crossdraft type manufactured by the British Gazogenes Ltd. and described in Reference 6 was rated number one. The second rated producer was the Swedish Swedlund downdraft producer manufactured by the Gas Generator Co. in Orebro and described in Reference 7.

248

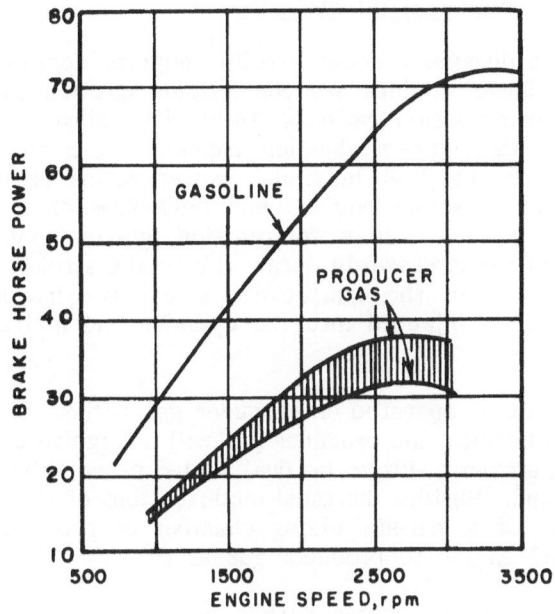

Figure 199. Power Output as a Function of Engine Speed (5).

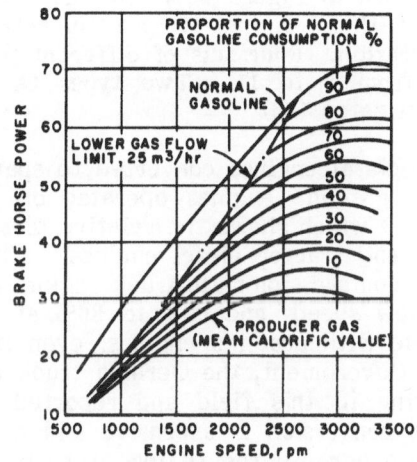

Figure 200. Power Output as a Function of Percent Gasoline Injected (5).

249

Conversion of diesel engine to producer gas:

Most of the previous discussion about gasoline engines applies to diesel engines as well. However, diesel engines are compression ignition engines and operate at a much higher compression ratio of 16-20 depending on whether they are direct injection chamber, pre-combustion chamber, four stroke or two stroke engines. Their piston speed at maximum power rating is about the same as industrial and tractor gasoline engines and only 70% of the piston speed of automobile gasoline engines. In a compression ignition engine, usually a full unthrottled charge of air is drawn in during the intake stroke. The temperature of the air near the end of the compression stroke is quite high. Just before top center, diesel oil is injected into the cylinder and ignited by the high air temperatures.

A diesel engine cannot be operated on producer gas without injection of a small amount of diesel oil because the producer gas will not ignite under the prevailing pressure. A diesel engine needs to be dual fueled or completely converted into a spark ignition engine. Besides the usual modifications of the induction manifold and the installment of a gas-air mixing chamber as previously described, one can convert a diesel engine to producer gas as follows:

1. Rebuilding of the entire engine with a new piston and new cylinder head and installment of electric ignition equipment. This kind of conversion is expensive and time consuming. Nevertheless it has been done successfully. Figure 201 shows the modified cylinder head of a 6 cylinder, 95 hp, truck diesel engine with an original compression ratio of 17 and 7.6 after conversion. Another design is shown in Figure 202. In this case, a 3 cylinder, 56 hp, tractor diesel engine with an original compression ratio of 16.5 was converted to spark ignition operation. Four sets of different pistons were tested with compression ratios from 9 to 11. Two types (A and B) of combustion chambers were investigated (26).

   The power drop in diesel engines converted to spark ignition operation is not as severe as in gasoline engines operated on producer gas. This is indicated in Figure 203 which shows the relative diesel power obtained with the converted truck and tractor diesel engines. Although the power drop is larger in the lower compression ratio truck engine, a relative power output of 70% to 85% at low speeds and 60% to 80% at high speeds is a result not readily obtainable with gasoline engines, even if they are dual fueled. Besides the Swedish Government, the German truck manufacturer MAN has done extensive testing in this field and reported similar results. It is emphasized that the conversion is expensive and does not always give the most favorable results because the fitting of a spark plug in the location previously occupied by the fuel injector nozzle may not be the best place in each case. Special attention should be paid to the spark plugs which are under an additional heat strain and need to be replaced by ones with lower heat values.

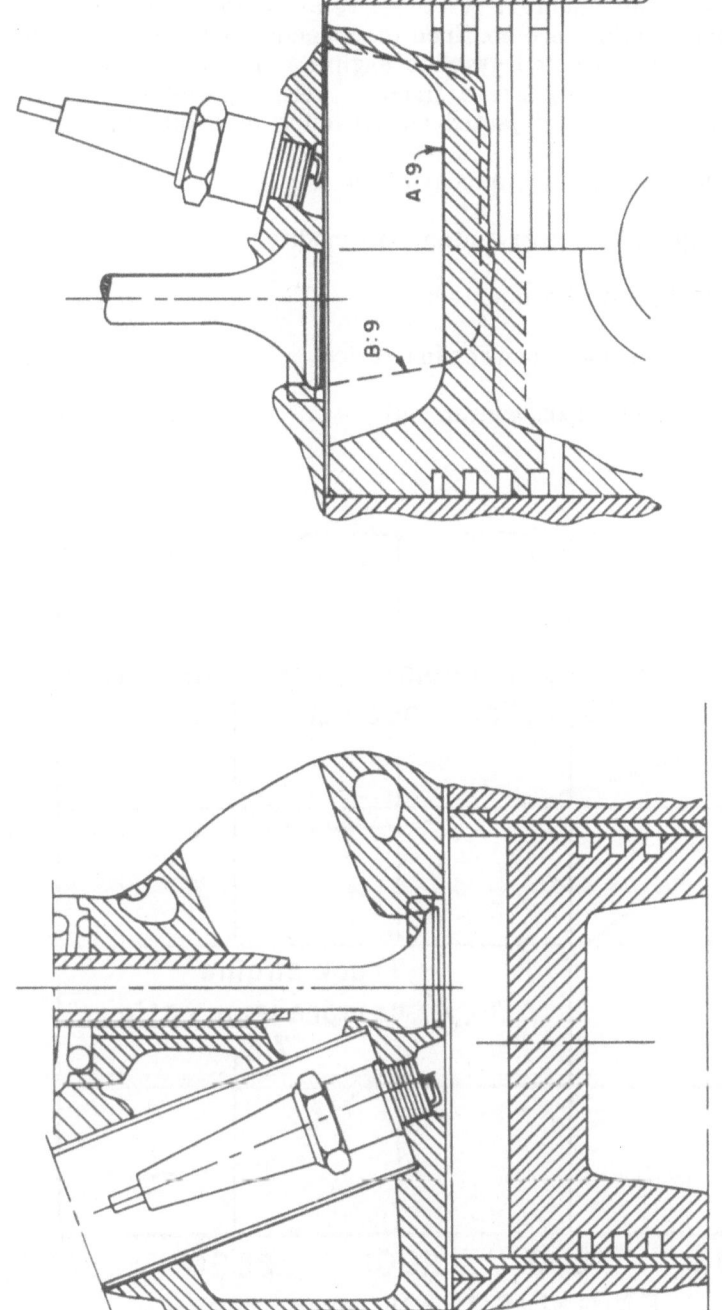

Figure 202. Converted Truck Diesel Engine (26).

Figure 201. Converted Truck Diesel Engine (26).

2. An alternative method of effecting diesel conversion for the use of producer gas is by retaining the existing compression ratio and arranging for dual fueling. In this case the fuel injection system is retained together with the original pistons and modifications are confined to a special induction manifold and a gas-air mixer as in converted gasoline engines. The injection pump needs to be modified to accommodate a fixed or variable amount of fuel injection smaller than the amount injected during idling of a diesel engine.

The main questions associated with such a conversion are:

1. What type of diesel engines are most suitable to modify?

2. How much diesel fuel needs to be injected?

3. Does knocking occur at the high compression ratios?

4. How severe is the power loss upon conversion?

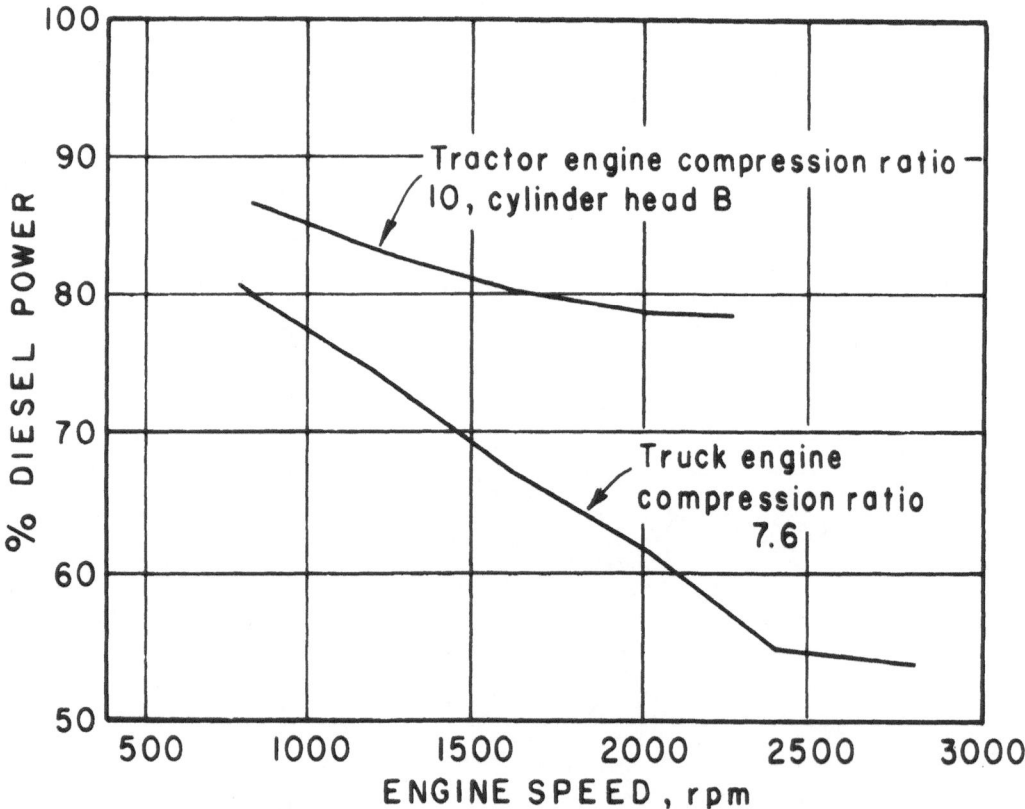

Figure 203. Power Output of Diesel Engines Operated with Producer Gas (26).

252

Not all diesel engines are suitable for this kind of conversion due to their compression ratios and the shape of the combustion chamber. Diesel engines are manufactured in three types: direct injection, turbulence chamber and ante-chamber engines. Direct injection engines, although they are working at high compression ratios compared to gasoline engines, are more suitable and do not require special low compression ratios as long as the compression ratio does not exceed 16 to 17. Ante-chamber and turbulence chamber engines are more difficult to convert. Their compression ratios are higher, up to 21, and need to be reduced to 16 or lower. Experiments conducted with unconverted engines of this type were very unsatisfactory and it was concluded, that a major rebuilding of the engine was necessary before they could be used for producer gas operation. The conversion to dual fueled engines would be as expensive and time consuming as a conversion into spark ignition engines (26).

The conversion of direct injection engines to dual fueled engines is well documented and the various test results are published in References 4,7,14,18,19,23,26,27,38. Figure 204 presents the power output and diesel oil consumption of a six cylinder diesel engine with a compression ratio of 16. The same engine converted to dual fuel at a compression ratio of 16 was then operated on producer gas with diesel pilot injections of various amounts. The results are shown in Figure 205 as a function of the heating value of the producer gas. Comparing Figures 204 and 205 one can conclude that a marginal power loss of 5% to 10% was reported, depending on the heating value of the producer gas. The pilot injection of diesel oil amounted to 16% to 28% of the original consumption or 10 mm^3 to 17.5 mm^3 per cycle.

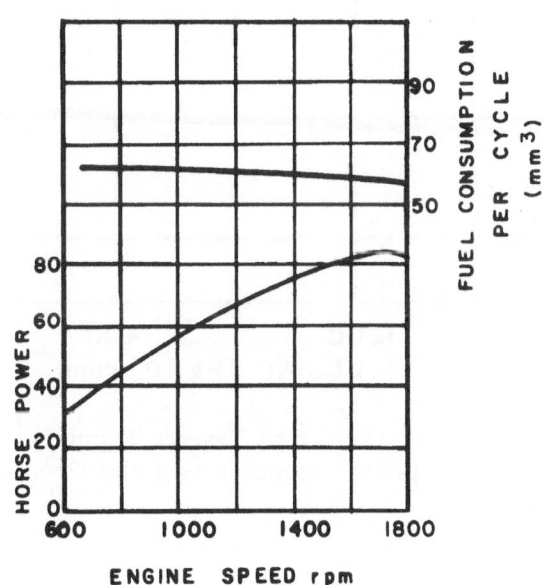

Figure 204. Performance of Unconverted Six Cylinder Diesel Engine (14).

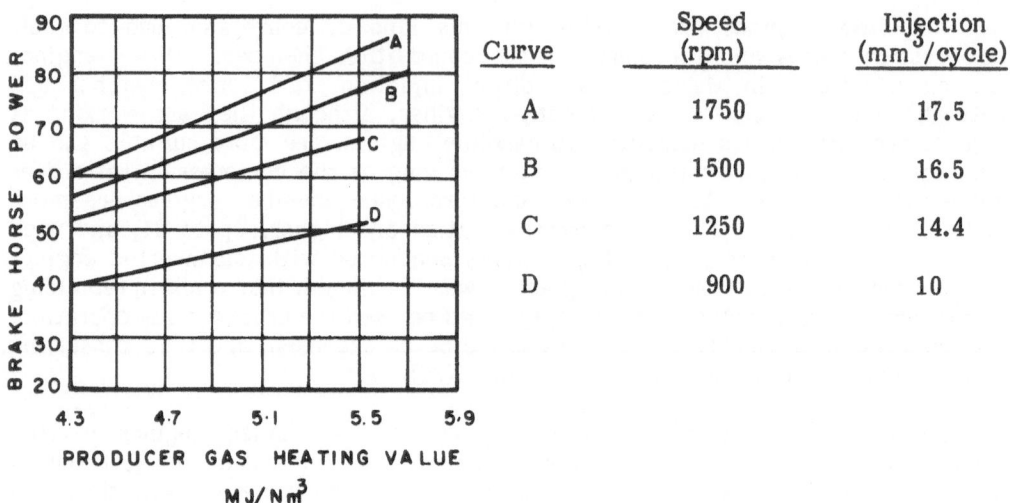

Figure 205. Performance of Six Cylinder Diesel Engine After Conversion (14).

Similar results were obtained in Swedish tests. Figure 206 shows the percent of the original diesel power obtained from a dual-fueled, 3.6 liter, tractor engine. The peformance of a 4 cylinder, 6.2 liter, truck engine is given in Figure 207.

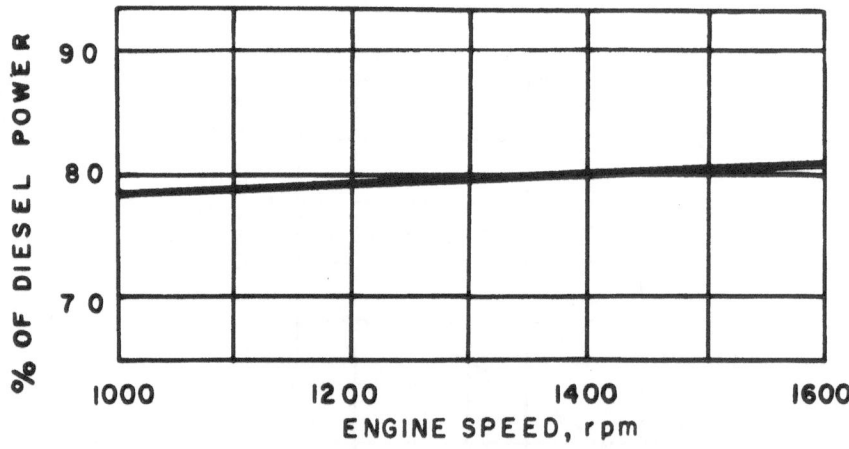

Figure 206. Power Output of Converted Tractor Engine with Diesel Oil Pilot Injection of 29 g/Nm^3 Producer Gas and Good Gas Quality of 5.4 MJ/Nm^3 (26).

All reports indicate that the power loss in dual-fueled diesel engines is by far much less than in dual-fueled gasoline engines, due to the higher compression ratios. It can be assumed that at least 80% of the original power can be restored.

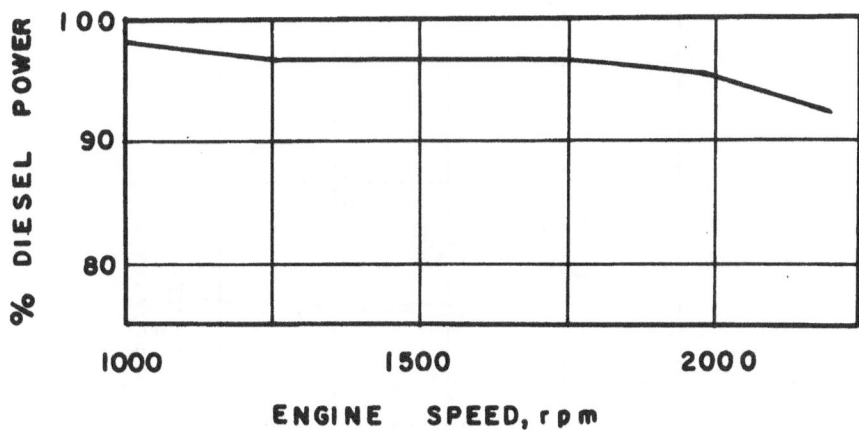

Figure 207.   Power Output of Converted Truck Engine with Diesel Oil Pilot Injection of 12-19 g/Nm^3 and Good Gas Quality of 5.5 MJ/Nm^3 (26).

It is of interest to point out that the amount of diesel oil injected can be very low. Going back to Figure 200, one can see that 70% of the original gasoline consumption must be injected in a dual-fueled gasoline engine in order to recover about 82% of the gasoline power at average engine speed. Surprisingly enough in diesel engines, the pilot injection of diesel oil is first done to guarantee the compression ignition of the producer gas-air mixture and stimulate a smooth combustion. Its effect on the power output is secondary to a large extent. The diesel amount necessary to guarantee ignition of the mixture and the amount injected for normal running conditions in a 3 cylinder, 3.4 liter, tractor engine is shown in Figure 208. The usual diesel oil consumption of this engine is 62 mm^3 per cycle. Consequently, only 8% to 16% of the original diesel oil is needed for satisfactory performance of the dual-fueled engine. Others reported somewhat higher numbers between 10% and 25% (4,14). For economical reasons it is best to inject only the amount of diesel fuel that is necessary for smooth operation of the engine. Additional amounts of diesel fuel do not have the desired effect of a significant increase in power at lower speeds. The better power output at higher engine speeds is also by no means proportional to the diesel fuel injection. In a particular case shown in Figure 209, the increase in pilot injection of diesel oil of 60% resulted only in a power increase of 7% at high speeds and 1% to 3% at partial load.

The difficulties with a proper injection timing in dual-fueled diesel engines are the same as with a proper ignition timing in producer gas operated gasoline engines. In either case, a fixed injection or ignition timing is just a compromise between bad combustion and rough running. All known reports reviewed agree

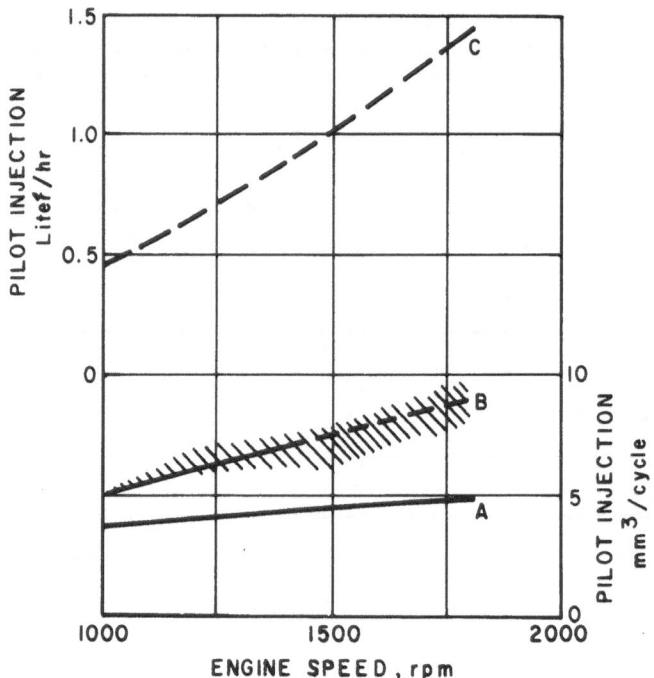

Figure 208. Amount of Pilot Injection for a 3 Cylinder, 3.4 Liter, Tractor Engine (26).

A: Necessary amount of guarantee ignition (mm^3/cycle).
B: Amount injected for normal running conditions (mm^3/cycle).
C: Amount injected in liter per hour under normal running conditions.

that the injection of the diesel oil must be advanced. There is little advantage in a variable injection time control because it complicates the entire system even more. Both past and recent experiments found an advanced injection timing of 30-35 degrees as a good compromise, and it should be emphasized that these numbers are only rough guidelines and the most proper timing must be found through trials in each particular case. One of the pecularities of dual-fueled diesel engines is their sensitivity to any change in injection timing when operated at high engine speeds. In these cases, misfiring, knocking and loss of power resulted. This is illustrated in Figure 210. At low engine speeds any change in injection timing does not have much influence on the power output, whereas at higher speeds a large advancement of the injection timing is necessary to obtain full power and this usually goes hand-in-hand with misfiring.

Modifications and the operation of an internal combustion engine fueled with producer gas are greatly simplified in stationary units which operate under constant load. In such cases, ignition and injection timing can be optimized and there are no difficulties to set the proper producer gas-air mixture. The gas producer can be kept in a semi-equilibrium state, only interrupted through batch feeding the unit. There will be some problems with a gas producer for transportation vehicles such as passenger cars, off-road vehicles, and vehicles operating under various loads or in difficult areas.

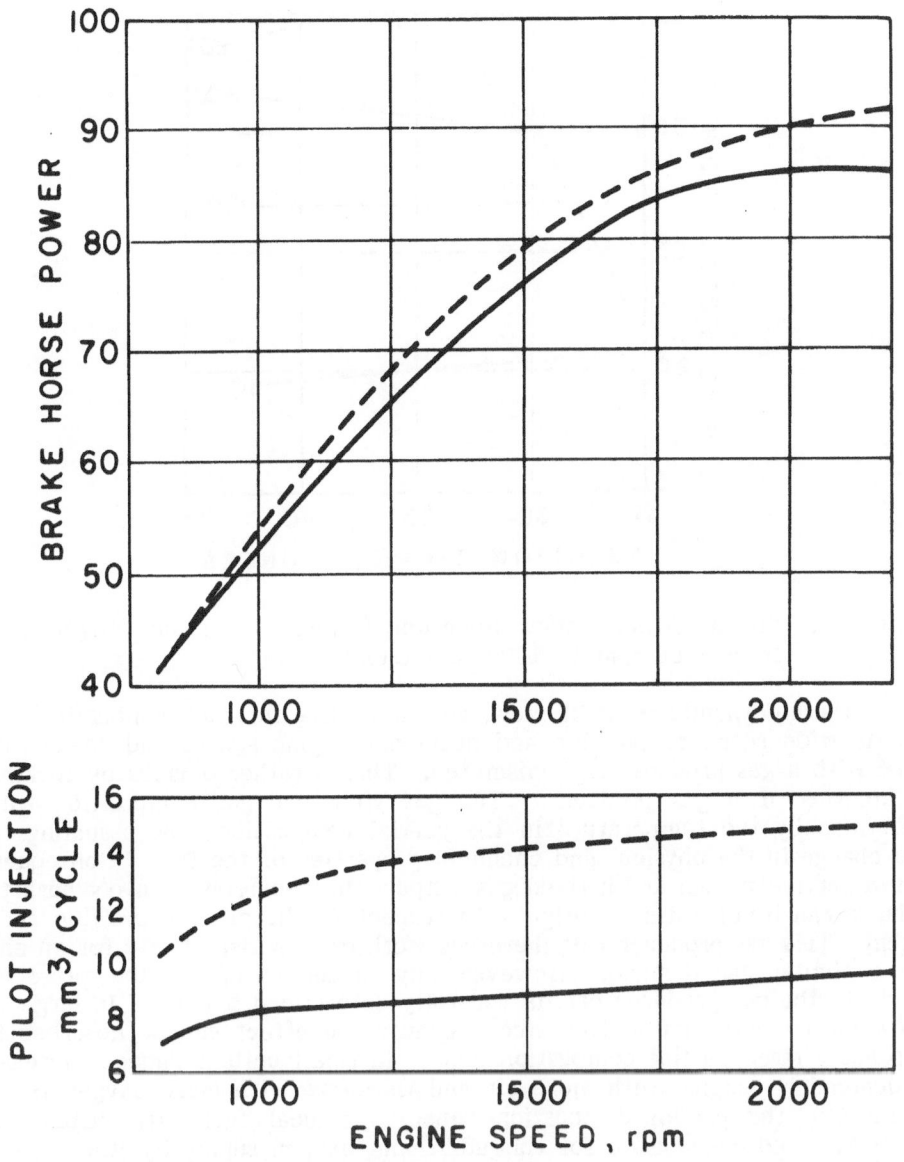

Figure 209. Effect of Increased Pilot Injection on Power Output at Various Engine Speeds (26).

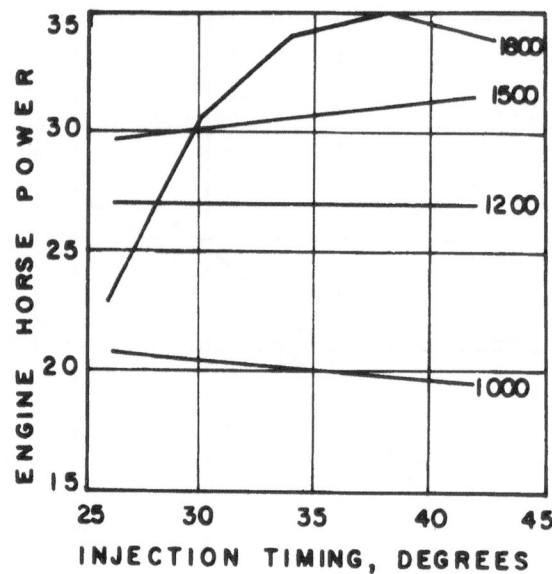

Figure 210. Engine Power Versus Injection Timing for Various Engine Speeds from 1000 rpm to 1800 rpm (26).

From a thermodynamic point of view, an automotive internal combustion engine with its wide range of possible and necessary engine speeds and power output fueled with a gas producer is a mismatch. This is rather obvious by looking at the kinetics of a gas producer. The gas yield and gas composition will be determined by the temperature in the partial combustion zone assuming there is no change in the physical and chemical properties of the fuel. Consequently, given a particular fuel and a fixed gas output, the geometry of a downdraft gas producer can be of optimal design with respect to the gas composition and tar content. This gas producer will therefore work quite satisfactorily for an engine with a similar gas demand. However, any sudden change in the gas demand will throw the gas producer off its carefully determined balance. In large units such a change does not matter much because the effect can be absorbed by a comparably large partial combustion zone. In small units a sudden increase in gas demand is coupled with more air and therefore with more oxygen per unit volume into the partial combustion zone. An ideal fuel with instantaneous reactivity could compensate for this additional oxygen supply by just expanding the partial combustion zone. In this case, there would be little change in the quality of the gas. The oxygen supply per unit volume remains constant, the reduction zone may be reduced depending on which side the fire will spread and most important, the temperature in the partial combustion zone stays the same. In a theoretical sense, the gas producer has only shifted to a different gear but not to a different equilibrium state. In practice this does not happen since there are no ideal reactive fuels.

Any sudden increase in the gas demand and therefore the oxygen supply to the partial combustion zone pushes the gasification process towards complete combustion either globally or locally if there is an inherent danger for channeling. A hot spot, where the useful chemical energy has shifted to useless sensible heat in the gas, will be the result. Any sudden decrease in the gas demand will be followed by cooling down the partial combustion zone instead of a shrinkage. Condensation and a low H_2O conversion will occur resulting in a wet gas and increased tar content. To make up for this, the ideal gas producer should change its important geometric parameters according to the gas demand. This is impractical. Some of the mentioned drawbacks can be partially eliminated through the use of specially prepared fuels such as charcoal or chemically treated fuels to increase the reactivity.

Highly controversial opinions exist concerning the engine wear, long-term effects on the engine and the engine exhaust. When dealing with these kinds of questions one should keep in mind the disorderly transition from fossil fuels to producer gas during WW II when most of the information was compiled.

Some drivers did not even make it out of their garage because the engine was totally clogged up with tar. Others ruined the engine within the first 50 miles of driving. Vice versa, there are reports about trucks operating up to 300,000 km on producer gas over a period of 4 years with less engine wear than with gasoline. The human element and frequent cleaning of the gas clean-up equipment seems to be the decisive factor concerning the engine wear and long-term effects. Table 55 lists average results obtained with various fuels.

Table 55. Cylinder Wear after 1000 km When Using Various Feed Materials (15).

Fuel	mm per 1000 km
Wood	0.003
Charcoal	0.006
Anthracite	0.009
Lignite coke	0.022
Coal coke	0.018
Peat coke	0.019
Lignite briquettes	0.03

The average cylinder wear of comparable gasoline engines during this time is given in Figure 211. It can be seen in this figure that the values do not differ much. Wear on engine parts is usually caused by abrasion and corrosion, the latter being predominant at low wall temperatures. Producer gas can contain a considerable amount of acetic acid, ammonia and sulfur compounds, depending on the fuel used and the mode of operation. In addition, the wall temperatures of engines are lower in producer gas drive due to the lower adiabatic flame temperature of producer gas. The wear by corrosion will, therefore, be greater than the wear by abrasion. The common gas cleaning equipment in automotive gas producers, although highly effective for solid particles larger than 5 microns, is not of much use to remove all the vapors in the gas. The gas will, therefore, reach the combustion chamber in a saturated state that depends on the pressure and temperature of the condenser. Water by itself does not cause undue corrosion since engines operated on hydrogen which yields a considerable amount of water as combustion product did not show any excessive cylinder wear. Consequently, there is a strong indication that the organic acids and others in producer gas are the main cause for excessive engine wear.

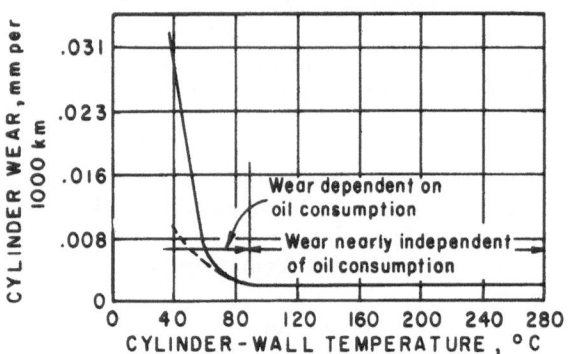

Figure 211. Cylinder Wear in Relation to Cylinder-Wall Temperature.

The question whether combustion of producer gas in an internal combustion engine will result in increased engine wear and shorter lifetime cannot be answered precisely. There is no such thing as uniform producer gas. The amount of mineral vapors carried into the engine have never been measured and their type is not known. Past experience indicated some problems with silica vapor usually found in the form of a fine white powder in producer gas. Silica, although predominant in most biomass fuels and coal, is not the only mineral evaporated in the process of gasification. In fact the alkalies start to evaporate at much lower temperatures than silica as outlined in Chapter VI.

The highly complex acetic acids generated in the distillation zone and the significant amount of ammonia and hydrogen sulfide associated with the gasification of coal and some biomass fuels are probably the main cause of engine wear, since their effect on the piping and condenser has been well demonstrated. This claim, however, cannot be backed by scientific data. Of course an efficient purification system will help to keep undue engine wear at a minimum but is certainly not a guarantee against ruined engines. Because of this uncertainty, some makers of automotive gas producers installed an oil drip feed to wet the incoming gas before it entered the mixer. Others employed a self-induced oil scrubber as the last cleaning stage and saturated the gas with a fine oil mist.

The same uncertainties apply to the quality of the engine exhaust gases. Until recently there was no concern about the engine exhaust and no data exists about the performance of the past units with regard to pollutants in the exhaust gases. The general awareness about the potential danger of engine exhaust gases has increased significantly during the last decade and standards have been set for the allowable percentage of hydrocarbons, carbon monoxide and the most dangerous nitrogen oxides in the engine exhaust. No data exists as to what extent an automotive gas producer could meet these standards. In theory, the combustion of producer gas should result in a less pollutant engine exhaust. This is due to the more complete combustion of the gaseous fuel and the lower adiabatic flame temperature which reduces the generation of nitrogen oxides and should reduce hydrocarbons and carbon monoxides in the exhaust. In practice however the difficulties with a fixed ignition timing and the rapidly changing gas composition will probably not result in any more complete combustion. One test conducted at UCD with a small 14 hp dual-fueled diesel engine showed a more favorable engine exhaust at high engine torque, Figure 212.

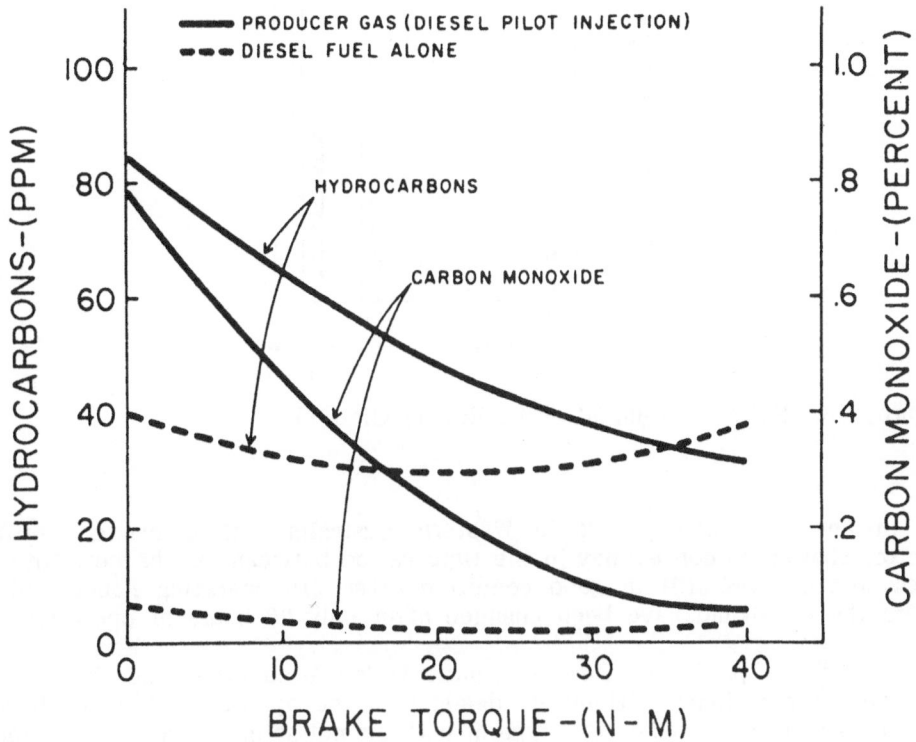

Figure 212.   Pollutant Emissions in Exhaust Gas (27).

The same argument applies to any effect on the lubrication oil. There are no data available that show consistent trends concerning the impact of producer gas operation on the lubrication oil. Due to the wide range of conditions in a gas producer and the chemical composition of the feed material, in particular the ash, any generalized statement would be misleading. In theory, the quality of an oil that governs its ability to lubricate two surfaces in sliding contact is its viscosity. The higher the viscosity the better the lubrication qualities. The dilution of the lubricant through unburned liquid fuels is absent in producer gas operation. Therefore, the viscosity of the oil should increase due to the burning off of the lighter fractions. The uncertainties are the mineral fractions in the raw gas and the efficiency of the gas cleaning equipment in removing it. The wide margin of the solid contamination of the oil that exists is illustrated in Figure 213 for filtered oil and various fuels.

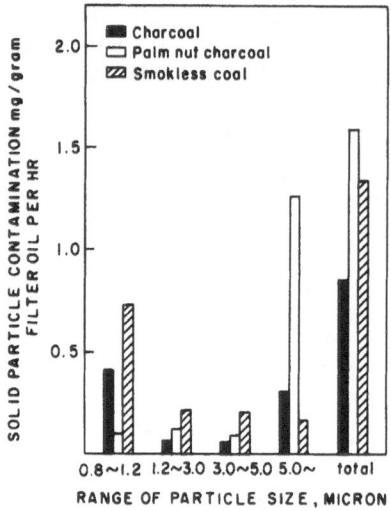

Figure 213. Solid Contamination of Engine Oil (36).

For instance, extensive trials in Western Australia with converted kerosene tractors showed no consistency in the time period between oil changes. In some cases, the oil was still in good condition after 210 operating hours while in others the oil should have been changed after only 20 hours of operation (10).

Automotive gas producers were the most widely used types and the reported data about operational difficulties, demand on the operator's skill and hazards involved are mostly based on this type of unit and not so much on small stationary units. The hazards and operational difficulties associated with gas producers must be seen in a larger context. Their broad introduction was always associated with some kind of emergency situation. World War II swamped Europe with at least one half million automotive gas producers within a short period of time. The collapse of the wheat prices in Australia during the year 1930 resulted in a rather hasty conversion of kerosene tractors to producer gas. The main motivation was to keep the fleet of essential vehicles such as trucks and tractors operating, as during WW II, or to offset the economical loss by cutting down the use of expensive kerosene as in Australia. The situation was worsened by many manufacturers trying to sell their unreliable equipment to a customer who knew almost nothing about automotive gas producers. It is obviously hard to decide whether this was done deliberately or the manufacturer not having much of an idea how to build a gasifier. The situation would not be much different today in case there is a need for the use of portable or stationary gas producers in the 5-100 hp range. Logistic problems combined with human failures of all kinds emerged and contributed much to the reported hazards, frustration and general displeasure with the new technology.

In order to do justice to the gas producer-engine system as an energy conversion system, one should distinguish between difficulties and hazards caused by the fuel, the gas producer itself and all the problems that have been created through human errors and insufficient knowledge about gasification.

The widespread belief that gas producers are energy conversion systems that can be operated with any kind of waste products in whatever form and physical condition has been most detrimental to this technology. It is hard for users of gas producers to understand that it will sometimes react rather drastically to changes in the fuel for which they were not designed. The misunderstanding about the flexibility of a gasifier with regard to its fuel is evident in brochures of a few manufacturers of small units. The functioning of a gas producer depends not only upon the moisture content of the fuel, but its ash fraction, size distribution and composition of the ash as outlined in Chapter V. In the past experience, the German and Swedish governments had to regulate not only the manufacturer of gas producers but also the fuel and its distribution. The average customer does not have the ability and knowledge to decide what fuel is best for a particular gas producer. The wide variety of available fuels made from wood, charcoal or coal was most confusing. It was soon recognized that minor incidents such as rough handling of a bag of charcoal or leaving the charcoal exposed to high humidity could severely decrease its suitability as a gas producer fuel. Any kind of dirt picked up during processing of the fuel may cause severe clinker formation in the gas producer. It was emphasized in Chapter V that the important key to successful gasification lies in the correct choice and the availability of the fuel.

The small-scale gas producer itself does not have inherent hazards or requires the attention of a specialist. Except for the very few supercharged gas producer-engine systems, all units are operated under suction. There is no danger of gas leakage during proper operation. However, if the engine is stopped or slowed down, pressure will build up due to continuing production of gas. This can lead to gas leakage if the gas producer is not properly sealed. High temperatures of 1300-1800 $^{\circ}$C are usually confined to a small zone in the gasifier and the walls of the unit are protected either by a fire lining or a layer of charred fuel. However, a breakdown in the cooling system of a crossdraft gas producer or a general overheating of the plant may lead to serious damage of the internal parts of the gasifier. The gas, because of its high carbon monoxide content, is toxic and the problems associated with carbon monoxide poisoning and general health hazards created by automotive gas producers in Sweden are thoroughly documented in Reference 7. A not so well documented hazard is the tar usually collected in tar traps and condensers of the unit. The phenols in the tar have been identified as a strong carcinogenic agent. Although an operator of a gas producer comes only in skin contact with this substance during cleaning of the plant, precautions such as wearing gloves should be taken. To what extent a gas producer operator is exposed to hazardous substances is not well documented.

The greatest hazard related to the operation of gas producer engine systems is the human being himself. It is certainly justified to ask the question, "How foolproof and simple to operate an automotive gas producer should be before it can be safely released to the public?" The picture may be different in stationary units, but the documented amount and type of accidents related to the use of an automotive gas producer shows clearly that the general public was neither informed nor capable of absorbing the fast transition to producer gas operation. There are no uncalculated risks in the operation of a gas producer such as high pressure lines or sudden rupture of particular parts. The reluctance of accepting the inconvenience of the automotive gas producer may have caused the most damage to people and the equipment. The list of accidents caused by negligence is long and special instruction booklets were published and schools had to be established to educate drivers on how to deal with the automotive gas producer.

The daily cleaning of the gas purification system was quite frequently not done and resulted in rapid wear of the engine or a breakdown of the cleaning equipment. The special safety filter incorporated in the gas line by some manufacturers was one way to protect the engine from damage and signal the driver to clean the main filter system. It was painfully recognized that driving with producer gas is an art requiring special skill and understanding of the overall process and much more and frequent attention to the gasifier and engine than with gasoline or diesel drive. A very detailed instruction manual on how to operate an automotive gas producer and recognize problems related to producer gas is given in Reference 32. Some parts of the Swedish and German instruction manuals have been translated into English and are published in References 7,21,30,31.

CHAPTER VII

1. Allcut, E. A. and R. H. Patten, **Gas Producers for Motor Vehicles, First General Report of the Subcommittee on Producer Gas,** National Research Council of Canada, Ottawa, Canada, 1943.

2. Allcut, E. A., Producer Gas for Motor Transport, Engineering Journal, v 25, n 4, 1942, pp 223-230.

3. Amin, H., Preliminary Evaluation of an Automotive Gas Producer, General Motors Research Publication GMR-3431, F&L-709, Warren, Michigan, September, 1980.

4. Anonymous, Converting Diesels for Producer Gas Working, Bus and Coach, January, 1943, pp 13-24.

5. Anonymous, Discussion on Tests on Transport Producer-Gas Units, Discussion in London on the paper by H. Heywood, Meeting of the Institution of Automobile Engineers, July, 24, 1942, Institution of Automobile Engineers, v 149, p 34.

6. Anonymous, Gas Producers, Automobile Engineer, November, 1942, pp 433-464.

7. Anonymous, Generator Gas The Swedish Experience from 1939-1945, Solar Energy Research Institute, Golden, Colorado, SERI/SP 33-140, January, 1979.

8. Anonymous, Producer Gas for Road Vehicles, Engineer, v 163, n 4248, 1937, pp 682-684.

9. Bainbridge, J. R., The Road Performance of Motor Vehicles Operated on Charcoal Producer Gas, The Modern Engineer, March 1942, pp 25-32.

10. Bowden, A. T., Bench and Field Tests of Vehicle Gas Producer Plant as Applied to Farm Tractors, Institute of Mechanical Engineering, University of Western Australia, v 146, 1941, pp 193-207.

11. Breag, G. R. and A. E. Chittenden, Producer Gas: Its Potential and Application in Developing Countries, Tropical Producers Institute, Report G130, London, England, 1979.

12. Egloff, G. and P. Van Arsdell, Motor Vehicles Propelled by Producer Gas, The Petroleum Engineer, v 15, n 4, 1944, pp 144, 146, 148, 150.

13. Eke, P.W.A., Gas as an Engine Fuel, Transactions of the Institute of Marine Engineering, 1970, pp 121-138.

14. Giffen, E., et al., The Conversion of Compression-Ignition Engines to Producer-Gas Operation, Engineering, August, 1944, pp 98-159.

15. Goldman, B. and N. C. Jones, The Modern Portable Gas Producer, Institute of Fuel, v 12, n 63, 1939, pp 103-140.

16. Goss, J. R., et al., Transient and Steady-State Temperature Fluctuation in a Downdraft Gas Producer, Meeting of American Society of Agricultural Engineers, Pacific Region, Hilo, Hawaii, March 1980.

17. Heywood, H., Loss of Power in Petrol Engines Running on Producer Gas, Engineering, January, 1941, pp 61-63.

18. Holmann, A., Erfahrungen mit MAN Fahrzeug-Dieselmotoren im Dieselgasbetrieb, Automobiltechnische Zeitschrift, v 44, n 8, 1941, pp 198-202.

19. Hurley, T. F. and A. Fitton, Producer Gas for Road Transport, Proceedings of the Institution of Mechanical Engineers, v 161, 1949, pp 81-97.

20. Kennedy, W. B., Mixtures of Producer Gas and Petrol, Institution of Engineers, Australia, v 12, n 9, 1940, pp 259-263.

21. Mellgren, S. and E. Anderson, Driving with Producer Gas, National Research Council of Canada, RP 15/43, Ottawa, Canada, 1943.

22. Meyer, W., The Supercharging of Internal Combustion Engine Plants Driven by Producer Wood Gas with Special Reference to Motor Vehicles, Brown Boveri Review, v 28, n 8, 1941, pp 206-208.

23. Michalski, W. and J. Spiers, Conversion of Compression Ignition Engines to Producer Gas Operation, Gas and Oil Power, v 39, n 468, 1944, pp 244-249.

24. Middleton, F. A. and C. S. Bruce, Engine Tests with Producer Gas, Journal of Research of the National Bureau of Standards, v 36, February, 1946.

25. Negretti, W., The Adaption of Turbo-Charged Producer-Gas-Operated Engines to Vehicle Running Conditions, The Brown Boveri Review, v 30, n 7, 8, July/August 1943, pp 184-187.

26. Nördstrom, Olle, Redogörelse för Riksnamndens for Ekonomisk Försvarsberedskap Försknings- Och Försöksverksamhet på Gengasområdet vid Statens Maskinprovningar 1957-1963, (from) Overstyrelsen for Ekonomisk forsvarsberedskap, Sweden, January, 1962.

27. Ogunlowo, A. S., Design Modification and Controls for the Operation of a Single-Cylinder Air-Cooled Naturally-Aspirated Diesel Engine on Producer-Gas Using Pilot Injection of Diesel Fuel, M.S. Thesis, Department of Agricultural Engineering, University of California, Davis, 1979.

28. Porter, J. C. and R. Wiebe, Gasification of Agricultural Residues, U.S. Department of Agriculture, Northern Regional Research Laboratory, AIC-174, 1948.

29. Renton, C., Producer Gas Tests in the Queensland Railway Department, Institution of Engineers, Australia, October, 1949, pp 274-278.

30. Ruedy, R., Mechanical Troubles and Remedies in the Operation of Producer Gas Vehicles, National Research Council of Canada, RP 18/43, Ottawa, Canada, September, 1943.

31. Ruedy, R., Sweden's Gas Producers, National Research Council of Canada, RP 16/43, Ottawa, Canada, September, 1943.

32. Ruedy, R., Wood and Charcoal as Fuel for Vehicles, National Research Council of Canada, n 1187, Ottawa, Canada, 1944.

33. Schläpfer, P. and J. Tobler, **Theoretische und Praktische Untersuchungen über den Betrieb von Motorfahrzeugen mit Holzgas,** Schweizerische Gesellschaft für das Studium der Motorbrenstoffe, Bern, Switzerland, 1937.

34. Spiers, J. and E. Giffen, Producer Gas: The Effect of Compression Ratio on Performance, Automobile Engineer, v 32, n 431, 1942, pp 523-527.

35. Takeda, S., Development of Gas Engine (I), The Bulletin of the Faculty of Agriculture, Mie University, Tsu, Japan, n 58, 1979, pp 137-141.

36. Takeda, S., Research on Gas Engine 2, Annual of Institute of Tractor Research and Testing, n 3, 1979, pp 19-36.

37. Taylor, C. F., **The Internal Combustion Engine in Theory and Practice, Volume I,** The Massachusetts Institute of Technology Press, Cambridge, Massachusetts, 1977.

38. Twelvetrees, R., Paving the Way for Producer Gas Operation, Bus and Coach, May, 1943, pp 104-107.

39. Williams R. O., et al., Development of Pilot Plant Gasification Systems for the Conversion of Crop and Wood Residues to Thermal and Electrical Energy, American Chemical Society, Symposium-Series, n 76, 1978, pp 142-162.

40. Williams, R. O., et al., Steam Raising with Low-Btu Gas Generators and Potential for Other Applications, Department of Agricultural Engineering, University of California, Davis, 1977.

41. Woods, M. W., An Investigation of the High-Speed Producer Gas Engine, The Engineer, v 169, n 4401, 1940, pp 448-450.

CHAPTER VIII: ECONOMICS

The introduction of small gas producer-engine systems into Third World Countries may not be much different from the present development in the United States and Europe. Two possible scenarios are most likely:

1. Providing a complete gas producer-purification-engine system manufactured in an Industrialized Country for a Third World Country.
2. Manufacture of the gas producer and purification system in a Third World Country and fitting the unit to suitable engines manufactured or in operation in that Country.

For various reasons, scenario 2. is most desirable; but, scenario 1. is more likely to take place at the initial development stage. The degree of difficulty to expect is lowest when single stationary units are introduced and highest for putting into service a whole fleet of automotive gas producers. The logistic problems with a reliable continuous supply of a suitable fuel will add to the complications in the latter case. It is therefore desirable to start at the lower end of the technology and test stationary gas-producer engine systems which are batch fed.

A small 30 to 50 hp unit with a simple cleaning and cooling system consisting of a cyclone, home made condenser, a fixed bed wet scrubber, and a tar extractor together with an engine and electric generator will fit on a trailer and occupy about 12 m^2. Larger units will be more complex and there will be a certain size beyond which batch feeding and hand removal of the ash will no longer be feasible.

For each horsepower-hour 1-1.5 kg of untreated biomass fuel must be supplied which amounts to 45 kg per hour for a 30 hp engine. If one desires to batch feed the unit in hourly intervals with olive pits only 0.07 m^3 of hopper space is required, whereas with wood 0.15 m^3 or with rice hulls 0.74 m^3 is necessary.

The batch feeding of large units above 100 hp is hardly practicable and a great deal of automatization with all the complications and expenses involved are necessary to run larger units.

Technical data on the material used in past automotive gasifiers and their approximate weights are listed below:

Empty Gas Producer	90-156 kg
Cooler	16-77 kg
Filter	20-81 kg
Entire Plant	135-301 kg

Material used for the manufacture of the gas producer, cooler and filter:

Copper	0-11 kg
Aluminum	0-7 kg
Brass	0-9 kg
Stainless Steel	0-29 kg

268

Alloy Steel	0-22 kg
Mild Steel	110-294 kg
Refractory	0-36 kg
Hopper capacity	22-80 kg

These data are based on 10 different units used on trucks with a rated output of 50 hp at 50 km/h and an actual achieved shaft horsepower of 16 to 31 hp with producer gas as the fuel. In addition, small amounts of high temperature gasket that can be made out of aluminum and perhaps fiberglass material for an improved filter system are necessary.

To what extent this material will be available in Third World Countries, either imported or domestically produced, is difficult to judge. For instance, a country like India is perfectly capable of producing all the necessary constituents for a gas producer and mass production of the entire gas producer-engine system. The opposite situation prevails in Afghanistan. All the needed material must be imported and no facilities are available to mass produce a gas producer. The situation of the individual person in terms of his monthly income and average living is much the same in both countries although India can be considered highly industrialized compared to Afghanistan.

Gas producer-engine systems completely assembled on trailers are manufactured in Europe and the United States for $700 to $1,200 per installed kW, depending on the size of the plant. Smaller units (10-30 kW) are more likely to cost $1,200/kW while larger units can be purchased under $1,000/kW. These units are usually fully automated. The search in 63 countries did not find manufacturers of small (less than 10 hp) off-the-shelf units. Some companies do have the know how and facilities to custom make and assemble complete units for the above price but are rather vague about the time required for manufacturing or guarantees for successful operation.

The gas producer itself can be manufactured for $2,000-$3,000 (1981 cost) in a size suitable to power a 30 to 50 hp engine. All these prices refer to custom made units and are based on quotes and manufacturer's information.

The installment of a 30 kW electrical generator fueled with biomass fuels would therefore cost about $35,000 when imported. Assuming a gasoline price of 60 cents per liter at the installation location, the purchase and installment price of the unit is equivalent to 58,332 liters of gasoline. On the other hand, a 30 kW gasoline generator requires 180 liters of gasoline when operated for 10 hours a day. This amounts to 49,140 liters a year assuming the unit is on duty for 273 days a year (75%). A gasoline engine driven generator of this size could therefore be replaced by a new imported gas producer generator unit. This transaction can be economical under some restrictions such as:

1. The maintenance and operational costs of the gas producer are comparable to the unit that has been replaced.
2. The fuel can be provided for a reasonably low price.
3. The reliability of the plant and the associated logistic problems with the fuel supply do not cut down drastically on the on duty time of the plant.

There are obviously many factors to consider in determining to what extent the above conditions can be met. A gas producer-engine system must be operated by a trained person that has to be on duty all the time for a batch fed unit. The gas purification system must be cleaned each day under such continuous operation. In many locations in Third World Countries the necessary technical personnel to operate the unit are available. The maintenance costs associated with a producer gas system will be higher due to the vulnerability of some internal parts of the gas producer and a frequent replacement of the gas cleaning filters. Consequently, any imported unit will depend on replacement parts such as fabric filters, tuyeres and refractory lining. How fast these parts will wear out and need to be replaced will depend on the skills and competence of the operators of the unit.

Annual maintenance and spare part costs for the gas producer and gas cleaning system is conservatively estimated to be 10% of the initial cost. Because money is not readily available in most Third World Countries and the difficulties to obtain loans for an unsecured project such as a gas producer-engine generator system, an interest rate of 15% with a necessary short capital recovery time of 4 years is assumed. The life expectancy of the plant will be about 4 years. This low number takes into account the unproven field experience, the uncertainties in engine wear and corrosion problems in humid climates. Consequently, $15,795 in annual capital cost and maintenance for the first four years are well covered by the annual savings of $29,484 in gasoline. How much of the annual savings of $13,689 needs to be spent to solve all the logistic problems with the processing of the new fuel depends very much on the local situation.

This example highly favors the gas producer unit but has some economical uncertainties. Assuming the fuel is wood and the amount used does not compete with other needs for wood or has any environmental impact and other negative long term effects, then only the cost of processing the wood needs to be taken into account. The project can then be viewed as economically and socially acceptable. Approximately 500 kg of wood are necessary to replace the daily amount of gasoline that would be used. This amount of wood as chips or blocks occupies about 2 m^3. It is realistic to assume that the wood will be hand sawed, cut and transported with locally available resources. Consequently, a large enough labor force is necessary to process at least 500 kg of wood a day. Though there is much labor available in most Third World Countries, this situation should not be exploited because of high unemployment. The assumption of $3 a day for unskilled farm and construction work is on the high side for most developing countries. A three man work force would require 23% of the annual savings ($90 per person a month). Whether it is economically and socially acceptable to use wood or any other fuel and to what extent the necessary low fuel moisture content for downdraft gas producers will require storage facilities and fairly advanced planning will be discussed later.

In case the imported gas producer-engine-generator replaced a diesel engine driven generator, the savings in fuel and money do not look favorable because of the higher efficiency of the diesel engine system and the lower diesel oil price. The 30 kW dual-fueled diesel engine needs at least 10 liters of diesel oil per hour of operation which amounts to an annual cost of $13,650 for a diesel oil price of 50 cents/liter. Assuming an annual interest rate of 15% and a necessary capital recovery time of 4 years, $12,259 must be paid back per year. This is almost equal to the savings in diesel oil and does not leave much for additional costs such as fuel processing and the higher maintenance and spare part costs for producer gas fuel. Both examples have deliberately overestimated the costs for switching to producer gas operation and underestimated the costs for the old gasoline or diesel engine units. However, under the first scenario it is economical when gasoline is replaced even with higher interest rates and high biomass fuel prices. The dual-fuel operation of a diesel generator cannot be economical without interest free loans. Buying a 30 kW gas producer, dual-fueled diesel engine generator set under the first scenario is even more uneconomical in this size range. Such a unit will still need 1.5 liters of diesel oil per hour and reduce the wood consumption to 260 kg a day. Taking into account the additional cost of $2,048 for diesel oil, the annual fixed costs will exceed the costs for a system fueled with gasoline.

The decisive factor in all such projects is the local gasoline and diesel oil price. In Third World Countries with underdeveloped infrastructure, the diesel and gasoline price can double and triple from one location to the other due to the transportation difficulties. This has led in some instances to greatly reduced duty times for generators and water pumps and even prevented some farmers from cultivating at all because the fuel was not available or too expensive.

The first scenario is undesirable for several reasons, although it can be justified in an initial stage for demonstration projects and pilot plants. One major point against such plants is the complexity of the unit and spare part supply. In a 30 kW and higher range, the unit will be well automated, probably have a rotating grate, a hopper vibrator to prevent bridging and several electric motors to support the automatic ash disposal, fuel feeding system and possibly an air blower. The energy to operate the automated plant will only consume 1%-2% of the gas producer fuel, an insignificant amount compared to what is lost when the gas producer has inadequate insulation. The plant is equipped with a small gasoline generator that provides the necessary power to start the unit before the generator can take over. In case no electricity is generated with the plant, this small generator must be operated all the time.

The second scenario is more desirable and will require further investigation to determine to what extent the conversion of existing gasoline and diesel engines and the manufacture of gas producers can be achieved on site. The possible costs involved will not exceed $5,000 which is about the price in the United States to convert a 30 kW diesel engine and install a gas producer-gas purification system which is not automated. However, the annual capital costs of $1,751 for interest and loan repayment and $500 for additional spare parts and maintenance costs in Third World Countries together with $2,048 for diesel oil are less than the diesel oil savings of $13,640 a year. The net savings of $9,351 a year will certainly cover the cost for 71 tons of biomass fuel with a heating value equivalent to wood.

Table 57 lists the three most technically feasible cases. All computations are based on the following assumptions:

a) Life of the entire plant is 4 years
b) Capital recovery in 4 years

c) Annual interest rate of 15%
d) Gasoline price of 60 cents per liter
e) Diesel oil price of 50 cents per liter
f) Fuel consumption of 1.25 kg per horsepower-hour for producer gas operation of a gasoline engine with a fuel equivalent to wood.
g) Fuel consumption of 0.65 kg per horsepower-hour for dual-fuel operated diesel engines.
h) Diesel oil consumption of 15% of the orignial amount used.
i) Annual maintenance and spare part cost of 10% of the capital investment in addition to what was spent for conversion. This high number is justified through practical experience with the costly and time-consuming procedure of obtaining spare parts for imported machinery in Third World Countries.

The case of an existing gasoline engine converted to gas producer operation has been excluded since the severe power drop of 40% or more may render such a conversion useless in industrial applications or at least cause serious inconvenience.

The most severe problems, however, are caused by the uncertain biomass situation in most Third-World Countries. In arid zone countries where irrigation requires the most energy, the wood situation is extremely grave. The trend of dwindling forests has been accelerated for many years and is at a level where most people cannot afford to have at least one hot meal a day since fire wood is simply too expensive. The traditionally used stickwood for cooking accounts for 80% of the monthly income of an unskilled laborer in Afghanistan and may explain why not many people can afford more than one hot meal a week. This situation has a huge impact on the general health of the individual in such countries and is the cause of wide-spread parasitic diseases which are usually transmitted by the consumption of uncooked food and unboiled water. The fuel shortage for household food preparation in most arid zones may rank second behind malnutrition in the cause of disease and very short life expectancy. Consequently, cutting or collecting wood as a gasifier fuel is neither practicable nor advisable in such countries.

In areas with higher annual rain fall which have a good supply of wood, the use of wood for gasification is practical and also economical. It is illustrative to express the amount of wood needed as fuel for a 30 kW gas producer in terms of trees per year. Table 56 gives the average yield of Douglas Fir trees as a function of age. The computed weight is based on the realistic assumption that the gasifier will be fed with seasoned, air dry wood. The impact of a 30 kW gas producer on the forest resource can be estimated. When the cutting of the trees is associated with effective reforesting of the area, about 110,000 m^2 of

Table 56. Douglas Fir Production and Utilization for Gasification

Age of Tree	Avg. Diam. cm	Useable Vol. m^3	Air Dry Weight kg	Trees Gasified Per Year For 30 kW
20	20	0.28	168	812
25	28	0.65	390	350
30	36	1.16	696	196

Table 57. Cost Comparison for a 30 kW (40 hp) Installation[1]

Scenario	Capital Investment $	Fixed Annual Costs Except for Biomass Fuel	Annual Fuel Consumption Tons of Wood	Money Available For Each Ton of Wood (Break Even Base) $
Replacement of gasoline engine generator by imported complete gas producer-engine unit	$35,000	$15,795[2]	137[3]	+100[4]
Replacement of diesel engine generator by imported complete gas producer engine unit	$35,000	$15,795	137	-16
Modification of existing diesel engine to a dual fueled-gas producer engine system	$5,000	$4,300	71	+132

[1] 45 hp input to generator shaft with 90% generator efficiency to produce 30 kW.

[2] Fixed Annual Cost: $(\$35,000)\dfrac{0.15}{1-(1.15)^{-4}} + (0.1)(\$35,000) = \$15,759$

[3] Annual Fuel Consumption: $\dfrac{(30\text{kW})(0.00125 \text{ tons/hp})(273 \text{ days})(10 \text{ hours/day})}{(0.746 \text{ kW/hp})} = 137 \text{ tons}$

[4] Money Available for Each Ton of Wood Fuel: $\dfrac{(0.6 \text{ \$})(18 \text{ liter/hr})(10 \text{ hr})(273 \text{ days}) - (\$15,759)}{137 \text{ tons}} = \$100 \text{ per ton}$

273

forest are needed as the renewable fuel supply for one single 30 kW gas producer unit. This number is based on a forest growth rate of 2100 m³ per square kilometer per year.

The use of waste wood from logging activities or sawdust will be practicable only in updraft gas producers. For any other type of gas producer careful screening recovery operations will be necessary because of the sensitivity to dirt and the large range in fuel size.

A major problem with the use of biomass fuels is their natural moisture content. It is not only desirable but essential to reduce the natural moisture content below 20%. This requires air and sun drying of wood, corncobs or other fuels for an extended period of time. The fuel cannot just be collected on a daily basis as needed. How well a reliable fuel supply can be assured in a particular social environment is an important factor in determining where the gas producer-engine system is to be located.

The example of a 30 kW plant has been chosen because it is most feasible to replace the many existing small gasoline engine generators by gas producer systems. The same applies to stationary water pumps of equal or smaller size. The operation of such a unit by a village will require totally different planning and evaluation than operation in industrialized areas.

Due to the normally inadequate infrastructure surrounding villages and their inaccessibility in Least Developed Countries, the fuel supply should come from within the village district. It may not be possible for a village to buy fuel from outside its district if it cannot be supplied from within the district. Long-term encounter with a highly industrialized, well infrastructured society usually rejects the thought that a sufficient amount of money cannot be exchanged for new needs such as 500 kg of corncobs or wood each day. Nevertheless, it cannot in most cases. For instance, villages in Afghanistan are autarkic, not by choice but by necessity. The food supply, water supply and usage of any kind of crop residue and biomass are carefully balanced. The replacement of the existing generator by a gas producer unit will most likely throw the system off balance because the gas producer needs fuel that has other uses. A classical example is the one or two man bakeries which supply the entire village with the daily bread and quite often cannot deliver because they either run out of fuel for their stoves or they are short of grain.

On the other hand, there are many cases where a gas producer-engine system has been and will be quite successful because of the abundance of biomass fuels. Gas producer-engine generators have been installed on tropical islands where because of the isolated situation fossil fuel is extremely expensive and high vegetation growth generates enough biomass to be used for operation of the gas producer.

The foregoing situation concerns only one portion of the huge pool of biomass that is available. Most efforts so far have only explored wood and wood charcoal gasification. There are many biomass fuels which are at least equal or even superior. For instance, most kinds of fruit pits and nut shells as well as corncobs are excellent fuels and available in appropriate form which requires very little processing. They can be available in large enough amounts and thus are a source for renewable energy. In addition, the gasification characteristics of many fast growing tropical woods such as eucalyptus, mahogany and bamboo are not known and should be explored. There are many crop residues such as rice husks, straw and cotton gin trash which are essentially a waste product in many parts of this world. To manage their gasification either in upgraded form or in their natural form could be highly beneficial for industrial plants processing these types of crops.

There is a certain point in the technological assessment of a gasifier where the "art" of gasification and its reputation could well benefit from the "science" of gasification. The final choice of a Third World Collaborator and a possible test site should be, therefore, mostly guided by the willingness of the Third World Party to carefully assess the social impact and the fuel situation. This will at least reduce the risk of the worst possible case, an installed gasifier that cannot handle the locally available fuel. This happened quite frequently during the Second World War and such a situation is usually followed by efforts to upgrade the fuel in order to make it suitable. One should not count on such an approach because upgrading a biomass fuel may either become too expensive or even worsen its already unsuitable gasification properties. The science of upgrading biomass fuels is young. Upgrading the fuel can be done with respect to its physical and chemical properties. The most simple and best known upgrading procedure is natural drying and screening, with the goal to reduce the moisture content and range of size of the fuel. Another method almost as old as the human race itself is charring the fuel. The end product of this process, charcoal, is in most cases a better fuel for a gasifier but about 60 percent of the energy in the raw fuel is lost. It does not always improve the fuel quality, because charcoal can be highly friable and so high in ash that its gasification will cause even more problems than the original material. Furthermore, the cost of densification is a major cost amounting to $9.00 to $14.00 per ton.

The production of charcoal, whether from wood or other biomass, is an art. The quality of the end product cannot always be controlled to the extent needed for gas producer fuel. In particular, the charcoal production on village levels is a gamble with respect to the suitability of the charcoal as a fuel for gasification. It is one of the most wasteful methods to produce gasifier fuels. The pyrolitic oils and gases are not recovered in almost all practical cases. As previously outlined, a 30 kW generator required about 812 trees each 20 years old a year as fuel. If this amount of fuel were supplied as charcoal, 2030 trees are needed annually. The situation is more economical and technically possible if some fast renewable crop residues could be carbonized on an industrial scale with the use for the pyrolitic oils and gases and quality control of the charcoal.

The recent increased demand for biomass fuels has stepped up efforts to densify biomass fuels to various shapes such as pellets, briquettes, cubes and cylinders. This is an expensive process but capable of upgrading all kinds of dry biomass such as grasses, leaves, sawdust, straw and rice husks to a uniform highly densified fuel. Depending on the fuel, it can be done with or without a binder that holds the shredded material together in the densified form. The energy input to run the equipment will range from 0.75 to 1.5 percent of the energy in the processed fuel.

There are many situations which may be unacceptable for the installment of a gas producer-engine system for one or more of the following reasons:

1. High ash content of the fuel.

2. Medium ash content of the fuel and unsuitable chemical composition of the fuel.

3. Insufficient storage facilities which cannot accommodate a few months supply of fuel.

4. Unsuitable processing procedures of the biomass which result in dust, dirt and a high fraction of small-sized particles.

5. Unsuitable fuel size reducing equipment that changes the physical properties of the biomass which may induce bridging. For example, it matters how a coconut shell is crushed.

6. Natural form of the fuel is too small or will pack too dense in a gasifier.

7. Flow properties of the fuel is poor because of its shape or the flowage worsens under thermal destruction due to packing together of the fuel or clinging to surfaces.

There is a large array of devices and techniques to cope with one or more of the above conditions. But hopper vibrators, fuel bed stirrers, devices to control the fuel bed temperature, heat exchangers to preheat the incoming air or even cogeneration require additional maintenance and trained personnel and can increase the overall costs considerably. Whether they are a cure for inferior fuel and unsuitable processing methods is a question which has been explored for the last decade at research institutes. Most of their research is based upon improving the physical properties of the fuel and creating more favorable conditions within the fuel bed by adding chemical slurries to the fuel. The little information gained so far does not permit a conclusive answer to be given for the technical and economical feasilibity of these methods.

Although it was our intention to present the collected data within a consistent framework of acceptable metric units, this goal could only be partially achieved. The still widespread use of English units and more convenient practical units did not in all cases allow transfer of the reported data to metric units. The internationally established gram (g), meter (m), second (s), and joule (J) system is therefore occasionally replaced by more convenient units which are more familiar to the reader.

All chemical equations include the energy balance on a one kg-mole basis that refers to the reactant appearing as first term in the equation. There is no consistency in the literature about how to report exothermic and endothermic reactions. We have adapted the policy of writing the net energy of the process together with the products and energy released by the reaction as viewed within the observer's control volume. In this context, an exothermic reaction is positive. The tendency of many researchers to report properties of producer gas without referring to the state of the gas makes it impossible to consistently report the data. The possible errors introduced in analyzing such data can be huge and is one major reason why so many conflicting opinions exist in this field.

The prefixes used in this report for mass (gram), length (meter), energy (joule) and power (watt) are as follows:

$$M \text{ (mega)} = 10^6$$

$$k \text{ (kilo)} = 10^3$$

$$c \text{ (centi)} = 10^{-2}$$

$$m \text{ (milli)} = 10^{-3}$$

$$\mu \text{ (micro)} = 10^{-6}$$

The conversion from one set of units to one more familiar to the reader is given in the following table.

Length:
    1 km = 1000 m
    1 cm = 0.3937 inch
    1 foot = 30.48 cm
    1 mile = 1609.344 m

Volume:
    1 liter = 0.0353 ft^3 = 0.2642 gal = 1000 cm^3
    1 m^3 = 35.3147 ft^3 = 1000 liter

Mass:
    1 lbm = 453.59237 g
    1 ounce = 28.3495 g

Pressure: 1 lbf/in^2 = 2.036 in Hg at 32 ^OF = 6894.76 newton/m^2

1 inch Hg = 0.0334 atm

1 atm = 14.696 lbf/in^2 = 760 mm Hg at 32 ^OF

$\qquad$ = 1.01322 x 10^5 newton/m^2

1 bar = 0.9869 atm

1 cm H_2O = 0.394 inch H_2O = 0.0289 cm Hg = 0.0114 inch Hg

$\qquad$ = 0.0334 atm

Energy: 1 Btu = 1055.06 joules = 4415.954 calories

1 cal = 4.1855 joules

Power: 1 watt = 1 joule/sec = 3.413 Btu/h

1 hp = 745.6 watt = 2545 Btu/h

Temperature: Degree Rankin (^OR) = 1.8 x Degree Kelvin (^OK)

Degree Celsius (^OC) = Degree Kelvin - 273.16

Degree Fahrenheit (^OF) = Degree Rankin - 459.69

Degree Fahrenheit = 1.8 x Degree Celsius + 32

Miscellaneous: 1 Nm^3 (one norm cubic meter) = one cubic meter of gas at

$\qquad$ 0 ^OC and 1 atm

1 scf (one standard cubic feet) = one cubic feet of gas at

$\qquad$ 77 ^OF and 1 atm (previously at 68 ^OF and 1 atm)

cfm (cubic feet per minute)

1 kg-mole of producer gas = 22.4 Nm^3 of producer gas treated

$\qquad$ as ideal gas.